Spatial Simulation

Spatial Simulation

Exploring Pattern and Process

David O'Sullivan and **George L. W. Perry**

*University of Auckland – Te Whare Wānanga o Tāmaki Makaurau,
New Zealand*

WILEY-BLACKWELL

A John Wiley & Sons, Ltd., Publication

Library of Congress Cataloging-in-Publication Data

O'Sullivan, David, 1966–
   Spatial simulation : exploring pattern and process / David O'Sullivan, George L.W. Perry.
     pages cm
   Includes bibliographical references and index.
   ISBN 978-1-119-97080-4 (cloth) – ISBN 978-1-119-97079-8 (paper)
   1. Spatial data infrastructures–Mathematical models. 2. Spatial analysis (Statistics) I. Perry, George L. W. II. Title.
   QA402.O797 2013
   003–dc23

          2012043887

A catalogue record for this book is available from the British Library.

Wiley also publishes its books in a variety of electronic formats. Some content that appears in print may not be available in electronic books.

Cover image: Photo is of Shinobazu Pond in the Ueno district, Tokyo, April 2012.
Photo credit: David O'Sullivan.
Cover design by Nicky Perry.

Typeset in 10.5/13pt TimesTen-Roman by Laserwords Private Limited, Chennai, India.

First Impression 2013

To
Fintan & Malachy
and
Leo & Ash

Contents

Foreword

Space matters. To answer Douglas Adams' question about life, the universe and everything, we need to detect and understand spatial patterns at different scales: how do they emerge, and how do they change over time? For most spatial patterns in social and environmental systems, controlled experiments on real systems are not a feasible way to develop understanding, so we must use models; but analytical mathematical models that address all relevant scales, entities, and interactions are also rarely feasible.

The advent of fast computers provided a new way to address spatial patterns at multiple scales: it became possible to simulate spatial processes and check whether or not our assumptions about these processes are sufficient to explain the emergence and dynamics of observed patterns. But then, as when any new scientific tool appears, results piled up quickly, perhaps too quickly, produced by thousands of models that were not related in any coherent or systematic way. There is still no general 'theory' of how space matters. Spatial simulation is thus similar to chemistry before the discovery of the periodic table of elements: a plethora of unrelated observations and partial insights, but no big picture or coherent theory.

This book sets out to change the current situation. The authors group spatial simulation models into three broad categories: aggregation and segregation, random walk and mobile entities, and growth and percolation. For each category, they provide systematic overviews of simple, generic models that can be used as building blocks for more complex and specific models. These building blocks demonstrate fundamental spatial processes and principles; they have well-known properties and thus do not need to be justified and analysed from scratch.

Using this book, model developers can identify building blocks appropriate for their own models. They can use the principles of model building and analysis that the book's opening and closing chapters summarize with amazing clarity. The authors demonstrate how all this works in a case study, in which a moderately complex spatial simulation model of island resource exploitation by hunter-gatherers is developed and analysed.

To facilitate the re-use of building blocks, this book itself is based on building blocks, using the free software platforms NetLogo (Wilensky 1999) and R (R Development Core Team 2012, Thiele and Grimm 2010, Thiele et al. 2012), which are widely used. All the models and R scripts are available on the internet as a model zoo, or model kit, for the spatial modeller. There (unlike at real zoos) you are invited to play with the zoo's specimens, and even to change and combine them!

This book helps establish a new culture of model building and use by reminding us that, as scientists, we need to see the forest through the trees and, with spatial simulation, the patterns in our explanations of spatial patterns. This book was badly needed. It makes us think, and play.

Volker Grimm,
Helmholtz Centre for Environmental Research – UFZ, Leipzig

Preface

This book had its genesis in three ideas.

The first, hatched at the 2006 Spatial Information Research Colloquium (or more precisely on the flight back from Dunedin to Auckland), is that for all their seeming variety, there are *really* only a limited number of truly different spatial processes and models of them. This modest proposal came up in conversation when we realised that we have both become good at anticipating what spatial models will do when they are described at conferences or workshops, and in the research literature. This perhaps debatable insight, followed by extended periods of omphaloscopy (often over a beer), eventually led us to the three broad categories of spatial model (aggregation, movement and spread) whose presentation forms the core of this book. As will become apparent, the dividing lines between these categories are hardly clear. Even so, we think the approach is a valuable one, and we have come to think of these as 'building-block' models. It is a central tenet of this book that understanding these will enable model builders at all levels of experience and expertise to develop simulations with more confidence and based on solid conceptual foundations. Belatedly, we have realised that the building-block model concept has much in common with the notion of 'design patterns' in computer programming and architecture.

The second insight, gleaned from several years' teaching a class where we require students to develop a spatial model on a topic that interests them, is that we (the authors) had picked up a lot of practical and usable knowledge about spatial models over the years, but that that knowledge is not easily accessible in one place (that we know of). This realisation took some time to dawn. It finally did, after the umpteenth occasion when a group of students seeking advice on how to get started with their model were stunned (and sometimes a little grumpy) when the seeming magic of a few lines of code in NetLogo (Wilenksy, 1999) or R got them a lot further than many hours of head scratching had. Examples *par excellence* are the code for a simple random walk in NetLogo:

```
to step ;; turtle procedure
  move-to one-of neighbors4
end
```

or for a voter model:

```
to update ;; patch procedure
  set pcolor [pcolor] of one-of neighbors4
end
```

It is easy to forget how much hard-earned knowledge is actually wrapped up in code fragments like these. Having realised it, it seemed a good idea to pass on this particular wisdom (if we can call it that) in a palatable form. The numerous example models at http://www.patternandprocess.org that accompany this book aim to do just that. A closely related motivation was the steady accumulation of fragments of NetLogo in a folder called 'oddModels' on one of our hard drives: 'Surely', we thought, 'there is some use to which all these bits and pieces can be put?!'

Finally, while the building-block models that are our central focus are in some senses simple, they are deceptively so. In that most understated of mathematical phrases, these models may be simple, but they are 'not trivial'. The thousands of papers published in the technically and mathematically demanding primary literature in mathematics, physics, statistics and allied fields attest to that fact. Unfortunately, this also means that students in fields where quantitative thinking is sometimes less emphasised—such as our own geography and ecology, but also in archaeology, architecture, planning, epidemiology, sociology and so on—can struggle to gain a foothold when trying to get to grips with these supposedly simple models. From this perspective we saw the value in presenting a more accessible point of entry to these models and their literature. That is what our book is intended to provide.

The jury remains out on just how many building-block models there are: our three categories are broad, but we are painfully aware that our coverage is necessarily incomplete. Reaction–diffusion systems and network models of one kind or another are significant omissions. The relatively demanding mathematics of reaction–diffusion models is the main reason for its absence. And because networks are about spatial structure, rather than spatial pattern *per se*, and to avoid doubling the size of an already rapidly growing book, we chose not to fill that gap, albeit with reservations. Perhaps as 'network science' continues to mature we can augment a future edition. At times, we also wondered if there might be *only one* spatial model (but which one?). In the end, we decided that there are already enough (near) incomprehensible mathematical texts staking claim to that terrain, and that our intended audience might not appreciate quite that degree of abstraction. Our more modest second and third ideas are, we hope, borne out by the book and its accompanying example models. Ultimately, how well we have delivered on these ideas is for readers to decide.

David O'Sullivan and George Perry
School of Environment and School of Biological Sciences
University of Auckland—Te Whare Wānanga o Tāmaki Makaurau

August 2012

Acknowledgements

Any book is the product of its authors and a dense web of supporting interconnected people and institutions. Professionally we have benefited from a supportive environment at the University of Auckland. Research and study leave grants-in-aid in the first half of 2012 for both authors were vital to the timely completion of the manuscript.

David O'Sullivan is thankful to colleagues at the University of Tokyo who hosted a Visiting Professorship at the Center for Spatial Information Science from April to June 2012. He extends particular thanks to the Center's director Yasushi Asami; to Ryusoke Shibasaki, Yukio Sadahiro, Ikuho Yamada, Yuda Minori, Ryoko Tone and Atsu Okabe for making the visit such a rewarding one; and to Aya Matsushita for ensuring that all the administrative arrangements were handled so efficiently.

George Perry was fortunate enough to receive a Charles Bullard Fellowship, which enabled him to spend seven months working at Harvard Forest; this was an immensely rewarding time, both personally and professionally, and gave him time and space to think about and develop material in this book. Thanks to all the people at Harvard Forest who helped make his time there so productive and enjoyable. Over the last 15 years GP has spent considerable time working with colleagues at the Helmholtz Zentrum für Umweltforschung (UFZ) in Leipzig on various projects. Interactions with people there, especially Jürgen Groeneveld, have coloured his thinking on models and model-making. GP has also received funding from the Royal Society of New Zealand's Marsden fund and from the NSF (USA), ARC (Australia) and the NERC (UK)—the research that this funding enabled has contributed to his views on models and modelling.

We have been ably assisted at Wiley by Rachael Ballard, Jasmine Chang, Lucy Sayer and Fiona Seymour. We are both grateful for supportive noises and comments from many colleagues, and particularly to Mike Batty, Ben Davies, Steve Manson, James Millington and students in the ENVSCI 704 class of 2012 for insightful comments on late drafts of various parts of the manuscript. Iza Romanowska spotted what would have been an embarrassing error at a late stage, for which we are grateful. We are grateful to Volker Grimm for his kind words in the foreword. His work is a constant inspiration, and if this book delivers on even a few of the claims he makes on its behalf, we will consider it a roaring success! Conversations with many colleagues and students have improved the text at every turn. Any errors that remain are, of course, our own.

And finally, to those who have borne the brunt of it, our families: from David, heartfelt thanks to Gill, Fintan and Malachy for putting up with an often absent(-minded, and in body) husband and father for so long; and from George, to Nicky, Leo and Ash, for tolerating my peripatetic wanderings as I finished this book, thank you, arohanui, and I look forward to getting back to building hardware models with you!

Introduction

On the face of it, the literature on 'spatial modelling'—that is models that represent the change in spatial patterns through time—is a morass of more or less idiosyncratic models and approaches. Likewise at first glance, the great diversity of observed spatial patterns calls for an equally wide range of processes to explain them. Ball (2009) provides a useful classification of the patterns in natural and social systems.

Our view, and the perspective of this book, is that just as there is a manageable number of spatial patterns, and it is useful to classify them, a relatively small suite of fundamental spatial process models are useful in exploring these patterns' origins. These models we consider to be 'building-blocks', which when thoughtfully combined can provide a point of departure for developing more complicated (and realistic) simulations of a wide variety of real-world phenomena. As Chesson (2000, page 343) notes, albeit in a different context, 'The bewildering array of ideas can be tamed. Being tamed, they are better placed to be used.' We aim to show how building-block models can help to accomplish this task in the context of spatial simulation.

Organisation of the book

Here we set out the general plan of this book to assist readers in making best use of it. Broadly, the material is organised into three parts as follows:

Preliminary material in Chapters 1 and 2 presents an overarching perspective on models, and the process of model-making in general, and spatial simulation models in particular, that informs our approach throughout.

Chapter 1 reviews what dynamic spatial models are, and why and how they can be used across the wide range of disciplines that comprise the contemporary social and environmental sciences.

Chapter 2 introduces the concepts of pattern and process that are central to our understanding of the importance of specifically *spatial* simulation models. We present these principles in the context of spatial point process models, at the same time showing how such static models can be made more dynamic, so providing a link to the kinds of models that this book focuses on.

Building-block models are introduced in Chapters 3, 4 and 5. Each of these chapters considers a particular class of spatial models in some detail, roughly grouped according to their qualitative behaviour and outcomes. Each chapter starts with some examples of the sorts of systems and contexts in which the models being considered might be applicable. The models themselves are then introduced, starting as simply as possible so that the chapters are reasonably self-contained and can be read as standalone treatments. Each chapter closes with an overview of how these simple models have been used as a basis for more complicated simulations in a range of fields.

Chapter 3 considers models that produce patchy spatial patterns most commonly referred to as 'aggregated' or 'segregated'. Such models can provide the stage on which other dynamic processes might operate.

Chapter 4 provides an overview of simple approaches to representing the movement of individuals and groups, starting with random walk models and developing these to look at foraging and/or search, and flocking behaviours.

Chapter 5 reviews a number of simple models of processes of spread or growth through spatially heterogeneous environments.

Applying the models in practice is considered from a number of perspectives, with a particular emphasis on spatial and temporal representation in Chapter 6, and model evaluation in Chapter 7. An example is presented in Chapter 8 to show how all these ideas can be combined.

Chapter 6 looks at some aspects of the representation of time and space in simulation models. While the building-block models of Chapters 3 to 5 do not necessarily raise many concerns in this regard, experience suggests that careful consideration must be given to these aspects when all but the simplest simulations are being developed.

Chapter 7 is the most technically demanding in the book, providing an overview of how models can be evaluated so that their outcomes can be used to make inferences about the real-world systems that they seek to represent. This chapter revisits many of the themes first introduced in Chapters 1 and 2 concerning the use of models in science.

Chapter 8 aims to bring together ideas from all of the preceding chapters through an extended worked example combining elements of all three building-block models. The model is implemented using some of the methods considered in Chapter 6, and its analysis deploys the tools of

Chapter 7. This example is more complicated than any in Chapters 3 to 5, and is comparable with examples that appear in the contemporary research literature. Even so, we believe that its complications are comprehensible viewed from the perspective of the earlier chapters, particularly the building-block models.

Finally, **Chapter 9** concludes with a brief summary of the book's main themes.

Using the book

This book could form the basis for several courses with different emphases in several fields. Our own jointly taught course 'Modelling of Environmental and Social Systems' has been taken by students of geography, ecology, environmental science, environmental management, biology, bioinformatics, archaeology, transport and civil engineering, psychology, statistics and chemistry, among others. While we believe that building-block models have a broad and general applicability, not all of the ground that we cover will be suitable for all these disciplines. Nevertheless, we hope that even if it is not the core text this book will be a useful addition to reading lists in courses in all of these fields. We also hope that the book will be a useful resource for those researchers making their first steps in spatial modelling outside the confines of a traditional course.

There are multiple pathways through this book depending on the context in which it is being read, and the reader's background and interests. In some situations, it might make sense to fill in the background in a little more detail than Chapters 1 and 2 allow, and also to cover Chapter 6 before approaching the building blocks themselves. Alternatively, it may make sense to just dive in and postpone the 'preliminaries' until after some models have been sampled in Chapters 3 to 5. For some readers the context will make more sense after some examples have been absorbed. Last, but certainly not least, Chapters 7 and 8 will invariably make most sense with all the other material already absorbed. For some readers Chapter 7 may be rather more demanding than the earlier material but we firmly believe that there is more to spatial modelling than appealing animations. If we are to make robust inferences about real systems using spatial models then we need to evaluate them rigorously, and this is what the tools introduced in Chapter 7 allow.

The literature on each of the building-block model areas is massive and growing rapidly, and it was not our intention to exhaustively review it. In a book of this size in places our treatment is necessarily limited, but we hope that any chapter of the book can provide a good point of entry to the literature for graduate seminars, reading groups or individual study.

Using the example models

This book is accompanied by many freely downloadable NetLogo (Wilensky, 1999) implementations of the models described in the text. The models can be found at http://patternandprocess.org. All of these models are released under the Creative Commons Attribution-NonCommercial-ShareAlike 3.0 License. Most of the figures were directly generated from these models using the excellent NetLogo-R extension described by Thiele and Grimm (2010). The analyses we present were all conducted using the freely available R environment (R-Development-Core-Team, 2012), some taking advantage of the RNetlogo library (Thiele et al., 2012) which allows NetLogo to run within R and so takes advantage of the latter's analytical capabilities.

How these models are used is up to the reader. We learned a lot writing them, and we trust that experimenting with them and examining the code will be a useful adjunct to the main text, and one that helps circumvent the difficulty of conveying inherently dynamic things on a static page. The point at which each model is discussed in the text is identified by a marginal turtle icon. We suggest that readers download the model in question when they first encounter it, and experiment with its behaviour while reading the associated text. It should be possible to reproduce patterns similar to those shown in the figures that accompany the discussion of each model. Examination of the code should clarify any ambiguity over the definition of the model. It should also provide ideas for how to implement the model, and its variants, in NetLogo or other platforms.

About the Companion Website

This book is accompanied by a companion website:

www.wiley.com/go/osullivan/spatialsimulation

The website includes:
- Powerpoints of all figures from the book for downloading
- PDFs of tables from the book

1
Spatial Simulation Models: What? Why? How?

It is easy to see building and using models as a rather specialised process, but models are *not* mysterious or unusual things. We routinely use models in everyday life without giving them much thought, if any at all. Consider, for example, the word 'tree'. We may not exactly have a 'picture in our heads' when we use the word, but we could certainly oblige if we were asked to draw a 'tree'. The word is associated with some particular characteristics, and we all have some notion of the intended meaning when it is used. In effect, everyday language models the world, using concrete nouns, as a wide variety of categories of thing: cats, dogs, buses, trains, chairs, toothbrushes and so on. We do this because if we did not, the world would become an unfathomable mess of sensory inputs that would have to be continually and constantly untangled in order to accomplish even the most trivial tasks.

If you are reading this book, then you are already well-versed in using models in the language that you use everyday. We define scientific models as simplified representations of the world that are deliberately developed with the particular purpose of exploring aspects of the world around us. We are particularly concerned with *spatial simulation models* of real world systems and phenomena. Our aim in this book is to help you become as comfortable with *consciously* building and using such models as you are with the models you use in everyday language and life.

This aim requires us to address some basic questions about simulation models:

- What are they?
- Why do we need them and use them?
- How can (or should) we use them?

Spatial Simulation: Exploring Pattern and Process, First Edition.
David O'Sullivan and George L.W. Perry.
© 2013 John Wiley & Sons, Ltd. Published 2013 by John Wiley & Sons, Ltd.

It is clearly important in a book about simulation models and modelling to address these questions at the outset, and that is the purpose of this chapter.

The views we espouse are not held by every scientist or researcher who uses models in their work. In particular, we see models as primarily exploratory or heuristic learning tools, which we can use to clarify our thinking about the world, and to prompt further questions and further exploration. This view is somewhat removed from a more traditional perspective that has tended to see models as primarily predictive tools, although there is increasing realisation of the power of models as heuristic devices. As we will explain, our view is in large measure a product of the types of system and types of problem encountered in the social and environmental sciences. Nevertheless, as should become clear, this perspective is one that has relevance to simulation models as they are used across all the sciences, and becomes especially important when scientific models are used, as increasingly they are, to inform critical decisions in the policy arena.

After dealing with these foundational issues, we briefly introduce probability distributions. Our goal is to show that highly abstract models, *which make no claim to realism*, may nevertheless still be useful. It is also instructive to realise that probability distributions are actually models of a specific kind. Understanding the strengths and weaknesses of such models makes it easier to appreciate the role of more detailed models that take realism seriously and also the costs borne by this increased realism. Finally, we end the chapter by making a case for the more complicated dynamic, spatial simulation models that are the primary focus of this book.

1.1   What are simulation models?

You may already have noticed that we are using the word 'model' a great deal more than the word 'simulation'. The reason for this will become clear shortly, but in essence it is because models are a more generic concept than simulations. We consider the specific notion of a simulation model in Section 1.1.5, but focus for now on what models are.

The term *model* is a difficult one to pin down. For many, the most familiar use of the word is probably with reference to architectural or engineering models of a new building or product design. Until relatively recently, most such models were three-dimensional representations constructed from paper, wood, clay or some other material, and they allowed the designer to explore the possibilities of a proposed new building or product before the expensive business of creating the real thing began. Such 'design models' are often built to scale, necessitating simplification of the full-size object so that the overall effect can be appreciated without the finer details becoming too distracting. Contemporary designers of all kinds generally build not only *physical models*

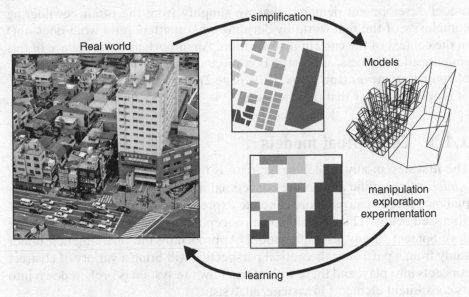

Figure 1.1 Schematic illustration of the concept of models. Models simplify the real world, enabling manipulation, exploration and experimentation, from which we aim to learn about the real world. Photograph from authors' collection.

but *computer models*, using computer-aided design (CAD) software to create virtual models that can be manipulated and explored interactively on screen. Design models then, are simplified representations of real objects that are used to improve our understanding of the things they represent. The underlying idea of model building of this kind is shown in Figure 1.1. An important idea is that more than one model is likely to be useful.

Scientific models perform a similar function—and follow the same general logic of Figure 1.1. Therefore, for our purposes, we define a scientific model as

> *a simplified representation of a system under study, which can be used to explore,*
> *to understand better or to predict the behaviour of the system it represents*

The key term in this definition is 'simplified'. In most scientific studies there are many details of the phenomena at hand that are irrelevant from the particular perspective under consideration. When we are tackling the transport problems of a city, we focus on aspects that matter, such as the relative allocation of resources for building roads relative to those for public transport infrastructure, the connectivity of the network and how to convince more people to car-pool. We do not concern ourselves with the colours of the cars, the logos on the buses or the upholstery on the subway seats. At the level at which we are approaching the system under study some components matter and others are irrelevant and may be safely ignored. The process of model development demands that we simplify from the often bewildering complexity of the real world by deciding what matters (and what does not) in the context of the current investigation. An important consequence of this simplification process, as George Box succinctly points out, is that, '[m]odels, of course, are never true' (Box, 1979, page 2). Luckily, as Box goes on to say, 'it is only necessary that they be useful'.

1.1.1   Conceptual models

The first step in any modelling exercise is the development of a *conceptual model*. *All* scientific models are conceptual models, and a particular conceptual model can be given more concrete expression as any of the distinct types discussed below. Thus, developing a conceptual model is fundamental to the development of any scientific model. Approaching the phenomenon under study from a particular theoretical perspective will bring a variety of abstract concepts into play, and these will inform how the system is broken down into its constituent elements in systems analysis.

In simple cases, a conceptual model might be expressible in words ('if parking costs more, fewer people will drive'), but usually things are more complicated and we need to consider breaking the phenomenon down into simpler elements. *Systems analysis* is a method by which we simplify a phenomenon of interest by systematically breaking it down into more manageable elements to develop a conceptual model (see Figure 1.2). A critical issue is the desired level of detail. In the case shown, depending on our interests, a forest might be simplified or abstracted to a single value, its total biomass. A more detailed model might break this down into the biomass stored in trees and other plant species, with a single submodel representing how both categories function, the difference between trees and other plants being represented by differences in attribute values. A still more detailed analysis might consider individual species and develop submodels for each of them. The most appropriate model representation is not predetermined and will depend on the goals of the model-building exercise. In this case, a focus on carbon budgets may mean that the high-level 'biomass only' model is most appropriate. On the other hand, if we are concerned about the fate of a particular plant species faced with competition from invasive weeds, then a more detailed model may be required.

In the systems analysis process, we typically break a real-world phenomenon or system down into general elements, as follows:

Components are the distinct parts or entities that make up the system. While it is easy to say that a phenomenon can be broken down into components, this step is critical and typically difficult. The components chosen will have an impact on the resulting model's behaviour so that these basic decisions

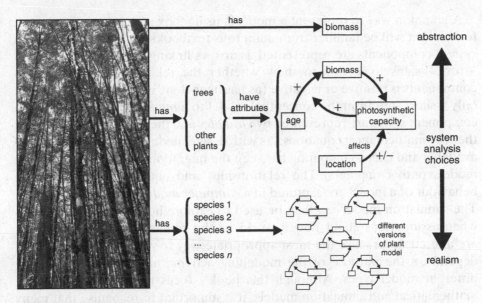

Figure 1.2   The systems analysis process. A real-world phenomenon is broken down into components, their attributes, how they interact with one another and how they change via process relationships. A particular phenomenon might be represented and analysed in a variety of ways, with the desired level of realism or, conversely, abstraction a key issue.

are of fundamental importance to how adequate a given representation will be. An important assumption of the systems approach is that the behaviour of the components in isolation is easier to understand than that of the system as a whole.

State variables are individual component or whole-system level measures or attributes that enable us to describe the overall condition of the system at a particular point in space or moment in time. Total forest biomass might be such a variable in Figure 1.2.

Processes are the mechanisms by which the system and its components make the transition from one state to another over time. Processes dictate how the values of the involved components' state variables change over time.

Interactions between the system components. In most systems not all components interact with each other, and how component interactions are organised is an important aspect of a system's structure. In many of the systems which interest us in this book, interactions are more likely and/or stronger between components that are near one another in space, making a spatially explicit model desirable.

Thus, a conceptual model of a system will consist of components, state variables, processes and interactions, and taken together these provide a simplified description of the phenomenon that the model represents.

A common way to represent a model is using 'box and arrow' diagrams, a format that will be familiar from numerous textbooks. Interactions between system components are represented as arrows linking boxes. We might add + or − signs to arrows to show whether the relationship between two components is positive or negative (as has been done in Figure 1.2). It is then only a short step from the conceptual model to a *mathematical model* where component states are represented by *variables* and the relationships between them by mathematical equations. As with design models, the advent of widely available and cheap computing has seen the migration of such mathematical models onto computers. The relationships and equations governing the behaviour of a model are captured in a *computer model* or *simulation model*. The simulation model can then be used to explore how the system changes when assumptions about how it works or the conditions in which it is set are altered. In practice the most appropriate way to represent a system will depend on the purpose of the modelling activity, and so there are many different model types. Although this book's focus is on spatially explicit mathematical and simulation models, it is important to recognise that many different approaches to modelling are possible. We consider a few of these in more detail in the sections that follow.

As we have already suggested, conceptual models can be represented in a variety of ways. Simple models may be described in words perfectly satisfactorily. Newton's three laws of motion provide a good example. The third law, for example, can be stated as: 'for every action there is an equal and opposite reaction'. Even very complicated models *can* be described in words, although their interpretation may then become problematic, and it is debatable how sensible it is for such models to remain as *solely* verbal descriptions. Nevertheless, in many areas of the social sciences, verbal descriptions remain the primary mode of representation, and extremely elaborate conceptual models are routinely presented in words alone (give or take the occasional diagram), at book length. The work of political economists such as Karl Marx and Adam Smith provides good examples. Partly as a consequence, the interpretation of these theories remains in dispute, although it is important to acknowledge that the subsequent mathematisation of economic theory through the twentieth century has not greatly reduced the interpretative difficulties: however we represent them, complicated models remain complicated and open to interpretation.

Even so, verbal descriptions of complicated conceptual models have evident limitations. As a result many conceptual models are presented in graphical form, with accompanying explanation. As we have noted, it is a short step from graphical representation to the development of mathematical models, and the success since Newton's time of those physical sciences which adopted mathematical modelling as a tool has been a persuasive argument in other disciplines for the adoption of the approach.

1.1.2   Physical models

We began by briefly touching on the three-dimensional scale models often used by design professionals. *Physical* or *hardware models* are also occasionally used in the sciences, and in engineering. Wave tanks can be used to simulate coastal erosion and wind tunnels to investigate turbulence around aerofoils or the aerodynamics of vehicles. Hardware models can provide guidance about a system's behaviour when that system is not sufficiently understood for mathematical or computational models to provide reliable guidance. Careful scaling of the model system's properties or extrapolation from the model to the target system is necessary for this approach to work well. For example, the grain size of sand in a hardware model of a beach must be carefully considered in combination with the wave heights and speeds if the results from a flume are to be applicable to real beaches. Rice et al. (2010) provide a good summary of the potential value of this approach in the specific context of river science, where they show how it can be used to relate stream hydraulics to stream ecology.

A highly specific kind of 'hardware' model is the use of laboratory animals in medical research, where mice or rats or fruit flies are considered 'model' organisms for aspects of animal biology in general. Similar to flumes or wind tunnels, this use of models recognises that our ability to mathematically model whole organisms remains limited at present, and for the foreseeable future, and these models provide an alternative. Even setting ethical issues to one side, the difficulties of generalising from findings based on such models are apparent.

An interesting hardware model of a different kind is provided by the hydraulic model of a national economy built by the economist Bill Phillips while he was a student at the London School of Economics (Fortune Staff, 1952). The so-called MONIAC (MOnetary National Income Analogue Computer) used a system of reservoirs and pipes to represent flows of money circulating in a national economy. Flow rates in different pipes in the system were adjusted to represent factors such as the rate of income tax. Several MONIAC machines were built and working examples are maintained by the Science Museum in London and by the Reserve Bank of New Zealand (Phillips was a New Zealander). While this may seem a strange way to model an economy, Phillips would have been well aware that his hydraulic model was a representation of an underlying conceptual and mathematical model.

1.1.3   Mathematical models

The MONIAC example points us towards *mathematical models*, the most widely adopted approach to scientific modelling. In a mathematical model, the state of the system components is represented by the values of *state variables* with equations describing how the state variables change over time. Newton's Second Law of Motion is a straightforward example, where the equation

$$F = m\frac{dv}{dt} \tag{1.1}$$

tells us that the velocity v of an object changes at a rate proportional to the net force on it, F, and in inverse proportion to its mass, m. Since Newton's time, many laws of physics governing the basic behaviour and structure of matter have been found to be well approximated by relatively simple mathematical equations not much more complicated than Newton's Second Law. Perhaps the best-known example is Einstein's celebrated $E = mc^2$ concerning the relationship between energy, E, mass, m, and the speed of light, c.

Such equations are simple mathematical models but taken in isolation they do not tell us much about the dynamic behaviour of systems. If we decompose a system into a number of interacting components, and equations representing the interactions can be established, then we have a mathematical model in the form of a system of simultaneous equations describing the system's state variables. This mathematical model can be used to explore how the state variables will change over time and through space, and how different factors affect overall system behaviour. How the necessary equations are determined depends on the nature of the system. The most productive approach over time has usually been the *experimental method*, where the relationships between different variables are determined by setting up laboratory experiments in which key system variables are carefully manipulated while others are held fixed, so that the independent effects of each can be determined. Experiments can be deliberately constructed to explore expected or hypothesised relationships between system variables. When the expected relationships are found not to hold, hypotheses are adjusted and new experiments designed, and over time a clear picture of the relationships between system variables emerges.

1.1.4   Empirical models

In an experiment, we artificially close the system under study so that only the single effect of interest is 'in play'. However, many systems, such as ecosystems or social systems, cannot be experimentally closed. In these situations, classic experimental methods are problematic. An alternative approach is to make empirical observations of such systems and to use quantitative methods to construct an *empirical model* of the relationships among the system variables. Such models are generally statistical in nature, with regression models the most widely used technique. An example of this approach is hedonic price modelling of real estate markets, where the sale price of housing is modelled as an outcome of a collection of other variables, such as floor area, date of construction, number of bedrooms and so on (Malpezzi 2008, provides a review of the approach). A statistical empirical model might show that other things being equal the sale price of larger houses is higher than that of smaller houses.

In empirical models, meaningful interpretation of observed statistical relationships can be challenging. In some cases, interpretations will be obvious and well-founded (for example larger houses attract higher prices because people are happy to pay more for extra space) but in others the mechanisms will be less obvious and interpretation may be controversial. It is particularly challenging to handle *interaction effects* among variables. In the case of house prices, an interaction effect would be an association between proximity to the central city and land parcel sizes. This might make it appear that small parcels are more expensive than larger parcels, when it is actually the more central location of smaller parcels that leads to the observed relationship between parcel size and price.

There are many examples of empirical models where interpretation is problematic, particularly in the social sciences. For example, what are we to make of findings relating different rates of gun-ownership in various US states to crime rates? Many interpretations are possible (and are offered), and the observed empirical relationships can only be linked back to underlying processes via long chains of argument and inference, which are not as readily tested as is possible in the experimental sciences. Such concerns have led to considerable dissatisfaction with empirical models (see, for example, Sayer, 1992, pages 175–203). The root of the problem is that empirical models do not consider the mechanisms that give rise to the observed relationships.

Nevertheless, in cases where the causal mechanisms remain poorly understood, empirical models can provide useful starting points for the development of ideas. We have a more or less stable empirical relationship: the question is why? As a result, and also due to the numerous statistical tools available for their development, empirical models are the most common type of model across the sciences, other than the ubiquitous conceptual model.

1.1.5 Simulation models

Whether experimentally, empirically or theoretically derived, quantitative relationships among the state variables in a system can be used to build a *simulation model*. In a simulation model, a computer is programmed to iteratively recalculate the modelled system state as it changes over time in accordance with the relationships represented by the mathematical and other relationships that describe the system.

The most widely used simulation models are *systems dynamics models*. Systems dynamics models consist of essentially three types of variable: stocks, flows and parameters:

Stocks are system variables that represent quantities stored in the system over time. Examples are the energy in a system, carbon in storage pools in the carbon cycle, money in the economy, population in a demographic model (as in Figure 1.3, which consists of linked population models) or workers in an international labour market.

Flows represent the movement of stock variables among different parts of the system. In a model of the carbon cycle, carbon moves from being stored in the ground into the atmosphere dependent on the rate at which fossil fuels are being used, among other pathways. In a population model, flows represent births 'flowing' into the population or deaths leaving the population.

Parameters (also called auxiliary variables) describe the overall state of the system and govern the relationships among stocks and flows, sometimes via several intervening parameters. In a carbon cycle model, the price of petrol, the propensity of consumers to drive and the rate of carbon taxation might be model parameters.

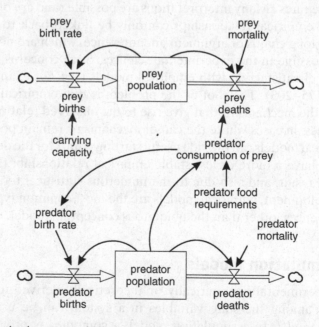

Figure 1.3 A simple systems dynamics model of the interaction between predator and prey species. Boxed variables are stocks, double-width arrows with 'valve' symbols are flows, clouds represent everything outside the system, and other named items are parameters. Arrows indicate where a relationship exists between variables.

Systems dynamics models are effectively computer simulations of mathematical models consisting of interrelated differential equations, although they often also include empirically derived relationships and semi-quantitative 'rules'. The predator–prey model in Figure 1.3 is a typical, albeit simple, example. Note that many of the probable relationships in a real predator–prey system are *deliberately* omitted from the model, for example the predator might have other potential food sources whose availability affects the level of predation. Phillips's MONIAC model referred to in Section 1.1.3 was a systems dynamics model of a national economy simulated using reservoirs and plumbing rather than a computer. The diagrammatic representation of systems dynamics models (see Figure 1.3) closely reflects the basis of such models in mathematical models of physical systems.

While systems dynamic models are among the most widely used type of simulation model, many other model types and variations are possible. Of particular interest in this book are *spatially explicit* models, where we decide that representing distinct spatial elements and their relationships is important for a full understanding of the system. Thus, a spatially explicit model has elements representing individual regions of geographical space or individual enities located in space (such as trees, animals, humans, cities and so on). Each region and entity has its own set of state variables and interacts continuously through time with other regions and entities, typically interacting more strongly with those objects that are nearest to it. These interactions may be governed by mathematical equations, by logical rules or by other mechanisms, and they directly represent mechanisms that are thought to operate in the real-world system represented by the simulation.

The simplest way to think of the relationship between models and simulations is that simulations are implementations, usually in a computational setting, of an underlying conceptual or mathematical model. Simulation is often necessary because detailed analysis of the model is difficult or impossible, or because simulation allows convenient exploration of the implications of the model. Because any simulation implements an underlying model, most of the time in this book we discuss models rather than simulations. In each case discussion of the models and understanding of their properties is only possible because we have built a simulation of the model.[*] Winsberg (2009a) provides a useful account of the role of simulation in science that clarifies both the distinction and the close relationship between models and simulations.

---

[*]Where the marginal turtle icon appears, as shown here, we have provided simulations of most of the models discussed in the text for readers to explore at http://patternandprocess.org

1.2 How do we use simulation models?

Some of the reasons for using simulation models have already been mentioned in passing. Here we expand on these ideas and present the perspective on the role of simulation models in research in which this book is grounded. The critical issue is that not all models are created equal. Figure 1.4 shows in schematic form how two critical aspects of any phenomenon, the data we have available and how well we understand it, affect our ability to use simulation models for different purposes.

Systems for which we have reliable, detailed data and whose underlying mechanisms we understand well are the ones most suitable for predictive modelling. Engineered systems such as bridges, cars, telephones and so on are good examples. We have a thorough understanding of how they work, and we have abundant and reliable data based on previous experience with using such systems. In this context, using models for prediction is worthwhile. However, in many cases we lack reliable, detailed observational data, even if we have a reasonable grasp of the underlying processes. For example, animal population dynamics are relatively well understood in general terms, but detailed rich datasets are scarce, especially for long-lived organisms. In this context, prediction is more problematic, and the best use for models may be to inform us about where critical data shortages lie. On the other hand, there are an increasing number of areas where we now have access to detailed data, but still have a poor understanding of how the system works, global financial markets being a prime example (May et al., 2008). Here, using models to help in developing theories may be appropriate. Finally, there are many fields where both the available data and current levels of understanding are poor. Here, the best prospect may be to use models as devices for learning, which

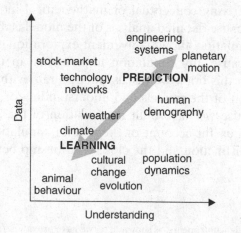

Figure 1.4 How data and understanding affect the use of models, with indicative fields of study (after Starfield and Bleloch, 1986). Note that the locations of different fields in the figure are suggestive only. See text for details.

may inform the development of theory and also steer us towards the areas of most pressing need for better data.

1.2.1 Using models for prediction

It is generally not long after a simulation model is presented to an interested audience before someone asks something like: 'So what is Auckland going to look like 25 years from now?' or 'So are polar bears going to survive climate change?' or any number of other questions about the future that are of interest. It is understandable that researchers, policy-makers and the public in general are interested in using simulation models to predict the future. After all, 'real' sciences make reliable predictions. Arguably, it was the advent of reliable predictions of the tides, seasons and planetary motion that began the human species' rise to dominance of life on Earth. Added to this, the ability of engineers to confidently predict the behaviour of the systems they design has enabled rapid technological progress over the last few centuries. Both kinds of prediction ultimately rely on scientific models. But there are good reasons to be wary of seeing models primarily as predictive tools.

We have already suggested that lack of data or of good understanding of the field in question are reasons for being cautious about prediction. If there are deficits in either dimension, it is quite likely that prediction will do more harm than good, although in such cases, it may not be prediction *per se* that is the problem. Placing unwarranted faith in inherently uncertain predictions is as likely to lead to ill-founded decision-making as having no predictive tools available at all. More circumspect approaches that consider alternative scenarios and possible responses to allow better-informed decision-making are usually a more appropriate way forward than blind faith in predictive models (see Coreau et al., 2009, Thompson et al., 2012). We consider aspects of the uncertainty around model outcomes in more detail in Chapter 7.

It is also important to be wary of the prospects for prediction in contexts where although individual elements are well understood, the aggregation of many interacting elements can lead to unexpected nonlinear effects. Perhaps the most important examples in this context are technological networks of various kinds. Here, individual system elements are designed objects, whose individual behaviour is well known. In many such networks, we also have good data on overall system usage and patterns. Nevertheless the nonlinear systems effects discussed in Section 1.3.2 have the capacity to surprise us, and there is an important role for modelling in improving our understanding of such systems (see, for example, O'Kelly et al., 2006).

1.2.2 Models as guides to data collection

Where the most pressing deficits in our knowledge of systems are empirical, so that we lack good observational data about actual behaviour, the most important role for models may be to help us decide which data are most critical to improving our understanding. Assuming that adequate models of the underlying processes in such situations can be developed, then we can perform *sensitivity analysis* (see Section 7.3.2) on the models to find out which parameters have the strongest effect on the likely outcomes. Once we know which data are most essential to improving our ability to forecast outcomes more reliably, we can direct our data collection energies more effectively.

Good examples of this sort of application of models are provided by animal conservation efforts. For endangered animals we may not have reliable data on their populations, but we do have good general frameworks for modelling population dynamics (for example Lande et al., 2003). Applying such models to endangered species with best guess estimates of critical parameters and the associated uncertainties may help conservation workers to target future data collection on the critical features of the system, and also to design improved management schemes. Unfortunately, population models applied to fisheries over recent decades provide depressing evidence of how important this is: arguably too much faith in the predictive capacity of models, and ignoring uncertainties in the models and the data used to inform them, has left many fisheries in a worse state than they may otherwise have been (see Clover, 2004, Pilkey and Pilkey-Jarvis, 2007).

1.2.3   Models as 'tools to think with'

In some cases there is no shortage of empirical data, but we have limited understanding of how a system works. The advent of sensor technologies of various kinds—remote-sensing platforms being the most obvious, but more recently including phenomena such as volunteered geographic information (Haklay et al., 2008)—has created a situation where we may have vast amounts of empirical data (Hilbert and López, 2011), but no clear idea of what it tells us. While data-mining methods may offer some relief in these situations (Hilbert and López, 2011) and provide at least candidate ideas about what is going on, science without theory is an unsatisfactory approach.

In this situation, simulation models can become 'tools for thought' (Waddington, 1977). In this approach we use simulation models that represent whatever working hypotheses we have about the target system and then *experiment on the model* to see what the implications of those theories and hypotheses might be in the real world (see Dowling, 1999, Edmonds and Hales, 2005 and Morrison, 2009). This turns the traditional hypothetico-deductive approach to science on its head. In deductive science, we make observations of the world and then evaluate hypotheses that could explain those observations. We then make controlled experimental observations to decide if our hypotheses provide adequate explanations of the observations. When we experiment on models by simulation, we effectively explore what the implications of various possible hypotheses would be. Observation of model behaviours and comparison of those behaviours with real-world observations of the target system may then allow us to dismiss some hypotheses as implausible and so refine our theories. Some authors refer to this approach as science *in silico* (see, for example, Epstein and Axtell, 1996, Casti, 1997).

This is not an easy approach. In particular, it is beset by the *equifinality* problem, which refers to the fact that large numbers (in fact an infinite number) of possible models, including the same model with different input data, could produce behaviour consistent with observational data. In other words identifying the processes responsible for a given pattern, on the basis of that pattern alone, is problematic. This makes problems of model selection and inference acute for anyone adopting this approach. We consider some of these issues in more detail in Section 7.6, where we consider *pattern-oriented modelling*, which may be particularly suited to making inferences with spatially explicit models (Grimm et al., 2005, Grimm and Railsback, 2012).

In sum, the proper role of simulation models in advancing our understanding of the world and the consequent variety of their uses is complicated and heavily dependent on the nature of the systems under study and on the current state of empirical and theoretical knowledge about them. As is often the case, it is sensible to be flexible and pluralistic in outlook, using models as one tool among many in a mixed methods approach.

1.3 Why do we use simulation models?

Simulation models are a relatively recent addition to the scientific toolbox, and the reasons for their widespread adoption call for some explanation. Broadly speaking, we believe there are three reasons that scientific research based on simulation modelling has become so prevalent in recent decades. The first is obvious and quickly dealt with: simply that this approach has now become possible. Cheap, widely available computing, although it appears commonplace from the vantage point of the early twenty-first century, is a historically recent development. The electronic computer was invented towards the end of the Second World War, and the desktop personal computer only appeared towards the end of the 1970s, and was not widely available until a decade after that. In the same way that widely available computing power has changed society generally, we might also expect it to have changed aspects of scientific practice, and simulation modelling is an example of such change.

However, given the success of science *before* simulation models, a better answer to the question 'Why use simulation models?' than 'Because we can!' is surely necessary. This leads us to two further reasons: the difficulty of experimental science in many fields of enquiry and the realisation that many systems are nonlinear.

1.3.1   When experimental science is difficult (or impossible)

The experimental method of science has proved to be extremely successful at systematically extending our understanding of the world. However, it is not always recognised that the success of the experimental method relies on certain features of the systems under study.

Some systems of interest to social and environmental scientists are not amenable to experiment for rather prosaic reasons. Archaeologists, anthropologists and palaeontologists are interested in systems that no longer exist. Climate scientists are interested in a system that is effectively the whole of the terrestrial, atmospheric and oceanic system, and hence not controllable in any meaningful sense. In many fields, such as forest ecology or geology, the time horizons of interest are too long for experiments to be practical.

Other limitations to experiment relate to the nature of the systems themselves. The most obvious consideration is that experiments only produce conclusive evidence when they are *closed systems*. Laboratory experiments go to great lengths to control all the factors that might have an effect on the phenomenon under study. For example, chemistry experiments must strictly control the temperature and pressure at which reactions are carried out, as either is likely to affect the outcome of any chemical reaction. In specific cases, many other factors will have to be controlled, such as the concentrations of the various reagents and the presence of even trace amounts of any impurities. The benefit of these efforts in experimental set-up is that it is clear when the experimenter varies a particular system input of interest that any response from the system is a result of that change. A laboratory experiment is thus a tightly controlled way to investigate the mechanisms in a system.

Such tight control is only possible for certain kinds of science. In many systems of interest in the environmental and social sciences system closure is impossible, impractical or unethical. A fragile ecosystem cannot easily be sealed off from external influences and have its climate controlled so that we can investigate the effects of a pesticide. It is generally not considered ethical to experiment with different education policies in different parts of a city to see which works, because the impact of potentially poor policy will have to be borne by some of the children going through the school system at that time. Even if such experiments were considered ethical, it is impossible to close off the education system from external variables such as the degree of home involvement in education or differences in the income, culture, language and ethnicity of the households in each school, any of which might have an equal or greater effect on outcomes than the policies being tested.

Drug trials are among the most carefully designed and constructed observational studies in the non-experimental sciences. Properly conducted trials involve an elaborate double-blind case-control methodology where a control group of patients are administered with placebo or 'sugar pill' treatment rather than the treatment under investigation. Neither the patients nor those administering the treatment know if they are part of the control group or the treatment group. Control and treatment groups are carefully selected to balance their characteristics in terms of the many demographic and physiological factors that might affect the outcome. Even so, interpretation of the results of such trials remains difficult and controversial.

It is rarely possible for field-based scientists to control the observational setting as closely as in medical trials. A fire ecologist may be able to set up (at great expense and with considerable difficulty) a study where control and treatment sites are matched and subjected to prescribed burns in order to investigate the effects of wildland fires (one example is the Kapalga savanna fire experiment in northern Australia described by Andersen et al., 2003). However, the number of observations that can be made in this way is severely limited, and the control and treatment sites are not identical—they are, after all different sites!—so that determining which differences in outcome are attributable to the treatment and which are variation in the outcomes that might have occurred anyway is extremely difficult. This is not to say that such experiments are not valuable, but that they are limited in fundamental ways that model-based approaches can possibly circumvent.

Occasionally *natural experiments* arise that provide brief windows when something like experimental results can be obtained for environmental systems. For example, when all air traffic over the United States was suspended for a period after the events of 11 September 2001, it provided an opportunity for climate scientists to collect data on the effect of aircraft contrails on the daily temperature ranges (Travis et al., 2002). But such opportunities are rare, they are not truly controlled and the observational sciences would make slow progress indeed if they relied solely on this approach.

Another possibility, rather than hoping to get lucky or trying to control the field *in situ*, is to deliberately set up laboratory experiments in the social or environmental sciences. The difficulty then becomes deciding how relevant to real systems are the findings from such artificial settings. Indeed, it can be readily argued that laboratory experiments in such cases are best understood as simplified models of the real world (see, for example, Diamond, 1983, Carpenter, 1996), so that the findings must be treated with caution.

In short, the inferential lot of observational field scientists compared with their laboratory colleagues is not a happy one. One option that simulation modelling opens up is to experiment not on real systems, but on virtual ones. If an adequate simulation model of the system of interest can be developed, then hypotheses about how the system behaves in different circumstances can potentially be investigated in a controlled manner. This approach is sometimes referred to as *surrogative reasoning* because the model is a surrogate for the real-world system. Of course, there are serious difficulties with this approach.

How can we know if the model is an adequate representation of the system? How can we map findings from experimenting with a model to conclusions about the real world? These are certainly not trivial considerations, and we consider them in more detail in Chapters 2 and 7. For now, it is sufficient to note that the potential of this approach is a major impetus for the growth of simulation modelling across various fields.

1.3.2   Complexity and nonlinear dynamics

The 'game of life' automaton   An informative way to approach complex systems and complexity is by means of one of the most celebrated examples: John Conway's 'game of life' cellular automaton. *Cellular automata* (CA) are a standard type of spatially explicit simulation model, and we will encounter many examples in this book (see particularly Chapter 3). A CA consists of a *lattice* of *cells*, which defines for each cell those other cells that are its *neighbours*. The most obvious spatial lattice is a two-dimensional grid of square cells, with neighbours defined either as the four immediately adjacent orthogonal neighbours or as the eight immediately adjacent neighbours, including the diagonals. Given the lattice structure, the full CA definition includes a set of *states* for each cell—most simply 'on' or 'off'—and a set of *transition rules* that determine how the state of each cell changes from one time step to the next, based on the current state of a cell and those of its neighbours.

A brief history of cellular automata is presented by Chopard and Droz (1998). CAs were proposed in the 1940s by John von Neumann in an attempt to devise a self-replicating machine—that is, a machine capable of creating copies of itself. Von Neumann's solution (described by Burks, 1970) was an elaborate CA with a lattice of around 200 000 cells, and 29 distinct cell states. Although von Neumann proved that a self-replicating machine was possible, his solution is so complicated that even up to the present it remains a historical curiosity that has not been implemented (although Pesavento, 1995, describes a partial implementation).

These unpromising beginnings changed dramatically with the advent of John Conway's game of life (as described in Gardner, 1970). Conway was intrigued by the question of how simple he could make the rules of a CA and still get 'interesting' behaviour. Thus, unlike the examples we consider in later chapters, the game of life is not a model of any specific system or entity. Rather, it is a mathematical system whose purely theoretical interest lies in the relationship between the intricacy of the rules that define a system's behaviour and the richness of that behaviour. The game of life CA is defined as follows:

- The lattice is a two-dimensional grid, theoretically infinite but in practice as large as needed.

- Cell neighbours are the eight immediately adjacent orthogonal and diagonal grid cells.
- Cell states are 'alive' or 'dead'.
- There are two transition rules:
  birth a dead cell is born if it has three live neighbours, otherwise it remains dead
  survival a live cell survives if it has two or three live neighbours, otherwise it dies.

The best way to appreciate the game of life CA is to experiment with a computer simulation. It does not take long to discover that the simplicity of the rules belies a rich array of dynamic behaviour. A flavour of this is provided by examining the small patterns in Figure 1.5 (see also Gardner, 1970, who presents a similar diagram). The smallest patterns, (a) and (b), unsurprisingly die immediately. Adding another live cell to produce the 'corner' pattern (c) results in a four cell block of live cells that is stable. Three live cells in a line (d) is a blinking pattern that switches each time step between a horizontal and vertical line. Adding one more cell to (d) to give the 'T'-shaped pattern (e) produces a sequence of nine time steps resulting in four copies of the three cell blinker pattern (d).

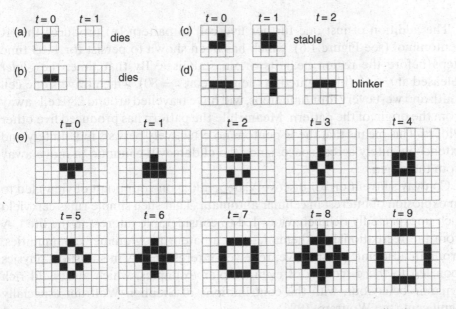

Figure 1.5 Some simple life patterns. Patterns (a) and (b) die immediately, pattern (c) produces a stable 'block', pattern (d) results in a 'blinker' that switches indefinitely every time step between two configurations and pattern (e) results in four copies of pattern (d) which blink in the same way.

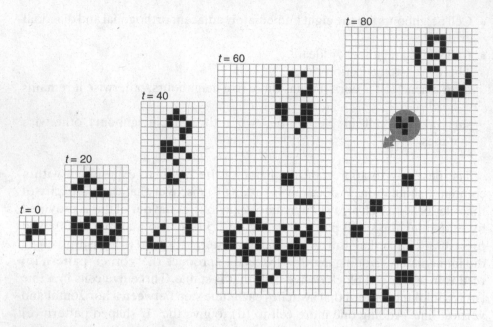

Figure 1.6 The first 80 time steps of the R pentonomino. The highlighted five cell pattern in the $t = 80$ snapshot is a *glider* which moves one grid cell diagonally in the direction shown every four time steps.

The addition of just one further live cell to pattern (e) produces the 'R pentomino' (see Figure 1.6), which has been shown to persist for 1103 time steps before the region near its origin stabilises. By that time, the *glider* released at $t = 69$ (highlighted in the image at $t = 80$), which travels one cell south and west every four time steps, will have travelled around 280 cells away from the origin of the pattern. Meanwhile the pattern has produced five other gliders. This means that the R pentomino effectively persists indefinitely and extends infinitely across space, since the gliders will continue to move away from the origin.

Conway's invention (or discovery, depending on your point of view) led to an explosion of interest in cellular automata. That such simple rules can yield such unexpectedly rich behaviour led to considerable interest in the life CA from the recreational mathematics and amateur programmer communities. From an academic perspective, renewed interest in CA in statistical physics soon led to the discovery of even simpler examples, which also yield rich dynamic behaviour, with the contributions of Stephen Wolfram especially significant (see Wolfram, 1986).

In Chapter 3 we describe other examples of the general category of *totalistic automata* of which Conway's life is an example. These examples further reinforce the important lesson of this case, that the behaviour of even simple *deterministic* systems can be unpredictable. In fact, it has been demonstrated that for life (see Berlekamp et al., 2004, pages 940–957) and other even simpler cases (see Cook, 2004) the only way to find out what state a particular system will be in after a specified number of time steps is to set up a simulation and run it!

Complexity science   It is tempting to see the game of life as little more than an obscure amusement. In fact, it is an exemplar of the sorts of systems studied in *complexity science*. Complexity science is unfortunately rather ill-defined, but roughly speaking it encompasses the study of systems composed of many basic and interacting elements (see Coveney and Highfield, 1995, page 7). On this loose definition a gas might be a complex system, so we need something more distinctive than this. In an early paper, Warren Weaver (1948) argued that the systems successfully studied by classical science were composed of either very few or very many elements. The former are usually analytically tractable, while the latter can be successfully studied using statistical methods, the statistical mechanics derivation of the gas laws being the most compelling example. Unfortunately for classical science, many real-world phenomena are 'middle-numbered' systems somewhere between these two extremes. Weaver calls these 'systems of organised complexity' (1948, page 539) where interaction effects are important—they don't simply average out as happens to molecules in a gas—and individual interactions between particular elements in one part of the system can unexpectedly scale-up to cause system-wide transitions. Weaver perceptively (and prophetically) suggested that the study of these systems would involve the use of the then recently developed electronic computers:

> [it seems] likely that such devices will have a tremendous impact on science. They will make it possible to deal with problems which previously were too complicated, and, more importantly, they will justify and inspire the development of new methods of analysis applicable to these new problems of organised complexity. (Weaver 1948, page 541)

Complexity science is a research programme along the lines envisaged by Weaver—a diverse one extending across many disciplines—which, as he anticipated, makes extensive use of computer models to understand otherwise intractable systems. It is impossible here to give a thorough account of the disparate findings of complexity science. General features of the structure and dynamics of complex systems have been characterised, including *path dependence*, *positive feedback*, *self-organisation* and *emergence*. Readers interested in learning more about complexity science should consult any of the large numbers of general and popular introductions to the field (see, for example, Waldrop, 1992, Coveney and Highfield, 1995, Buchanan, 2000, Solé and Goodwin, 2000) or more formal treatments in the academic literature (Simon, 1996, Allen, 1997, Byrne, 1998, Cilliers, 1998, Holland, 1998, Manson, 2001, O'Sullivan, 2004, Batty, 2005, Solé and Bascompte, 2006, Érdi, 2008, and Mitchell, 2008).

Nonlinear dynamics   The complexity of the life CA is an example of the broad category of *nonlinear dynamics*. Like the game of life, many nonlinear systems exhibit surprising shifts in behaviour in response to seemingly minor changes in their initial states. Linear systems, by contrast, behave more predictably, with small changes in inputs producing small changes in the response, and large changes in inputs required before large changes in the response will occur. While the systems studied in what Weaver (1948) dubbed classical science were linear, those that have increasingly come to preoccupy scientists are nonlinear. In fact, we now recognise that linear systems are a convenient idealisation of the real world that enables us to build mathematical models. In many situations, linearisation is a good approximation and a useful one (remember: *models are never true . . .*), but this is not always the case, and many systems require nonlinear models to represent them adequately. The key point to note is that nonlinear systems, because of their structure, are often more conveniently analysed by means of computer simulation models than by more traditional mathematical methods.

Chaotic systems provide another example of nonlinear behaviour and demonstrate that deterministic systems need not consist of large numbers of elements for unpredictability to emerge; deterministic systems with only a few elements can also be mathematically intractable and unpredictable (Gleick, 1987, Schroeder, 1991). Given *perfect* information about the current state of a deterministic system, *in principle* its state at any particular future time is calculable and knowable. In practice, however, chaos and complexity prevent such predictability. In chaotic systems the difficulty is that even minor perturbations—or minor inaccuracies in the data recording the initial state—can lead to arbitrarily large differences in the outcome at some later time, rendering them unpredictable for practical purposes. In fact a commonly used definition of a chaotic system is one that shows extreme sensitivity to initial conditions. In a complex system the lack of predictability arises from the unexpected ways in which, via positive feedbacks, self-organisation, lock-in and other mechanisms, local interactions among system elements scale up to cause system-wide outcomes and effects.

Beckage et al. (2011) expand on the implications of these characteristics in ecology, and similar conclusions are warranted for social, economic and cultural systems. The essence of their argument is that such systems are *irreducible*, that is, we cannot easily reduce their behaviour to aggregate rules of thumb or predict the precise outcome of a given starting configuration *even if the systems are completely deterministic*. The easiest way to think of this is that many nonlinear systems lack the quality of 'predictability'. This makes simulation an essential tool for understanding and exploring their behaviour.

1.4 Why dynamic and spatial models?

1.4.1 The strengths and weaknesses of highly general models

Many typologies of models have been developed and these tend to emphasise dichotomies, such as 'complicated' *versus* 'simple', in model design. Although such black and white dichotomies are invariably simplistic, they provide a useful way of thinking about the trade-offs that must be made when developing appropriate and manageable abstractions or representations of real-world phenomena. In a classic paper on the strategy of model building, Levins (1966) argues that model building requires the modeller to make trade-offs between generality, precision and realism. A single model cannot have all three and so the modeller has to sacrifice generality for realism and precision, precision for generality and realism, or realism for generality and precision (Perry, 2009, provides a schematic illustration of these trade-offs). In other words, a highly detailed simulation model of a specific system at a specific place and time may be realistic and useful for understanding that particular system, but will be difficult to transfer to other systems and so it is not general. On the other hand, a general framework, such as von Thünen's land-use model of urban spatial structure, may lack realism when applied to specific city systems because it glosses over the geohistorical details of the growth of particular cities, but it can inform us about the behaviour of cities in general terms.

To explore the idea of trade-offs between generality and realism more closely, in this section we discuss how we might model observational data using probability distributions such as the normal or lognormal and the highly general mechanisms that underpin them.

An additive process: the normal (Gaussian) distribution A familiar way to model observed data is to use the Gaussian (or normal) distribution, the 'bell-shaped' curve that underpins many data analysis methods (Figure 1.7). In many systems *approximately* Gaussian distributions arise as the result of a series of independent *additive* processes, as explained by the (additive) central limit theorem. The statistician Sir Francis Galton described a machine called a 'Galton board' or quincunx, which demonstrates how Gaussian distributions arise from repeated additive processes (see Figure 1.8). On the Galton board a ball is dropped from the top and on hitting a pin moves, with equal probability, either a unit to the left or a unit to the right before

1.2

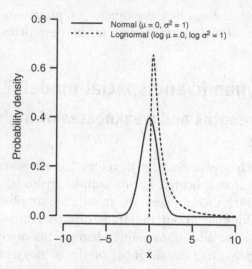

Figure 1.7   Normal and lognormal probability distributions. The symmetrical normal distribution arises from multiple independent *additive* processes, whereas the right-skewed lognormal distribution arises from multiple independent *multiplicative* processes. The normal (or Gaussian) distribution is defined by the mean (μ—the location parameter) and the standard deviation (σ—the scale parameter).

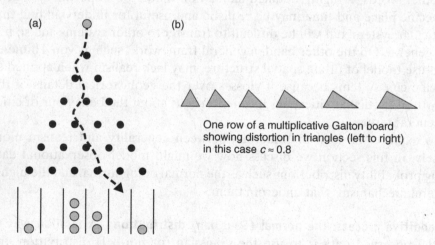

Figure 1.8   (a) The Galton board described by Sir Francis Galton. Balls are dropped and move through the system additively, eventually being collected. The heights of the piles of balls in each unit at the bottom of the board follow a Gaussian distribution. (b) One row of the hypothetical multiplicative Galton board described by Limpert et al. (2001).

eventually being collected at the bottom. The sequence of movements that each ball makes constitutes an independent additive process (there is a direct relationship between this model and a random walk in one dimension—see Chapter 4). Our interest here is in the distribution of the final position of the balls on the board, that is, the relative sizes of the piles that collect below the board. After many balls have been dropped, the size of the piles under the board approximates a binomial distribution, which, according to the central limit theorem, will tend to a normal distribution if a sufficiently large number of balls are dropped, as shown in Figure 1.9(a).

Multiplicative processes: various skewed distributions We might expect that a different distribution will arise in systems dominated by multiplicative processes, and this is indeed the case. For processes comprising many independent but *multiplicative* events, a right-skewed approximately lognormal distribution can arise (also known as the Galton–McAlister, Kapteyn or

1.3

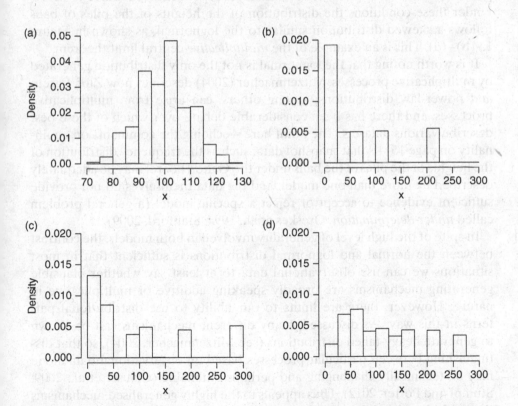

Figure 1.9 Outcomes of simulations of (a) additive and (b)–(d) multiplicative *in silico* Galton boards, with $c = 0.7$, 0.8 and 0.9 in (b)–(d), respectively. The right-skewed distributions produced by the multiplicative process become more obvious as c decreases and the strength of the multiplicative process increases. Note axes are scaled differently on (a) than on (b)–(d).

Gibrat distribution, see Koch, 1966). Thus, while the Gaussian distribution is widely used, many (perhaps most) processes in environmental and social systems are multiplicative rather than additive and so, as Heath (1967), Koch (1966), Limpert and Stahel (2011), Limpert et al. (2001), and others have argued, in many cases a right-skewed distribution, such as the lognormal, is a more appropriate model. For example, population growth is typically multiplicative (Limpert and Stahel, 2011), as is the growth of firms (Simon and Bonini, 1958) and many other things besides.

Limpert et al. (2001) describe a modification to the Galton board (Figure 1.8) that demonstrates the outcome of multiple independent multiplicative processes. In their modified Galton board, shown in Figure 1.8(b), the isosceles triangles are replaced by scalene triangles. Each triangle has a skew to the right such that at each successive level the ball moves to either $x \times c$ or x/c, where x is the current position on the horizontal axis and c is the skew to the right in the relevant triangle. This modified process is multiplicative rather than additive, but it is still independent from move to move. Under these conditions the distribution of the heights of the piles of balls follows a skewed distribution similar to the lognormal, as shown in Figure 1.9(b)–(d). This is an example of the *multiplicative* central limit theorem.

It is worth noting that the lognormal is not the only distribution generated by multiplicative processes. Mitzenmacher (2004) describes how Zipf, Pareto and power-law distributions, among others, can arise from multiplicative processes, and there has been considerable debate over which of these best describe various datasets. The point here—echoing the comments on equifinality on page 15—is that snapshot data, such as the frequency distribution of the heights of the piles of the balls under the Galton board, may be adequately described by more than one model, and the data themselves do not provide sufficient evidence to accept or reject a specific model (a general problem called *under-determination*, Oreskes et al., 1994, Stanford, 2009).

In spite of the high level of generality involved in both models, the contrast between the normal and lognormal distributions is sufficient that in most situations we can use observational data to at least say whether plausible generating mechanisms are broadly speaking additive or multiplicative in nature. However, there are limits to our ability to use distributional patterns in this way. As discussed, many different mechanisms can be shown to generate heavy-tailed distributions (see Mitzenmacher, 2004), so that distinguishing between particular processes based on observational data alone may be technically challenging and perhaps untenable (Clauset et al., 2009, Stumpf and Porter, 2012). Thus appeals to the highly generalised mechanisms that underlie probability distributions are problematic. Many different mechanisms yield similar distributional outcomes. Furthermore, the mechanisms themselves are so generalised that determining the most suitable distribution to describe a particular case can provide only limited information about systems and the processes driving them. This points to the potential value of more detailed models incorporating more specific mechanisms, although, keeping in mind Levins's (1966) trade-off, we might expect any advantages in specificity to come at a cost in generality and precision.

1.4.2 From abstract to more realistic models: controlling the cost

As we have seen, probability distribution functions are highly abstract and highly general—if we have right-skewed data, with all values greater than zero, then the lognormal distribution is (possibly) an adequate description of those data regardless of context, and we should consider how multiplicative processes might drive the system. If, however, we are considering specific patterns in specific environmental or social systems and want models that represent in more detail the processes that determine and are influenced by those patterns, such generality will be lost. Does this mean that more detailed models of particular systems cannot also teach us more general lessons about how the world works?

The key idea behind this book is that if we are alert to the similarities between patterns across diverse domains then perhaps the loss of generality as we move from abstract to more specific models can be mitigated to a degree. As an example, in many cities populations with different socio-economic characteristics are segregated with respect to each other, and conversely aggregated with respect to themselves. Likewise in temperate forests competitive interactions often result in individuals of different species being segregated from each other, and conversely they show strong conspecific aggregation. No one would suggest that the same causal mechanisms are driving these patterns, but is it really the case that there are no general ways of thinking about the processes that produce 'segregation' across the very different contexts in which it appears? Likewise, visitors to an art gallery looking for a celebrated work of art and caribou searching for food in the tundra are clearly very different, but is there any commonality in these 'foraging' processes that means we can, at least in the first instance, think about their representation in the same way?

We believe that many ostensibly different spatial patterns, arising in very different systems, can usefully be represented using similar frameworks (see also Ball, 2009). This does not mean that the processes are the same (of course they are not!) but rather that it may be useful to use similar models to understand the ways in which very different processes can produce similar patterns. Our goal is to make building and interpreting spatial models easier by showing how a relatively small number of models of spatial processes can generate many of the types of patterns seen across a wide range of social and environmental systems. Such models will not be as general as the independent additive and multiplicative processes that underlie the normal and lognormal distributions, but, with some care, we can identify more specific spatial processes that can account for diverse spatial patterns commonly observed in a wide variety of systems.

Central to our approach is the view that some broad categories of spatial outcomes, namely aggregation, movement and spread, can be accounted for by a relatively small range of relatively simple dynamic spatial models. These 'building-block' models, which are covered in Chapters 3, 4 and 5, can in turn be combined to create more complicated, detailed and realistic models of particular real-world systems (see Chapter 8).

Developing detailed, dynamic, spatial models comes at some cost in generality and interpretability, but buys us realism and the ability to represent specific processes in specific contexts. If we are willing to see the similarities in patterns and processes across many different domains, then some of those costs are offset by making the models themselves easier to understand and less daunting to work with. But before we tackle the building-block models themselves, we must first consider more closely two key concepts: *pattern* and *process*, and these form the subject matter of the next chapter.

2
Pattern, Process and Scale

The processes that are dominant in a system leave distinctive fingerprints in the form of dynamic spatial patterns. As Meentemeyer (1989, page 168) argues 'it is from spatial form that most processes are "discovered"'. That processes can be expected to leave distinctive patterns suggests that we might be able to identify process *from* pattern, and this is the (often unstated) holy grail of spatial analysis and modelling.

Unfortunately, as we saw in Chapter 1, things are not so simple. While the careful characterisation of pattern in a system can help us to identify candidate driving processes or, perhaps more realistically, rule out less important ones, it cannot unequivocally isolate the processes responsible for the observed patterns. There are also difficult issues related to the scales at which we detect and perceive patterns, how patterns change with scale and the inferences we can make at scales different from those at which we have made observations. Many of the issues surrounding scale, scaling and the inference of pattern from process apply not only to models and modelling, but also to empirical and experimental investigations, and are of wide scientific interest. Developing appropriate representations of specific processes in spatial models and then making inferences from those models requires that we take such issues seriously, and that is the subject of this chapter.

We will start by considering what we mean by 'pattern' and 'process', and the difficulties in effectively disentangling them, especially in light of the issues of scale and scale-dependence. Using a simple model of the spatial dynamics of a forest we will then explore issues surrounding how to describe patterns. As we shall see, meaningfully describing patterns requires the use of *null models*, which tell us the patterns that we might expect given some prior assumptions about the spatial character of the processes that produced them. That developing meaningful descriptions of patterns requires close consideration of process demonstrates just how closely interrelated these

Spatial Simulation: Exploring Pattern and Process, First Edition.
David O'Sullivan and George L.W. Perry.
© 2013 John Wiley & Sons, Ltd. Published 2013 by John Wiley & Sons, Ltd.

concepts are. This chapter concludes with a brief look at some of the pitfalls of inferring process from pattern. In particular, we look more closely at the difficult issues of equifinality and under-determination, which bedevil such efforts and which are considered in more depth in Chapter 7.

2.1   Thinking about spatiotemporal patterns and processes

2.1.1   What is a pattern?

'Patterns are regularities in what we observe in nature; that is, they are 'widely observable tendencies'' (Lawton 1999, page 178). Lawton goes on to argue that patterns in natural and social systems arise from the fundamental physical principles that control all systems, such as the laws of thermodynamics, Darwinian selection and so forth. In short, patterns are the discernible outcomes or signatures of the processes operating in a given system. In the more specific context of simulation models, Grimm et al. (2005, page 987) define patterns as '. . . defining characteristics of a system and often, therefore, indicators of essential underlying processes and structures. Patterns contain information on the internal organization of a system, but in a 'coded' form.' Similarly, Unwin (1996, page 542) defines pattern as 'that characteristic of the spatial arrangement of objects given by their spacing in relation to each other'. These definitions are very broad and span data including time series, maps, frequency distributions, rates of change, single numbers and many more, as shown in Figure 2.1.

In this book we are specifically interested in spatial patterns that change over time, that is, they are *dynamic*; nevertheless, it is easier to consider the intricate relationship between pattern and process by first considering static patterns. Ball, who is also particularly focused on *spatial* patterns, notes that the notion is difficult to rigorously define, but suggests that 'a pattern is a form in which particular features recur recognizably and regularly, if not identically or symmetrically' (Ball 2009, page 20). For our purposes regularity is not necessarily present but it is important that the features of the pattern are explicitly related to one another in space (as in Unwin's definition). This does not restrict the spatial patterns we are considering to geographical spaces. They might, for example, be on social networks. It also means that a pattern encompasses both the space in which the entities are embedded (or located) and the entities themselves.

Spatial patterns can be represented using a range of data models (see Section 2.2 and Chapter 6). Banerjee et al. (2003) distinguish point-referenced (geostatistical), areal and point process data, but the utility of a spatial pattern for informing model development is not related to the types of spatial objects it contains. In this chapter we focus on point patterns for convenience, but

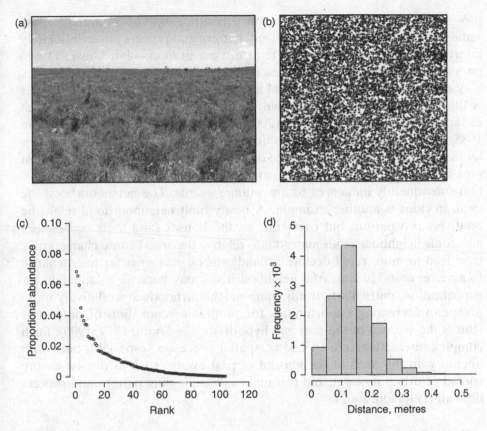

Figure 2.1   Examples of patterns from a 30 × 30 metre plot in a high-diversity shrubland in Western Australia containing more than 100 species and more than 12 000 individuals. The plot is depicted here as (a) a photograph of the site, (b) a map of the (x, y) location of the individual plants in the plot, (c) the relative abundance of each species in rank order (the species–abundance distribution) and (d) a frequency histogram of the distances from individuals to their nearest neighbours.

the general themes that we emphasise hold across all forms of spatial data. Fortin and Dale (2005), O'Sullivan and Unwin (2010), Gelfand et al. (2010) and many others provide more background in the area of spatial analysis.

2.1.2   What is a process?

As Meentemeyer's quote opening this chapter emphasises, the spatial and temporal patterns that we observe in the systems we are investigating are the outcomes of the processes operating in those systems. For our purposes a *process* is any mechanism that causes a system to change its state, and so potentially to produce characteristic patterns. Scientific inquiry tends to emphasise how processes generate patterns, but it is important to recognise that there are feedbacks in both directions. Patterns are the outcomes of processes, but they also act, in their turn, to either suppress or amplify those same processes (Turner, 1989). Indeed, pattern and process are so intimately intertwined that their definitions tend to be a little circular. Unwin (1996) goes so far as to suggest that process is, in fact, a component of pattern!

A good example of the reciprocal interaction between pattern and process is the spread of wildfires, as shown in Plate 1. Large fire complexes, such as those that burned more than $3\,200\,km^2$ in Yellowstone National Park in 1988, create a heterogeneous mosaic of vegetation that is burned more or less severely. Sites that experienced different severity burning show different post-fire establishment patterns, leading to a heterogeneous forest structure that subsequently influences future wildfire events. The neighbourhood life cycle in cities is another example. A newly built neighbourhood might be relatively prosperous, but over time as the houses (and their occupants!) age, some neighbourhoods may go into relative decline. Tenure changes may then lead to more rapid decline as landlords neglect or defer maintenance to save on costs. In time, the neighbourhood may become a candidate for gentrification, particularly if adjoining neighbourhoods are relatively more prosperous, creating opportunities for profitable renovation of properties (this is the essence of the rent gap hypothesis; see Smith, 1979, 1996). Such complex interactions are typical of spatial processes, especially ones with 'memory' in the form of the imprint of past events, such as disease or fire spread or urban renewal, and this makes disentangling pattern and process difficult (Peterson, 2002).

2.1.3 Scale

Any discussion of pattern and process and their interactions requires us to think carefully about the vexing issue of *scale* (Lock and Molyneaux, 2006, Levin, 1992, Urban, 2005). Scale colours any consideration of pattern and process—it affects how we perceive pattern, how we represent process and how each affects the other.

The meaning of scale depends on the context in which the term is used (Meentemeyer, 1989, Englund and Cooper, 2003). Schneider (2009) distinguishes between measurement scale, which relates to the nature of a variable; cartographic or geographic scale, which is the ratio of distances on a map to distances in the real world; and spatial/temporal scale, which refers to the resolution of data in space/time relative to the extent of those data. He then goes on to adopt as a general definition, 'scale denotes the resolution within the range or extent of a measured quantity' (Schneider, 2009, page 22). When referring to scale we will roughly follow Schneider's definition, although we note that, confusingly, what a geographer or cartographer calls large scale is what in many other disciplines would be referred to as small (or fine) scale and *vice versa*. Irrespective of semantics, the concept of scale has risen to prominence across the social and environmental sciences over recent decades (Lock and Molyneaux, 2006, Schneider, 2009), although in geography concerns about scale, alongside efforts to adequately conceptualise space and time, have long been prominent (Meentemeyer, 1989).

In the context of spatial analysis and modelling, scale can be decomposed into *grain* and *extent*. Spatially, grain refers to the resolution of the data. The pixel size in remotely sensed imagery is an example. Grain defines the finest patterns we can perceive. If we collect data at a grain of (say) 30×30 m, then we cannot say anything about spatial heterogeneity at resolutions smaller than that. Spatially, extent refers to the total area that the dataset spans. Temporally, grain is the frequency of the data, generally controlled by how often measurements were taken, while extent is defined by the duration over which the data were collected. The spatial and temporal extent place bounds on our models, or on our data, and affect our ability to generalise from them. If we collect data over a decade for one city it will be difficult or impossible to use them to make statements about change over centuries for cities in general.

The crux of the scaling problem for environmental and social scientists is that 'nature'—the real world—is *both* fine-grained *and* of large extent. For example, when considering problems such as climate change we need to consider and integrate across spatial scales from the cellular to the global and temporal scales from the millisecond to the multi-millennial. It is logistically challenging, even in an age of remote (high-resolution) data capture, to collect very fine-grained information over large extents. This forces us to trade-off grain against extent, so that we can only have fine-grained information over small extents or, if we want to work over large extents, then we must accept the loss of detail that comes with coarser-grained data (Urban, 2005).

With access to burgeoning computing power and software tools, it is fair to ask whether developing highly detailed, fine-grained simulation models that cover broad extents provides a way of circumventing the grain–extent trade-off. To some degree this has started to happen. Parker and Epstein (2011), for example, outline the framework for an agent-based model (EpiSIM) that simulates the progress of epidemics through the global human population at the *individual* level, hence potentially considering billions of people. Certainly this is an impressive achievement from a technical standpoint. On the other hand the amount of data produced by such models will be overwhelming, and will almost certainly require aggregated macroscopic analyses for the results to be interpretable. This begs the question of how useful or necessary such detailed models really are. It also suggests that we run the very real risk of substituting the 'black box' of the system we are trying to *simplify* for another, in the form of a complicated model that we find equally difficult to understand (but that we know to be 'wrong' in that it omits many aspects of the real system). In our view, advances in computing power and software engineering *do* promise new avenues for simulation modelling, but it is not obvious that 'solving' scale–grain–extent dilemmas simply requires the use of bigger computers!

Scale-dependence and patterns   Pressing environmental problems such as elevated nitrogen deposition, ocean acidification and urban sprawl are global in extent, but typically the data we have describing them are local (catchment level or smaller) and limited in their temporal extent (decadal or less). Thus there is often a significant disconnect between the scale of the processes of interest and the scale of the observational data (Blöschl and Sivapalan, 1995). In such cases we would like to be able to extrapolate, or interpolate, data from one scale to another, whether for the purposes of description or inference. Moving information between scales in this way is called *scaling* or *rescaling*. As Englund and Cooper (2003) point out, the underlying premise of experimentation, which typically occurs at scales smaller than the systems being investigated, is that we can effectively scale data.

Unfortunately (see Figure 2.2), scaling is beset by a number of difficulties and challenges (Meentemeyer, 1989, Levin, 1992, Blöschl and Sivapalan, 2.1 1995). As we zoom in and out of a given pattern, the details that we see change. This phenomenon is termed *scale-dependence*, and means that the patterns that we perceive in a given system are a function of the space–time scale at which we view it (Levin, 1992, Englund and Cooper, 2003). For example, the human perspective on change in environmental and social systems is strongly determined by the fact that we are, on average, about 1.7 m tall and live for 70 years or so. An important corollary of scale-dependence is that there is *not* a single 'correct' scale for data collection or model construction (Levin, 1992, Urban, 2005). We are free to choose the scale(s) at which we wish to operate, and the appropriate scale depends on the questions we want to ask with our model(s).

As an aside, it is worth pointing out that some patterns do *not* change with scale—that is, they are not scale-dependent (Figure 2.3). Such geometric constructs are termed *scale-invariant, self-similar, self-affine* or *fractal*, wherein 'each portion can be considered a reduced scale portion of the whole.' (Mandelbrot, 1967, page 636). Fractal objects of one sort or another have been described in a wide range of ecological (Brown et al., 2002), chemical/physical (Meakin, 1990) and social systems (Goodchild and Mark, 1987, Batty and Longley, 1994). Gouyet (1996) provides an overview of the concepts underpinning the geometric properties of fractals and their measurement, and Halley et al. (2004) provide a list of the conditions under which they might arise in natural systems. Although it is not an issue that we focus on in depth, many of the models introduced later in the book produce approximately fractal patterns (in space and/or time) and considerable interest has centred on describing and estimating their fractal properties.

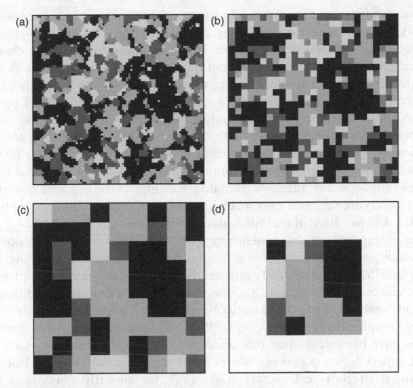

Figure 2.2 Examples of how changing grain and extent affect pattern. (a) The original pattern on a lattice of 81 × 81 cells with five distinct types (the grey shadings). (b) and (c) The effects of coarsening the grain by factors of three and nine, respectively, using a majority rule such that in the coarser grains cells take on the most common value of the original pattern. (d) A change in extent, in this case just the inner 5 × 5 cells of (c) are shown. Note how as we move from (a) to (c) more and more of the fine spatial detail is lost, and some of the smaller patches disappear altogether.

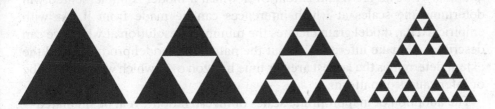

Figure 2.3 Four iterations of the Sierpiński triangle (or gasket), a well-known pure fractal object with a structure that repeats at all scales and so does not show scale-dependence. Many real-world objects, such as coastlines, mountain ranges, drainage systems and cities, can be shown to be approximately fractal.

Scale-dependence and processes　In the same way that patterns are usually scale-dependent, so too are processes. A particular process will likely reveal its effects most strongly across a limited range of scales, and so as we change the scale at which we look at a system the dominant suite of processes will shift. Unfortunately, scale-dependence in patterns does not necessarily easily map onto the scale-dependence in processes, and as Urban (2005) points out we have a much better understanding of how patterns change with scale than we do processes. Processes that occur rapidly but whose effects are felt much more slowly are particularly difficult to pin down. Schneider (2009) discusses the example of genetic mutation, which at the mechanistic level happens almost instantaneously (the sudden change in the genomic material) but its effects are revealed over much longer (up to macroevolutionary) time scales. Likewise, Blöschl and Sivapalan (1995) note that some hydrological processes have effects that potentially span time scales from sub-second to supra-millennial. Similarly, in his account of the development of Los Angeles, Davis (1990b) shows how small events, in the context of larger economic forces and tendencies, can, over time, produce outcomes at a dramatically different scale, in the shape of one of the world's largest metropolises. To simplify some of this complexity, at least conceptually, tools such as space–time domain figures have been developed that attempt to isolate the scales at which we might expect to see a process operate and the organisational level of the patterns it affects (see Delcourt et al., 1983, for an early example in the context of Quaternary vegetation dynamics).

Scaling, scale-dependence and inference　Thinking about scale-dependence in the design of (field) experiments and when collecting empirical data is crucial (Schneider, 2001, 2009). However, even if scale-dependent effects are more frequently considered in the context of data collection and analysis, decisions about scale are no less important in the context of model design and analysis. The grain and extent over which a model is implemented will determine the scales at which inferences can be made from it. As with empirical data, model grain dictates the minimum resolution at which we can describe and make inferences about the patterns a model produces, and the extent determines the largest area or time horizon over which we can describe or infer causes of pattern.

　　The adoption of inappropriate scales of investigation—real or simulated—can have damaging consequences. In a survey of 16 multi-scale experiments looking at trophic interactions, Englund and Cooper (2003) found that ten showed strong evidence for scale-dependence such that the results of the studies varied with scale. Such outcomes suggest that transferring experimental, empirical or simulation results and associated understanding across scales will rarely be easy, and points to the importance of scale-related decisions in model conceptualisation, building and analysis. It also highlights the necessity to be clear about why the particular spatial and temporal scales in a model were chosen, and to be honest about the limits that these choices place on subsequent inferences about the system under study. Considerable effort has been put into developing a theory (or theories) of scaling, with the goal of enabling prediction of how patterns change with scale. Such a predictive framework would allow the (re)scaling of experimental, observational or modelling outcomes from the scale of investigation to any desired scale. Heuristic rules of thumb have been suggested (see, for example, Wiens, 1989), but generalities are elusive other than for specific cases such as allometric scaling (Enquist and Niklas, 2002, Schneider, 2009) and even then they tend to be descriptive rather than underpinned by a mechanistic understanding (Urban, 2005).

Meentemeyer (1989, page 168) identifies three broad types of inferential error commonly made when transferring information (and/or understanding) across scales. While he is speaking broadly in the context of empirical data these concerns are equally valid for models and their interpretation:

individualistic fallacy is the inappropriate use of finer grain data to make macro-level inferences (problems of interpolation or up-scaling)

cross-level fallacy is the inappropriate use of data from one part of a population to make inferences about another (problems to do with transferring information between systems at the *same* scale)

ecological fallacy is the inappropriate use of macro-level data to make inferences about finer grains (problems of extrapolation or down-scaling).

Such concerns are highlighted by considering a very simple example, shown in Figure 2.4. Scale-dependent effects are often both nonlinear and nonadditive, which means that we should not take data at one scale and simply multiply or divide it to estimate some parameter of interest at another scale. Imagine we collect a map of plants in a plot—or households in a city block, or archaeological artefacts at an excavation, the difficulties are the same. We then repeat the exercise in another plot for comparative purposes. Thinking about just one of the plots, heterogeneity and scale-dependence mean that an estimate of density in a sub-plot is not necessarily applicable to the whole plot in which it is contained. This makes it problematic to estimate the number of individuals in the larger plot from the smaller sub-plot, and is an example of nonlinearity. On the other hand we can be certain that the number of individuals (plants or households or artifacts) in one plot added to the number in a second plot will be the sum of the total in both plots because such counts are *additive*. On the other hand, the 'richness', that is the number of plant species (or household types or artifact classes) contained in the two plots, is not necessarily the sum of the types contained in the two plots because this is a *nonadditive* measure.

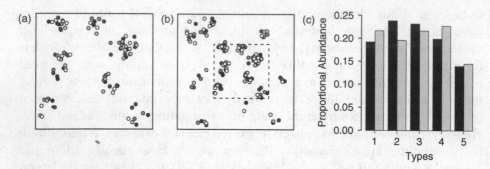

Figure 2.4 Data collected at one grain and extent are not always easily transferred to other scales. Here we have a map of trees from five different species (represented by the different shades of grey) collected in a forest. (a) is a sub-window 25% of the area of the larger plot in (b). The spatial density in (a) is more than twice that in (b), an example of nonlinear scale-dependence, and this means that estimating the total number of individuals in the larger plot by multiplying the number in the smaller area by four may not be sufficient. Both (a) and (b) contain five species, although their rank abundance does change (c). This is an example of non-additivity.

Thus, even operations as seemingly simple as up-scaling from a smaller to a larger plot or comparing one plot to another of the same size are rarely trivial. Actual scaling operations are usually much more complex than this rather simple case; for example they may involve taking ecophysiological measurements, such as photosynthetic rates from a small area, and hoping to deduce landscape or regional consequences in terms of (say) total carbon sequestration. While such attempts may be the best we can do, we must be aware of the difficulties.

In short, decisions about scale are absolutely critical for making appropriate decisions about model representation ('What processes shall I include?') and also about interpretation of model outcomes ('What patterns am I seeing, and what do they tell me?'). Models are thus subject to scale-dependence in how processes are represented, and also in the patterns of model inputs and outcomes (see Heuvelink, 1998, for an extended discussion). Spatio-temporal scale needs to be considered at all stages of model development, *particularly at the outset*.

2.2 Using models to explore spatial patterns and processes

When thinking about how patterns are represented in models, a useful distinction can be made between phenomenological and mechanistic models, although as with most dichotomies this one is not rigid. According to Bolker (2008, page 8) *phenomenological* models, 'concentrate on observed patterns in the data, using functions and distributions that are the right shape or sufficiently flexible to match them'. Phenomenological approaches emphasise reproducing observed patterns and not the mechanisms responsible for generating them. *Mechanistic* models, on the other hand, 'are more concerned with the underlying processes, using functions and distributions based on theoretical expectations' (Bolker 2008, page 8). Despite phenomenological models lacking mechanism, their careful and effective use can guide us towards the mechanisms responsible for the patterns operating in a given system.

In the following section we make use of statistical point process models to provide a more concrete example of the phenomenological approach. While point process models are typically static they usefully illustrate: (i) the use of phenomenological models and (ii) how static 'building-block' models can be used to develop richer and more dynamic ones.

2.2.1 Reciprocal links between pattern and process: a spatial model of forest structure

In (some) species-rich forests, the strength of local aggregation of a species is negatively correlated with its abundance (Condit et al., 2000). In other words, rarer species seem, for reasons that are not well understood, to be more aggregated than more common ones. Building on this observation Flügge et al. (2012) argue that snapshot patterns of the spatial distribution of a species might provide clues to its demography, and they developed a spatial model to explore this hypothesis. The model presented by Flügge et al. (2012) is deceptively simple. At the outset n individuals of a single species are placed at random in the model space. One model time-step (or generation) consists of selecting n pairs of individuals at random. The first member of the pair dies and is removed. The second member of the pair is a 'parent' and reproduces by adding a new individual to the population at a random distance and direction from itself, with the distance being drawn from a negative exponential distribution.

Following Condit et al. (2000), Flügge et al. (2012) measured aggregation as the neighbourhood density function (NDF) at a short distance $\Omega_{a,b}$. This is the density-normalised average number of individuals within an annulus of inner radius a and outer radius b away from the focal individual (Perry et al., 2006). We will refer to distance b in the following as the 'neighbourhood' size and we will usually assume that $a = 0$. The NDF quantifies the average pattern that an individual tree might experience in its neighbourhood. For a random pattern the NDF will be close to 1.0, with values above 1.0 suggesting aggregation and those below 1.0 suggesting segregation of the individuals.

Despite its simplicity, the model reproduces the negative correlation between abundance n and local aggregation $\Omega_{a,b}$ (Figure 2.5). After around

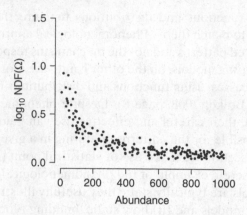

Figure 2.5 The relationship between aggregation Ω and abundance n, based on 200 model simulations each lasting 200 generations with abundance a random deviate from a discrete uniform distribution U[2,1000], mean dispersal distance of five and neighbourhood radius (b) of ten. The model space is 100×100 units.

200 generations the model settles down to a stable spatial pattern. Figure 2.6 further suggests that, over time, the virtual forest becomes more aggregated than its initial random pattern. Heuristically, this dynamic is easily understood. Starting from a random pattern individual trees are selected and removed, and are replaced by a new plant whose location is likely to be near another individual, as controlled by the parameterisation of the negative exponential distribution used to describe seed dispersal. Given many such birth–death events the pattern will become more aggregated before stabilising when all (or most) of the existing individuals are products of the the model's birth process.

2.2.2   Characterising patterns: first- and second-order structure

Given the array of spatial dynamics and patterns that even a simple model can produce, we need to be able to effectively describe and synthesise these patterns if we wish to make inferences about the processes underpinning them. There is a rich conceptual literature and a wide range of associated quantitative tools with which to characterise spatial patterns (see O'Sullivan and Unwin, 2010, for a survey), particularly point patterns. In this section we outline how these tools can help us to understand the interaction between patterns and processes, drawing on the model presented above. Although they are static, point patterns provide an entry point into the simulation of dynamic spatial patterns and processes, which is the goal of our treatment here. This discussion also shows how effective description of static patterns is an important step in developing dynamic spatial models. Note that although

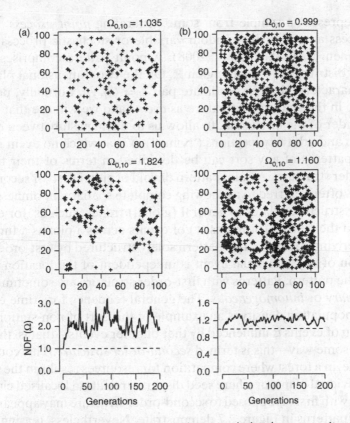

Figure 2.6 The relationship between aggregation Ω and abundance n as it emerges from the spatial birth–death process, with mean dispersal distance of five and a neighbourhood radius of ten in a 100×100 model space, for populations of (a) $n = 200$ and (b) $n = 800$ individuals. Graphs are, from top to bottom, the initial spatial pattern, the spatial pattern after 200 generations and change in Ω over time. Note that the vertical axis scaling differs between the lowest pair of figures. Dashed lines on the lower figures indicate the expected value of the NDF for a completely random pattern.

point patterns and their analysis are central to this section we do not intend to provide a comprehensive introduction to point pattern and process analysis and modelling (see Diggle, 2003, Illian et al., 2008, Gelfand et al., 2010, for this). Rather we are using point pattern analysis methods as a vehicle for outlining a general analysis strategy.

Point patterns are one of the simplest, and among the most frequently collected and encountered, forms of spatial information (O'Sullivan and Unwin, 2010), and provide a simple *data model* of the spatial processes that they represent. Points can represent objects as diverse as trees in a forest, cell-phone tower locations, cases of disease, the location of groundwater bores, fire ignition points in a city, earthquake epicentres, crime locations, archaeological sites, galaxies and so on. Statistically speaking, observed point patterns represent a sample from some generating *point process*, with the point process an underlying random variable or stochastic process (Stoyan and Penttinen, 2000, Illian et al., 2008). A point pattern comprises a set of points $\{p_i\}$ distributed in some region $\mathbb{R}$. On the two-dimensional plane each point is characterised by a coordinate pair $s_i = (x_i, y_i)$. Formally, points are referred to in the statistical literature as *events* and we will use that term for the remainder of the discussion, to allow us to distinguish between events in the pattern and arbitrary locations ('points') in $\mathbb{R}$ at which no event is found.

Spatial patterns of any sort can be described in terms of their first- and second-order structure. A point pattern devoid of either first- or second-order structure is often referred to as showing complete spatial randomness (CSR). *First-order* structure relates to general (global) trends, such as, for example, variation in the density (or intensity) of events across a plot as a function of some underlying covariate. In a pattern solely structured by first-order effects the location of any individual event is independent of the location of other events in the pattern. Patterns with first-order structure are sometimes called *non-stationary* or *inhomogeneous*. The general tendency for crime events to vary with population density is an example of this sort of non-stationarity. If the location of events *is* influenced by that of other events—that is, the events interact in some way—this is termed *second-order structure*. This could arise, for example, in a forest where competition for resources results in the segregation of individual plants or where seed dispersal results in localised clustering.

Patterns with first- as opposed to second-order structure may appear similar, as the two patterns in Figure 2.7 demonstrate. Nevertheless, teasing the two effects apart, while difficult, is important because the types of mechanisms generating the two effects are different and so the inferences we might draw about generating process are also different. For example, localised aggregation of species in a forest could arise from first-order effects driven by

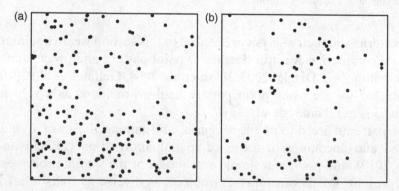

Figure 2.7 Examples of aggregated patterns. (a) First-order structure, with intensity (λ) varying across the plot from south-west to north-east. (b) Second-order, but not first-order, structure, with points 'attracted' to each other and forming distinct clusters.

heterogeneity in microhabitat suitability or second-order effects arising from dispersal or, most likely, a combination of the two. If we can characterise a pattern by separating out first- and second-order components, then the direct link to the *types* of processes responsible might allow us to rule out some possibilities even if we remain unable to isolate the specific generating mechanisms responsible for the pattern.

Many summary statistics have been developed to describe spatial point patterns (Diggle, 2003, Illian et al., 2008), ranging from the very simple, such as mean nearest-neighbour distances (Clark and Evans, 1954), to more complicated multi-scalar measures such as the NDF. As summaries, all such measures have their strengths and weaknesses in how they capture the information embedded in the spatial pattern, and in a formal analysis it is often a good idea to use more than one (Perry et al., 2006). In a general sense, first-order features of a pattern are characterised by the intensity of the underlying point process, which can be *estimated* as the number of events divided by the area. Second-order (covariance) structure can be described using a summary statistic such as Ripley's K-function (Ripley, 1977)

$$\hat{K}(r) = \frac{1}{\lambda n} \sum_{i \neq j} I(r) \qquad (2.1)$$

where n is the number of events in the process, r is the separation distance being considered, and I is an indicator function, such that

$$I = \begin{cases} 1 & \text{if } d_{ij} \leq r \\ 0 & \text{if } d_{ij} > r \end{cases} \qquad (2.2)$$

where d_{ij} is the distance from event i to event j. Under complete spatial randomness, $K(r)$ is given by

$$K(r) = \pi r^2 \qquad (2.3)$$

If $\hat{K}(r) > K(r)$ then it suggests that the pattern is more aggregated than expected under CSR up to distance r and *vice versa*. The NDF introduced in Section 2.2.1 is in essence a noncumulative form of the K-function.*

2.2.3 Using null models to evaluate patterns

As the previous section suggests, we often want to know how well our data match what we would expect if a given null model described the data adequately. For example, if we collect some point data in the field we may want to ask whether they depart from a null model such as CSR. We can address this question by comparing some quantitative summary of the

---

*The NDF is also sometimes referred to as the pair correlation function (PCF; Stoyan and Penttinen, 2000) but following Flügge et al. we will refer to it as the NDF (Ω).

observed data with that expected under our chosen null model. For simple summary statistics (for example, mean nearest-neighbour distances) we may be able to answer this question analytically (see O'Sullivan and Unwin, 2010, for details), but for measures such as Ripley's K-function we must resort to simulation-based approaches as follows (see Figure 2.8):

(i) Calculate the summary function (e.g. the K-function) for the observed data.
(ii) Simulate n patterns following CSR and compute the summary function for each. n should be at least 99, but the more simulations the better.
(iii) Use the n simulations to generate confidence envelopes at each distance by ranking the simulated values and using those that correspond with the desired error rate. Thus, if $\alpha = 0.05$ and $n = 999$ we would use the 25th and the 975th ranked values (assuming a two-tailed test).
(iv) Assess where the observed summary function lies in relation to the simulation envelope. If it falls outside the envelope then we may conclude that the pattern departs from the null model (subject to some subtleties that we need not consider here, but see Loosmore and Ford, 2006, Perry et al., 2006).

For our example data (Figure 2.8) we start by estimating the K-function on the basis of the observed data. We then produce (simulate) many realisations of patterns following a homogeneous Poisson (CSR) process and calculate the K-function for each of these. Finally, we compare the K-function of the observed data with the envelope of values produced by the simulations. Repeating this analysis for the spatial outcomes of Flügge et al.'s model and using the NDF as a summary statistic (Figure 2.9) we can see that the pattern is more aggregated than we would expect under a null model of CSR, especially at shorter distances—this is due to the localised dispersal dynamic.

This general strategy is known as *Monte Carlo simulation* (see also Section 7.3.3) and is not restricted to point pattern (or even spatial) analysis. If we can develop a null model, which might be one that simply leaves out the process of interest, we can use the Monte Carlo approach to assess how closely observed (or simulated) data match the expectations of that null model.

Deciding on an appropriate null model is important: in spatial analysis a model of CSR is often used but this is a limited approach (Diggle, 2003). For example, rejection of CSR in the direction of aggregation does not allow us to separate first- and second-order effects, since *neither* is present in CSR. If we want to tease such effects apart we need to use a null model that includes first- but not second-order structure. Rejection of that null model would allow us to evaluate the extent to which first-order structure alone accounts for any observed aggregation. A second key step in this procedure is deciding on a summary measure of the data to compare with the null model. Ideally we want

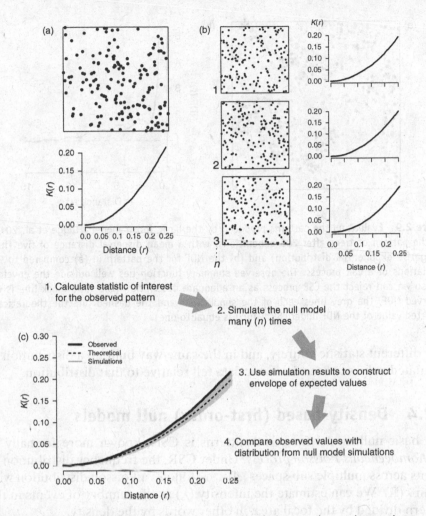

Figure 2.8 A schematic framework showing how simulation can be used to estimate how well a given pattern follows some expectation or null model, in this case complete spatial randomness. (a) The observed pattern and its K-function, (b) the simulations of CSR used to estimate the confidence envelope, with $n = 999$, and (c) the simulations compared to the observed data. In this case, the observed pattern's K-function lies outside the range of the simulated results shown by the grey lines: it is higher than all but one simulation of CSR, and so we might conclude that the CSR model does not adequately describe these data.

a statistic that effectively summarises all the detail in the data—a so-called *sufficient statistic* (Wood, 2010, Hartig et al., 2011)—but inevitably some information is lost in the summarisation process. Nevertheless, the use of a summary measure is necessary because it is impractical to compare all the information we have when using complex simulation models that generate dynamic spatial outcomes (see Chapter 7 for more on this). In the example discussed in Section 2.2.1 we used the NDF, but we could have used some derivative of it,

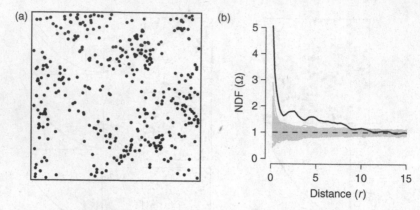

Figure 2.9 Evaluating the patterns produced by the basic forest model (Flügge et al. 2012): (a) the pattern of trees after 200 generations, with a mean dispersal distance of five (from a negative exponential distribution) and (b) the NDF for the pattern in (a) compared to 999 simulations of a CSR process. The observed summary function lies well outside the envelope and so we can reject the CSR process as an adequate description (the solid black line is the observed NDF, the grey lines each of the simulations and the dashed line the theoretically expected value of the NDF under CSR, which is equal to one).

or a different statistic entirely, and in the same way built up a distribution of simulated values to see where our data fell relative to that distribution.

2.2.4   Density-based (first-order) null models

The basic null model for point patterns is CSR, known more formally as the *homogeneous Poisson process*. Under CSR, the frequency distribution of events across multiple sub-spaces $|A| \subset \mathbb{R}$ follows a Poisson distribution with mean $\lambda(\mathbb{R})$. We can estimate the intensity (λ) as the number of events in the pattern divided by the focal area, in other words by the density.

Simulating a realisation of the homogeneous Poisson process is straightforward: simply draw an (x, y) coordinate pair for each event from a uniform distribution. In general, however, we expect the underlying intensity of the process to vary across space. For example, plants may respond to variability in soil nutrient and water availability so that their density varies in space even if they are not directly interacting, or non-contagious diseases may vary in localised prevalence as a function of the local population density. This spatial structure may take many forms from simple linear gradients to more complex structuring. We can simulate realisations of inhomogeneous Poisson processes reasonably easily as follows:

(i)   Start by calculating λ_{max}, that is, the maximum value of the desired intensity function $\lambda(x, y)$ at any location $\mathbf{s} \in \mathbb{R}$.

(ii) Generate a homogeneous Poisson process with $\lambda = \lambda_{max}$.

(iii) *Thin* the pattern by selecting events at random and deleting them with probability $p_{del}(\mathbf{s}) = (\lambda_{max} - \lambda_{\mathbf{s}})/\lambda_{max}$.

This algorithm will generate a different number of events each time it is run. The total number of events in the simulated pattern will be random with expected value $\mu = \lambda \times |\mathbb{R}|$. An example is shown in Figure 2.10(a).

2.4

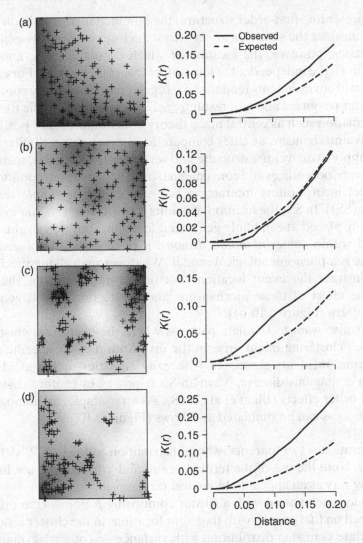

Figure 2.10   Examples of realisations of (a) inhomogeneous Poisson $\kappa(x,y) = \lambda(x,y) = 300$ $\exp(-5xy)$, (b) SSI, $d = 0.5$, $n = 100$ events, (c) homogeneous Thomas $\kappa = 20$, $\sigma = 0.04$, $\mu = 10$, and (d) inhomogeneous Thomas $\kappa(x,y) = \lambda(x,y) = 50\exp(-5xy)$, $\sigma = 0.04$, $\mu = 5$ processes overlain on their intensity surfaces shaded from low (light) to high (dark) on left, with their corresponding K-functions (right). Note the variable scaling of the vertical axes in the right-hand column.

This model is a phenomenological simulation of an underlying process which produces first-order trends. Thus, the algorithm does *not* directly represent any underlying mechanism, but is deliberately intended to produce a pattern with some desired statistical properties.

2.2.5   Interaction-based (second-order) null models

While representing first-order structure, the inhomogeneous Poisson process does not consider the structure arising from the interactions between events. In most cases, however, the location at which events occur is known and expected not to be independent of location or from other events. For example, shrubs in arid environments tend to show regular patterns as they compete for scarce water resources and as a result 'repel' one another. Classic theories of urban formation such as central place theory (Christaller, 1966) posit similar competitive mechanisms as cities compete for market share. Similar considerations apply to many infrastructure and service network nodes, such as cellphone towers, post offices and convenience stores in dense metropolitan areas.

 2.5
A model incorporating interaction between events is *simple sequential inhibition* (SSI). In SSI the location of events is constrained so that events are sequentially placed at randomly generated locations, but new events cannot be placed within some *inhibition distance d* of any already placed event. Again this is a phenomenological model. Whatever underlying mechanisms act to constrain the event locations are not represented, but the model mimics the effect of those mechanisms and the algorithm will generate a regular pattern (Figure 2.10(b)).

 2.6
We may also want to simulate patterns where the events are clustered in space. Such clustering might arise in the distribution of conspecific trees in forests, crime incidents associated with gang activities or the incidence of cases of a contagious disease. Neyman-Scott processes capture these kinds of second-order effects (Illian et al., 2008). As an example, an homogeneous Thomas process can be simulated as follows (Figure 2.10(c)):

(i) Generate a set of 'parents' whose distribution follows CSR with intensity κ. Note the use of the terms 'parent' and 'offspring' is not intended in any way as an appeal to biological mechanism.

(ii) Replace each parent with a cluster comprising μ points (the offspring) centred on that parent, with the event locations in the cluster following a bivariate Gaussian distribution with variance σ^2 (other Neyman–Scott processes, such as the Matérn, use different bivariate distributions).

(iii) Remove the parent events.

The basic Neyman–Scott process provides a point of departure for simulating more complicated spatial patterns. An obvious modification is to include

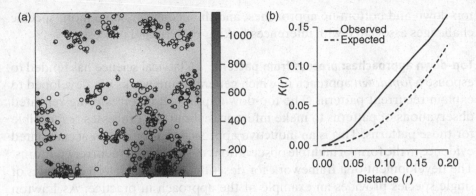

Figure 2.11 An example realisation of (a) a double Thomas process, final $n = 300$, $\kappa = 20$, $\sigma_1 = 0.05$, $\sigma_2 = 0.01$ with (b) its corresponding K-function. The large open circles are the parent process, the smaller open circles the first generation and the crosses (+) the second generation.

first-order structure such that the density of the clusters across the plot varies, as shown in Figure 2.10(d).

Another possibility is to generate multi-generational processes. For example, a double Thomas process (Wiegand et al., 2007) uses the first generation of offspring as cluster centres for a second generation, with the different generations possibly varying in the diffuseness of their clustering (Figure 2.11). Again, such processes could be modified to include first-order structure. The double Thomas process suggests how the types of static processes considered in this section can be made dynamic. There is clearly a close link between these static multi-generational processes and the forest model described by Flügge et al. (2012).

2.2.6 Inferring process from (spatio-temporal) pattern

Having collected snapshot data, such as the spatial distribution of an outbreak of disease or crime hotspots, we typically want to infer the processes or mechanisms that can explain those data. In other words we want to relate any observed patterns to the processes that generated them. The null model strategy introduced above allows us to take some tentative steps in this direction. Alternatively—and more commonly in this book—we may have some understanding of how the components of a system work, and we want to know how they interact to produce broader patterns of outcomes. So, a fundamental question is how can we link patterns to processes? This question is particularly important for dynamic spatial models where pattern and process are invariably closely intertwined. We begin this section by considering two broad frameworks for linking pattern and process, so-called top-down and bottom-up approaches, and then consider some more specific challenges associated with inferences linking pattern and process.

Top-down approaches: process from pattern   Classical science has tended to espouse a *top-down* approach in which general frameworks are developed to explain recurrent patterns. The top-down approach involves using repeated observations of patterns to make inferences about the processes responsible for those patterns. This is an inductive approach that builds on accumulated evidence in the form of multiple observations of similar and recurrent patterns. The development of a framework for describing the population dynamics of single species provides an example of the approach in practice. As Lawton (1999, page 179) notes, '[t]here are not ten million kinds of population dynamics; rather there are a multitude of essentially trivial variations on a few common themes.' A top-down approach to understanding population dynamics starts by trying to isolate Lawton's 'few common themes' and then developing a generally applicable framework to describe them. For example, many populations show a pattern where starting from low abundance they increase rapidly before stabilising at (or around) some higher abundance, as shown in Figure 2.12. This dynamic has been seen in populations of plants, animals and micro-organisms (perhaps humans too, although this remains open to debate), suggesting that it may be a general pattern.

The logistic (or Verhulst) mathematical model provides a general description of this population growth dynamic. It includes two terms, r and K, where r is the instantaneous *per capita* growth rate in the absence of density contraints and K is the *carrying capacity*, the maximum attainable population. Population growth is reduced by a factor proportional to the remaining available

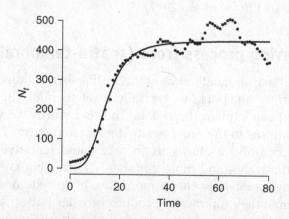

Figure 2.12   Typical dynamics of a single species exhibiting logistic growth. Starting from a low abundance the population rapidly increases before settling around some stable point (the *carrying capacity, K*). Here the points are 'observed' data and the solid line a best-fit logistic model.

carrying capacity as N increases, that is, with $1 - \frac{N}{K} \rightarrow 0$.

$$\frac{dN}{dt} = rN \left(1 - \frac{N}{K}\right) \qquad (2.4)$$

In terms of the discussion in Section 1.4.1, this is a highly abstract model that does not concern itself with the details of any particular system, but in return it is broadly applicable. Analysis of the logistic model shows that it has two stable points when $\frac{dN}{dt} = 0$, at $N = 0$ and $N = K$, and also that *per capita* population growth is most rapid when $N = K/2$. These observations are held to be true for *any* population following the dynamics in Equation 2.4. It does not matter whether we are considering galaxids, kauri trees, beavers, lichens or humans. The top-down framework aims to develop a *general* understanding of broad classes of systems and to make high-level inferences from this understanding.

Bottom-up approaches: pattern from process   Often, we want to use our understanding of fine-scale processes to predict the broad-scale (macroscopic) patterns that might emerge from them. In such cases, we are more likely to use a *bottom-up*, or atomistic, framework.

Individual-based and agent-based models, the subjects of much recent interest, are good examples of this approach (Grimm and Railsback, 2005, O'Sullivan, 2008). An agent-based model of pedestrian activity in a streetscape, for example, focuses attention on how phenomena such as streams or linear bands of movement arise from the decisions made by individual pedestrians in the absence of any central organiser (Helbing et al., 2001). Such macroscopic patterns, arising from individual, micro-level behaviours, are sometimes called *emergent* or *self-organising*, and are major themes in the complexity literature (see pages 18ff.). Another example of the approach is provided by individual-based forest gap models (reviewed in Perry and Millington, 2008). In such models the recruitment, growth and death of individual trees in a forest is simulated, the interest being how individuals interact with each other and their environment to produce changes in species composition (that is, succession) and ecosystem properties (such as biogeochemical cycling).

There is considerable current interest in the bottom-up approach to modelling as it provides a way to handle heterogeneity among individuals in their reciprocal interactions with complex environments and with each other. This development has been facilitated by advances in computer power and software architectures (O'Sullivan et al., 2012). Individual- or agent-based models have become frequently used in many disciplines, including sociology (Macy and Willer, 2002), land-use change (Parker et al., 2003), ecology (Grimm and Railsback, 2005), economics (Tesfatsion and Judd, 2006), geography (O'Sullivan, 2008) and archaeology (Costopoulos and Lake, 2010).

It may appear as if top-down and bottom-up approaches are diametrically opposed, and it is true that they do represent quite different ways of thinking about how systems behave and how best to understand them. In this book we are largely focused on bottom-up models. However, an approach that draws on both approaches is likely to be more useful and successful for theory development and learning about systems (Grimm, 1999, Vincenot et al., 2011). And this, after all, is the primary motivation for any modelling activity.

Linking pattern and process: some inferential traps Making inferences about processes from snapshot patterns and *vice versa*, while appealing, is difficult and must be considered carefully. A first problem is that the same process might generate many different patterns. A second is that different processes might result in the same pattern. Furthermore, a system can arrive at the same endpoint via many different trajectories even from similar starting conditions. These problems together constitute *equifinality* (Beven, 2002).

The equifinality problem is closely related to and difficult to separate from the issue of *under-determination* or *nonuniqueness* (Oreskes et al., 1994, Stanford, 2009). Stanford (2009, page 1) defines under-determination as being the problem arising when 'the evidence available to us at a given time may be insufficient to determine what beliefs we should hold in response to it' and notes that in the scientific context it is closely related to the old adage that correlation does not imply causation. In the context of models and model evaluation, under-determination means that just because our model reproduces some observational data we can not say that our model is 'correct'. Equifinality implies that more than one model structure, or even the same model with different parameterisations, could reproduce the same dynamics. This is clearly a problem for model evaluation methods (see Chapter 7) that emphasise comparing a model's predictions with observed data (Kleindorfer et al., 1998, Perry et al., 2006). Arguably such concerns are more serious when considering phenomenological models that deliberately do *not* represent mechanism, but the difficulty is a general one. These issues make drawing inferences about generating processes from snapshot patterns, whether they are empirical or model-derived, treacherous ground. Based on snapshot information we cannot unequivocally identify the mechanisms and dynamics operating in a system from observing it. The data do not provide enough information to draw the conclusions we might wish.

While efforts to directly link *single* patterns to generating mechanisms may be ill-founded, the careful use of patterns, spatial, statistical or otherwise, can help us to develop and evaluate appropriate models. Recently, for example, pattern-oriented modelling (Grimm et al., 2005, Grimm and Railsback, 2012), has been advocated as a way of using *multiple* patterns to guide model development and calibration, and to select between competing models. We discuss these issues further in Chapter 7. The careful use of spatial and temporal patterns thus remains central to model conceptualisation, implementation and evaluation. We often have only limited information to guide model development, and we must take care not to throw the baby out with the philosophical bath water!

2.2.7   Making the virtual forest more realistic

Armed with the knowledge we now have regarding how to confront model outcomes with specific null expectations and also some of the difficulties involved, we can return to Flügge et al.'s (2012) forest model. The Thomas process introduced in Section 2.2.5 appears a good candidate description of the patterns we might expect in this virtual forest. We can use the Monte Carlo approach laid out in Figure 2.8 to test if this is the case. First, we must parameterise a Thomas process using the data, and then simulate many patterns following this process before confronting our observed pattern with these simulations. Figure 2.13 suggests that the homogeneous Thomas process is an adequate description of the patterns depicted in Figure 2.9. We could also compare the pattern against the Matérn process in which the distances of offspring in the cluster from the parent is uniform. Assessing the fit of the observed pattern to *different* null models of aggregation might help us to home in on the nature of the dynamic driving the pattern that we see.

The basic form of model that we have experimented with up to this point is simple, and it is tempting to try to make it more realistic. One modification might be based on the observation that plant mortality is usually density-dependent. Trees with more other trees close to them may be at

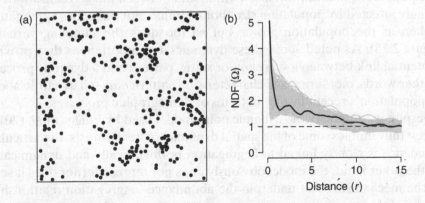

Figure 2.13   Evaluating the patterns produced by Flügge et al.'s forest model: (a) the pattern of trees after 200 generations, with a mean dispersal distance of five (from an exponential distribution) and (b) the NDF for the pattern shown in (a) compared to 999 simulations of an homogeneous Thomas process (with estimated $\kappa = 0.008$ and $\sigma^2 = 8.37$). The pattern in (a) is the same as that in Figure 2.9.

greater risk of mortality due to competition. We can, following Flügge et al. (2012), add density-dependence to the model by computing the strength of competition experienced by each individual tree. Competitive intensity in the model is a function of the distances to competing trees. We estimate competitive intensity for each tree i by summing, across other trees j within a specified competitive range, the density of a standard normal distribution ($\mu = 0$) at a z-score given by $z = d_{ij}/\sigma$, where d_{ij} is the distance to each competing tree. As $\sigma \to 0$ the relative impact of closer trees increases so that this parameter controls how competition varies with distance between trees. Having determined the competitive pressure on each tree, individuals are selected for mortality based on their ranking in the distribution of competitive pressure across the whole population. For example, if the mortality percentile is set to 50% an individual in the upper 50% of values is selected at random to die. As this percentile increases up to 100 the strength of the density-dependent effect increases. We represent density-dependence in a somewhat different way to that described by Flügge et al. (2012), but our approach is similar in spirit and produces very much the same outcomes. Under even weak density-dependence the trees exhibit a random or even regularly spaced pattern, with the variation between model runs and over time during a single run considerably reduced (see Figure 2.14).

2.2   Finally, we can look at how aggregation changes as populations grow or decline. This population dynamic is represented by adding (at a random location) or removing (at random) individuals from the model at the required rate. Flügge et al.'s model suggests that for a given abundance, populations that are growing may be more aggregated, and those that are declining less so, and that these changes in spatial pattern lag behind changes in abundance. It is also apparent that the variance (within and between model realisations) is strongly affected by population size. For example, not only does aggregation weaken as the population grows, but so too does the sampling variance (Figure 2.15). As noted above these dynamics are important as they provide a potential link between a snapshot, or static, pattern and a dynamic process. In other words, measuring and characterising pattern *may* tell us a little about the population's recent history in terms of demographic processes.

Despite its simplicity the dynamic point model used by Flügge et al. (2012) successfully mimics some of the spatial dynamics of 'real' forests. In particular, abundance is closely linked to aggregation both statically and dynamically. On the other hand, the model obviously does not represent (nor does it seek to) the mechanisms that underpin the abundance–aggregation relationship, even if does mimic them. This returns us to earlier themes (see Section 2.2.4) concerning the dangers of assuming that because a model reproduces an observed pattern it must be 'correct' or 'true'. As Millington et al. (2011) point out *mimetic accuracy*—the ability of a model to reproduce an

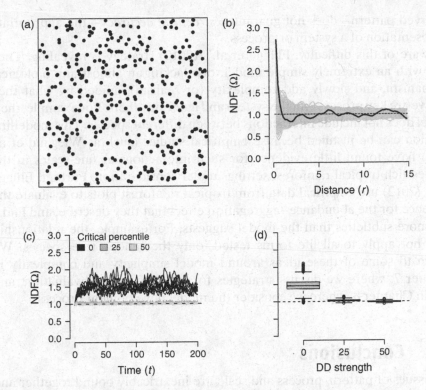

Figure 2.14 The relationship between aggregation (Ω) and abundance as it emerges from the spatial birth–death process in the presence of density-dependence. Here abundance n is fixed at 400 individuals and the mortality percentile (see text) is set at 0%, 25% and 50%, where 0% represents no density-dependent effect. Mean dispersal distance is five and the neighbourhood has a radius of ten. (a) Typical regular pattern produced under density-dependence (here 25%), (b) comparison of the pattern in (a) against a null model of CSR, (c) time-series of Ω under different density-dependent strengths (the dashed line is the value expected under CSR) and (d) the distribution of the Ω values after 200 generations as a boxplot.

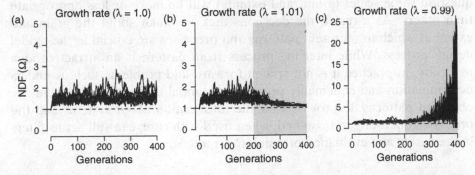

Figure 2.15 Changes in aggregation Ω as a function of changes in abundance starting from 400 individuals over 400 generations. (a) Population is constant, $\lambda = 1$, (b) population grows at rate $\lambda = 1.01$ from generation 200 and (c) population declines at rate $\lambda = 0.99$ from generation 200. The shaded areas are the periods when the population is changing. Note the different scaling of the vertical axis in (c).

observed pattern—does not guarantee *structural accuracy*—the appropriate representation of a system or process.

Aware of this difficulty, Flügge et al.'s (2012) strategy is revealing. They start with an extremely simple model, one lacking in almost any biological mechanism, and slowly add complexity (or realism), guided by what they believe are key processes in the system and ignoring others. For example, their model does not include interactions between different species. This modelling decision can be justified because empirical studies (such as Wiegand et al. 2012) have found little evidence for such interactions at fine-scales in the species-rich tropical rainforest settings under consideration. Further, Flügge et al. (2012) use empirical data from tropical rainforest plots to evaluate the evidence for the abundance–aggregation effect that they describe, and find a few more subtleties than the model suggests. For example, the relationship does not apply to all life forms tested, only the canopy tree species. We return to some of these ideas around model simplicity and complexity in Chapter 7, where we discuss strategies for effective model evaluation, and also in Chapter 8, where we consider the model development process.

2.3   Conclusions

The issues of pattern, process and scale are inextricably bound together and continue to challenge researchers across many disciplines. We can see patterns as the signatures or fingerprints of the processes that generate them, and this suggests that pattern can be used to infer generating process. However, both patterns and processes are scale-dependent so that the way we perceive a system will be strongly conditioned on the scale at which we choose to view it in space and/or time. Scale-dependence also means that there is no single 'correct' or 'best' scale at which to view a system, although for a given question some scales (grains and extents) will be more or less appropriate than others. As a result, the decisions that we make about the grain and extent at which to represent patterns and processes are crucial in the model design process. While inferring process from pattern is an attractive idea in theory, in practice it is not straightforward and problems such as under-determination and equifinality present substantial difficulties. Nevertheless, observed patterns do provide valuable information about systems and the processes operating in them and, when used with care, can still act as filters in the design and evaluation of simulation models.

3
Aggregation and Segregation

The most salient feature of many spatial patterns is summarized in Waldo Tobler's 'first law of geography: everything is related to everything else, but near things are more related than distant things' (Tobler, 1970, page 236). Interestingly, Tobler's law arises in the context of simulating urban growth, from the necessity of simplifying the system to be modelled from one where 'everything is related to everything else' (Tobler, 1970, page 236), a situation that presents obvious difficulties when it comes to developing a simple model.

The status of Tobler's premise as a *scientific* law has been questioned. In a forum in the *Annals of the Association of American Geographers* (Sui, 2004), two authors take exception to its status as a law (Barnes, 2004, Smith, 2004). Others are more comfortable with the idea (Phillips, 2004), asserting its usefulness as a guiding principle or useful null model when we approach the study of any spatial system. We should *expect* to find that near things are more related than distant things and evaluate empirical data in the light of that expectation. Miller (2004) and Goodchild (2004) even suggest that Tobler's law ought to be preceded by the even more general statement that spatial heterogeneity exists—that everywhere is not the same. Given this (seemingly) self-evidently true statement, Tobler's law describes the equally self-evident (but not inevitable) truth that the heterogeneity in any phenomenon is likely to be less at a local level than it is globally. We can probably agree that whether or not it is a 'law', Tobler's statement is generally true, but perhaps not all that original. In his response to the forum, Tobler (2004, page 309, note 5) points to a much earlier statement by Fisher, noting 'the widely verified fact that patches in close proximity are commonly more alike [. . .] than those which are further apart' (Fisher, 1966, page 66).

Spatial Simulation: Exploring Pattern and Process, First Edition.
David O'Sullivan and George L.W. Perry.
© 2013 John Wiley & Sons, Ltd. Published 2013 by John Wiley & Sons, Ltd.

3.1   Background and motivating examples

So what does this null model of spatial heterogeneity lead us to expect? Roughly speaking, we expect the world to be *patchy*. Conditions locally are more likely to be similar to one another than they are when we expand the scope of our study. We observe such patchiness in many different settings. The physical relief of Earth's surface, its geology, soils and climate all vary less in local areas than they do globally, leading to distinctive regional mixes of these attributes. Egmont (Mt Taranaki) National Park in New Zealand provides an example, shown in Figure 3.1(a). Ecosystems are commonly conceptualised as providing patches of different habitat types, each of which offers niches for different species, resulting in a patchy distribution of plant and animal species. 'Tiger bush' or *brousse tigrée* vegetation, which responds to subtle changes in the availability of water in areas of low relief, provides a striking example, as seen in Figure 3.1(b).

In urban settings, different socioeconomic, cultural and ethnic groups tend not to be homogeneously mixed. Rather, there is a mosaic of social, cultural and ethnic neighbourhoods: we find gay neighbourhoods, student neighbourhoods, a Little Italy and a Chinatown, like the one illustrated in Figure 3.1(c). None of these is exclusively occupied by members of the community with which we identify them, but in each that community's presence is sufficiently prominent for there to be general agreement about the designations. Similarly, commercial districts in cities tend to be functionally aggregated with commercial, retail and administrative areas distinctly spatially organised (Figure 3.1(d)). Such urban patterns help to explain the success of the geodemographic targeting methods employed by marketing professionals (see Harris et al., 2005), which rely on an assumption that 'birds of a feather flock together'.

From a simulation modelling perspective the interesting challenge is to identify process models that generate patchy landscapes (see also Chapter 5). Tobler's statement is a bald verbal description of a general, empirically observable tendency. To make the statement more useful for understanding processes that might give rise to patchiness, in this chapter we explore simple, spatially explicit models which produce patchy patterns. The key feature of all of these models is that they favour the local aggregation of similar things. This aggregation may express itself in a process such that model elements change to become more like neighbouring elements, but an important insight is realising that aggregation and segregation are two sides of the same coin, so that aggregation may also arise from a segregative mechanism favouring the separation of dissimilar entities.

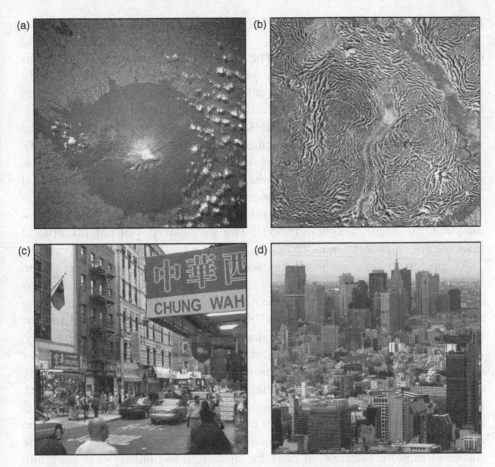

Figure 3.1 (a) Egmont (Mt Taranaki) National Park, New Zealand, (b) *brousse tigrée* vegetation patterns in Niger, (c) Chinatown in New York City and (d) the Shinjuku skyscraper district (background) with lower-rise retail and residential Aoyama in the mid-ground and another high-rise office district (Akasaka) in the foreground, seen from Roppongi, Tokyo. *Sources*: (a) image STS110-726-6, taken by Space Shuttle crew on 9 April 2002, courtesy of the Image Science & Analysis Laboratory, NASA Johnson Space Center, http://eol.jsc.nasa.gov/, (b) image from the declassified corona KH-4A National Intelligence reconnaissance system, 31 December 1965, http://commons.wikimedia.org/wiki/File:Tiger_Bush_Niger_Corona_1965-12-31.jpg, (c) photograph by Derek Jensen, 15 June 2004, http://en.wikipedia.org/wiki/File:Chinatown-manhattan-2004.jpg and (d) authors' collections.

3.1.1 Basics of (discrete spatial) model structure

Most of the models in this chapter adopt the cellular grid structure that we have already encountered in Section 1.3.2, which rests on ideas about how to represent time and space. These ideas are considered in more detail in Chapter 6, but it is appropriate to briefly summarise them here.

Broadly, time is represented by cells determining, and iteratively updating to their next state (see Section 6.1), while space is represented as a grid of square cells (see Section 6.2). The grid of cells is often referred to as a *lattice* of *sites*. In many cases, the grid 'wraps' around from left to right and from top to bottom, so that cells in the extreme left-hand column have as neighbours those in the extreme right-hand column as well as their immediate neighbours. This simulates an infinite space and avoids model artifacts due to edge effects, and is called *toroidal wrapping* (see Section 6.4). Another important decision concerns the neighbourhood structure used, that is, what is adjacent or connected to what? The most common neighbourhood definitions are either the four orthogonal immediate neighbours (the *von Neumann* neighbourhood) or the eight adjacent neighbours (orthogonal and diagonal, the *Moore* neighbourhood). The various types of neighbourhood possible are considered in more detail in Section 6.3.2.

At any given time each cell has some *state*, generally a number, often restricted to integer values when cell states are *discrete*. Change occurs iteratively, by updating cell states in accordance with transition rules governing how the current state and the state of neighbours in the grid lead to new cell states. The timing of state changes may occur either *synchronously*, when all cells determine their next state, and are updated simultaneously, or *asynchronously*, when one cell at a time is updated (see Section 6.1). In the asynchronous case, it is useful to distinguish between *ticks* of the model clock and *generations*. During each tick, one cell may change state. A generation is a number of ticks equal to the number of grid cells, so that during a generation every cell may have an opportunity to update its state, although due to the (usually) random selection of cells for update, it is unlikely that each and every cell will change state during a generation. Under synchronous update, ticks and generations can be considered equivalent.

3.2 Local averaging

3.1 The most obvious way to make places similar to their neighbours is to repeatedly apply a local averaging procedure so that each location's current value moves towards the mean value of its neighbours. Local values that are high relative to neighbouring values are dragged down by this averaging process, while values low relative to neighbouring values are pulled up. If we denote the value at time t of some attribute at a location i by $z_i(t)$, and the Moore neighbourhood of i as N_i, then the update process at each time step is

$$z_i(t+1) = (1-w)z_i(t) + \frac{w}{8}\sum_{j \in N_i} z_j(t) \tag{3.1}$$

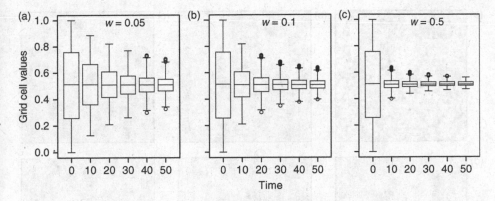

Figure 3.2 Increasingly rapid convergence of the local average values as the weight parameter w increases. Boxplots show the distribution of all cell values on a 50×50 grid over time.

where the w parameter sets the proportion of the new z value that results from the local average, with the remainder $1 - w$ derived from the current z value. If $w = 0$ then there will be no change in cell values. If $w = 1$ then the new z_i value will be the mean of the neighbouring z values. As w increases from 0 to 1 the rate at which the grid cell values converge to a uniform value everywhere increases, as shown in Figure 3.2.

Figure 3.3 shows how this plays out over space. An initially random surface (a) moves rapidly towards homogeneity as local averaging is applied. With the proportion of the new z value attributable to the local average $w \geq 0.5$ as in panels (d)–(f), the process results in a near uniform surface, although as the contours show, surface structure remains albeit over a reduced range of values. The point to note is that the averaging process changes the pattern from completely random to one that is separated into distinct regions of high and low values. Whereas for the random starting configuration in Figure 3.3(a) it is impossible to draw meaningful contours, after only a few time steps of local averaging, the z values are sufficiently spatially coherent or *autocorrelated* for contours to be drawn.

In Figure 3.4 the evolution of the pattern over time is shown for the $w = 0.1$ case. Here we have changed the colour scheme for z values at each time step so that the pattern remains clearly visible even though the range of values is reduced. As the number of steps increases, the averaging at each location progressively extends to greater distances across the model space, causing the scale of the pattern to increase. For cases where $w > 0.1$, the grain of the pattern increases at the same rate, but the degree of similarity of adjacent cells increases more rapidly. This is because it is the grid cell neighbourhood that governs the rate at which regions of similar values can grow, not the w parameter.

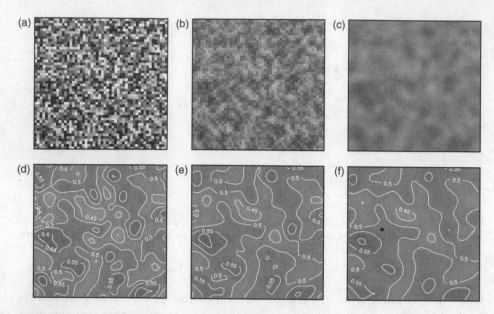

Figure 3.3 Results of the local averaging process applied to a 50 × 50 grid of cells starting from the random initial state shown in (a) and with the state after ten generations shown for (b) $w = 0.1$, (c) $w = 0.3$, (d) $w = 0.5$, (e) $w = 0.7$ and (f) $w = 0.9$. Contours at intervals of 0.05 show that the patterns are similar from case to case, and that convergence continues to increase as w increases from 0.5 to 0.9, as is evident from the diminishing range of values in the surface.

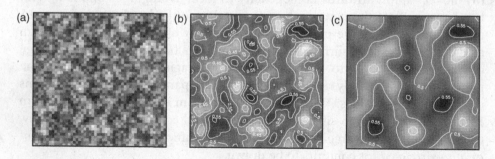

Figure 3.4 Results of the local averaging process over time with $w = 0.1$, showing snapshots at (a) $t = 10$, (b) $t = 40$ and (c) $t = 100$. The 50 × 50 grid wraps around at each side here, and as a result once $t > 50$ the scale of the pattern produced does not change. Note that the colour scheme is different in each panel.

Local averaging provides a simple model of the uniform diffusive spread of material because the averaging guarantees that the total $\sum z$ across the whole model remains constant. We look more closely at spatial aspects of spread in Chapters 4 and 5. For now, we note that the eventual effect of processes of spread in the absence of any mechanism to reinforce or 'lock in' differences between locations is for the landscape to become uniform, as is clear from Figure 3.3(d)–(f).

3.2.1 Local averaging with noise

It is straightforward to include a random element in the local averaging process, but its effects can be difficult to detect against the global smoothing dynamic. A model that simplifies the idea dramatically so that this aspect becomes clearer is presented by Vicsek and Szalay (1987). Their model is again grid-based but uses a von Neumann neighbourhood. The initial state of the model assigns $z = 0$ to every grid cell. Subsequently, synchronous updates occur with every cell changing its state to

3.2

$$z_i(t+1) = \frac{1}{5}\left[z_i(t) + \sum_{j \in N_i} z_j(t)\right] + \epsilon_i(t) \qquad (3.2)$$

where ϵ is randomly chosen with equal probability to be -1 or $+1$. In the first generation this is equivalent to randomly initialising each cell to $z_i = \pm 1$. Subsequently, where we might expect the averaging component of Equation 3.2 to produce the convergence we see in simple local averaging, the relative weight of the averaging and random elements is such that the system remains in a constant state of flux. Vicsek and Szalay (1987) add an additional dynamic (or rule), such that if at any time the z value in a grid cell exceeds a threshold z_c then that cell is permanently switched 'on'. Typical patterns resulting from this model with a variety of thresholds are shown in Figure 3.5. It is apparent that these 'random' data feature considerable structural complexity. This complexity arises because extreme high (or low) values in the underlying surface must necessarily be near one another, since reaching extreme values is dependent on having other extreme values nearby.

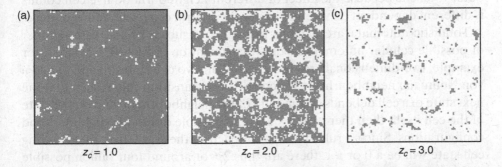

Figure 3.5 Typical outcomes of the Vicsek and Szalay (1987) local averaging model with noise. These examples were run for 50 iterations on a 100×100 grid which wraps both horizontally and vertically. Cells with z greater than the threshold state value z_c shown in each panel are coloured grey.

Vicsek and Szalay (1987) propose this system as a simple model of galaxy formation (no less!), and Batty (2005) suggests that it is a reasonable starting point for a model of the long-term development of urban regions. The idea is that the model embodies two essential features—aggregation and diffusion—which together drive urban (and many other) growth processes. Aggregation is represented by the summation of the averaging process, while diffusion results from the iterated summation across a neighbourhood, which over time spreads the total amount of 'development' around. The random component 'seeds' the system with latent spatial structure which averaging locks in over multiple iterations. This captures the idea of positive feedback as a critical driver of urban growth. A few 'lucky breaks' (a series of +1 random draws) in a particular area, early in the model evolution, are likely to lead to a long-term elevation in the state values in that part of the system.

It would be absurd to claim any great degree of realism for this model. Nevertheless, it is suggestive that repeated iteration of a simple localised process produces such a richly patterned structure and this is a common feature of many of the models in this chapter.

3.3   Totalistic automata

3.3

Arithmetic averages are based on summation, and so we can consider local averaging as closely related to a broad general category of cellular automaton models, the *totalistic automata*. Totalistic automata are cellular automata (CA) where the rules governing cell state changes are expressed solely in terms of the *sum* of the state values of neighbouring cells. Where the next cell state also depends on the current state of the cell itself (and not solely on the local neighbourhood sum of state values) the automata is sometimes referred to as *outer totalistic*. Conway's life automaton (introduced in Section 1.3.2) is outer totalistic because the effect of different neighbourhood live cell counts is different dependent on whether or not the central cell itself is alive or dead.

Totalistic automata are a more manageable subset of the vast number of possible cellular automata models that might be proposed. Consider, for example, two-dimensional grid-based CA with two cell states (0 and 1) and a von Neumann neighbourhood structure. If the transition rule is such that the next state of a cell depends on the state of the neighbourhood and on the state of the cell itself, then there are $2 \times 2^4 = 32$ possible different neighbourhood configurations. Since a rule specifies for each of these whether the resulting cell state will be a 0 or a 1, there are thus 2^{32} or around four billion possible different automata. On the other hand, if the CA is outer totalistic then the neighbourhood sum is an integer from zero to four, giving 2×5 possible neighbourhood states and thus only 2^{10} or 1024 possible different rules—a much reduced range of possibilities. A further advantage is that totalistic rules are guaranteed to be *isotropic*, meaning that they do not 'care' about the directional orientation of cell states, so that, for example, the neighbourhood configurations

$$
\begin{array}{ccc}
1\ 1\ 0 & 0\ 1\ 1 & 0\ 0\ 0 & 0\ 0\ 0 \\
1\quad 0 & 0\quad 1 & 1\quad 0 & 0\quad 1 \\
0\ 0\ 0 & 0\ 0\ 0 & 1\ 1\ 0 & 0\ 1\ 1
\end{array}
$$

are equivalent, which is likely to be desirable in many cases.

Although restricting our interest to totalistic automata substantially reduces the number of distinct cellular automata rules, there is still a wide range of possibilities. We have already seen Conway's life; we now consider some other transition rules.

3.3.1 Majority rules

A majority rule specifies that where one state is in the majority in a neighbourhood, the next state of the central cell will be the local majority state. For a four-cell (von Neumann) neighbourhood, with two states, this rule is:

3.3

$$
z_i(t+1) = \begin{cases} 1 \text{ if } \left(z_i(t) = 0 \text{ and } \sum z_j > 2\right) \text{ or } \left(z_i(t) = 1 \text{ and } \sum z_j \geq 2\right) \\ 0 \text{ otherwise} \end{cases}
$$

(3.3)

For the eight-cell (Moore) neighbourhood case this becomes:

$$
z_i(t+1) = \begin{cases} 1 \text{ if } \left(z_i(t) = 0 \text{ and } \sum z_j > 4\right) \text{ or } \left(z_i(t) = 1 \text{ and } \sum z_j \geq 4\right) \\ 0 \text{ otherwise} \end{cases}
$$

(3.4)

Similar transition rules can be defined for other neighbourhood sizes.

This rule produces the outcomes we might expect: as shown in Figure 3.6, the model rapidly segregates into two regions, one in each of the two available states. The symmetry of the rule is apparent in the snapshots, and even more so when viewed dynamically. The state 1 regions (dark) and state 0 regions (white) rearrange themselves around one another until no further state changes are required because all central cells are in the same state as the majority of their neighbours. Relatively rarely for the Moore neighbourhood, but more commonly for the von Neumann neighbourhood, small regions of the model do not stabilise but instead flip continuously back and forth between states. For larger neighbourhoods this behaviour does not occur.

The segregation behaviour of the majority rule is sensitive to the initial state of the system. In particular, if the proportion of either of the two states in the initial pattern is low, there is a high probability that the final system state will consist solely of grid cells in the initial majority state. The response in terms of the final proportion of grid cells in state 1 for random set-ups

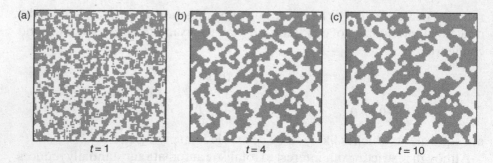

Figure 3.6 Results of a simple majority rule automaton with the rule from Equation 3.4 starting at $t = 0$ from a random configuration with 50% of cells in each state. The grid is 101×101 cells, wrapped both horizontally and vertically and using a Moore neighbourhood structure.

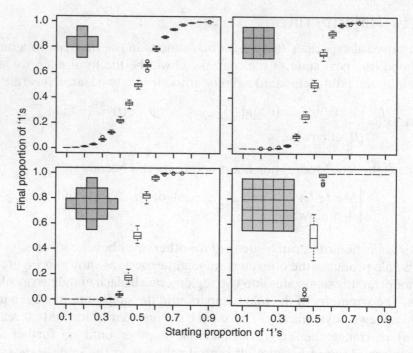

Figure 3.7 The dependence of the final proportion of state 1 on the initial proportion for different neighbourhood sizes. The neighbourhood shape is shown in each panel. As neighbourhood size increases, the response becomes more sharply defined as a result of the increasing size of the minimal stable configurations (see Figure 3.8).

with varying starting proportions of cells in state 1 is shown in Figure 3.7 for different neighbourhood sizes. The higher the initial proportion of cells in state 1 the more likely it is that the final configuration will consist entirely of 1s (the same is true for the 0 state). This behaviour varies depending on neighbourhood size. For small von Neumann neighbourhoods, because

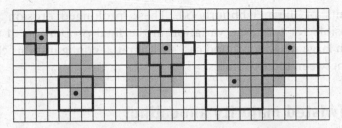

Figure 3.8 The minimal stable configurations of grid cells in state 1 (shown in grey) for various different neighbourhoods. The neighbourhood in each case is shown in heavy black outline, with the black dot at its centre. More tightly 'curved' boundaries than those shown are unstable for the neighbourhoods shown.

some small configurations are stable (see Figure 3.8), and these are likely to form spontaneously in a low density random configuration, the final system configuration does not necessarily end up being entirely 0s or 1s. As the neighbourhood size increases, the minimal stable clusters become larger and are very unlikely to form spontaneously except in higher density configurations. The result is that for increasing neighbourhood size, the final configuration is increasingly likely to be either all 0s or all 1s depending on which state is in the majority initially.

A somewhat surprising corollary of these effects is that the time taken for the system to arrive at a final stable configuration increases as neighbourhood size increases. This is evident in Figure 3.9, where a sequence of snapshots for a majority rule automaton with a circular neighbourhood of radius 2.3 is shown. Long linear features and narrow 'peninsulas' of either cell state are slowly eroded back to stable shapes consistent with the curvature

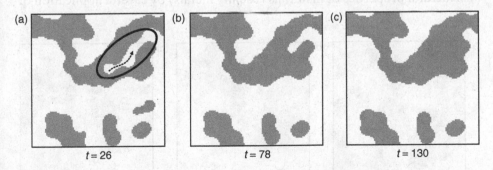

(a) $t = 26$ (b) $t = 78$ (c) $t = 130$

Figure 3.9 Erosion of elongated regions in a majority rule automaton. The initial proportion of cells in each state is 0.5. The neighbourhood in this case is the 20 cell area within a radius of 2.3 of the central cell. This leads to the initial formation then gradual 'erosion' of long peninsula regions such as the white area indicated in the first snapshot. The overall effect is longer elapsed times before a stable final configuration is reached, and the potential for stable configurations with an uneven split between the cell states.

requirements of the neighbourhood. These prolonged erosion processes are responsible for the large variation in the final proportion of 0s and 1s observed in large neighbourhood cases with initial configurations close to an even split between the two states.

3.3.2   Twisted majority annealing

 3.3

A variation on the majority rule, attributed to Vichniac (1984), introduces a 'twist' in the majority. For example, in the eight-cell neighbourhood case, we consider a majority to exist when the sum of the local cell states, including the cell itself, is one of $\{4, 6, 7, 8, 9\}$, omitting the actual majority total of five and including the erroneous majority of four. This minor change has surprisingly far-reaching effects.

By making convex regions such as those in Figure 3.8 unstable, the twisted majority rule forces the boundary lines between regions of 0s and 1s to become as close to straight lines as possible. Figure 3.10 shows a typical sequence. This process takes a long time: the 'wiggly' region in the example shown eventually becomes a near-horizontal east–west 'stripe' after more than 3000 generations. The only difference between this example and that of Figure 3.6 is the minor twist in the rule—they both use the same neighbourhood definitions.

Although it is not an obvious candidate for representing any real-world social or environmental processes, the Vichniac twisted majority rule holds wide interest. First, it again demonstrates the capacity of simple models to produce surprising results whose explanation requires careful consideration. Second, it is of considerable interest in solid-state physics where the model behaviour is analogous to *annealing* processes, which promote the refinement of structural properties of materials (usually metals) by careful application of

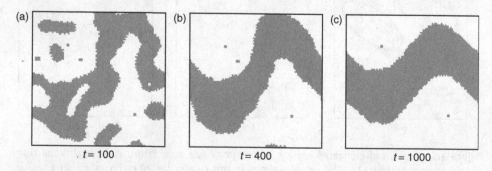

Figure 3.10   Time evolution of Vichniac's (1984) twisted majority rule. Here we use an eight-cell neighbourhood and the central cell transitions to state 1 if the sum of the focal and neighbouring cell states $z_i + \sum z_j \in \{4, 6, 7, 8, 9\}$. Compare this sequence with Figure 3.6. Particularly note the much longer time period in this case.

heat causing localised reorganisation of the fine-scale structure. In a similar way, the twist in the majority rule causes localised reorganisation of the boundary region between cell states. While there is no direct analogy with annealing in social and environmental systems, allowing a system to make minor localised changes either deterministically (as in this case) or randomly can have significant effects on system-wide patterns.

3.3.3   Life-like rules

An interesting question, given the widespread interest in Conway's life, is how unusual a case is it? That is, are there many transition rules and automaton configurations that produce such rich behaviour, or is the life automaton a rarity? In keeping with the first publication of the life automaton in a magazine column about recreational mathematics (Gardner, 1970), a lot of work in this area has been relatively informal computer hacking. Even more formal treatments are primarily phenomenological—empirical searches of suspected life-like transition rules, with weak theoretical underpinnings. In their early survey of two-state two-dimensional automata, Packard and Wolfram noted that among several thousand randomly sampled two-dimensional automata transition rules examined, 'no propagating structures were seen' (Packard and Wolfram, 1985, page 936)—in other words no 'gliders' (see Figure 1.6)—except in simple variants of life. An example, sometimes called *higher life*, allows births when dead cells have either three or six live neighbours but leaves the rule otherwise unchanged (see Figure 3.11(a)).

One obvious way to search for life-like behaviour is to focus on the idea that the life transition rules lead to birth or survival when local densities are neither too sparse nor too dense. Some success in the digital 'search for life' has been reported and generalised to *Larger than Life* (LtL) automata, where it is assumed that critical parameters are the lower and upper limits on the number of live neighbours for birth or survival, with these denoted β_1, β_2, δ_1

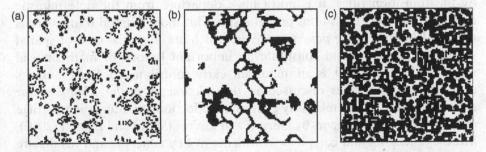

Figure 3.11   Snapshots of life-like automata: (a) higher-life which admits births when a dead cell has six live neighbours, (b) the Larger than Life (LtL) automaton {4, 22, 31, 28, 34} and (c) LtL {1, 1, 6, 2, 7}.

and δ_2, respectively (see Evans, 2003). If the neighbourhood range ρ is also recorded (assuming a Moore neighbourhood) then an LtL automaton can be summarised by $\{\rho, \beta_1, \beta_2, \delta_1, \delta_2\}$ and life is denoted $\{1, 3, 3, 3, 4\}$. Adopting this framework, Evans (2003) reports mobile self-replicating configurations with a variety of transition rules, many with very large neighbourhoods. An example is shown in Figure 3.11(b). Other cases have been found (Figure 3.11(c)) where the interest is not in gliders but in the unceasing activity of the automaton over time, such as LtL $\{1, 1, 6, 2, 7\}$ first noted by de la Torre and Mártin (1997). The birth–survival structure of these life-like transition rules can be stretched into a loose analogy with classical population models (Begon et al., 1996), where population growth is most rapid when populations are at mid-range densities. For example, Evans (2003, page 43) suggests that 'LtL is a spatial version of nonlinear population dynamics in which both space and time are discrete'. It is at least interesting that such behaviour is found in this space.

While the excitement around Conway's life and the discovery of its universal computational properties (see Berlekamp et al., 2004, pages 940–957; see also Section 1.3.2) injected considerable energy and interest (life, even!) into the study of CA, it can also distract us from the considerable potential of these simple systems as models of natural phenomena, which should be apparent from all of the examples in our brief coverage. Nevertheless, it is also clear that we need to ground automaton rules more empirically for them to be relevant to improving our understanding of real-world systems, rather than our understanding of artificial (albeit interesting) systems.

3.4   A more general framework: interacting particle systems

Cellular automata are deterministic systems. Given its current state, and the current state of its neighbours, the next state of each cell is exactly specified by the transition rules. It is not a huge departure from the deterministic case to make state transitions probabilistic. This more general framework is provided by *interacting particle systems*, which are a special case of spatial stochastic processes and constitute an important branch of mathematical statistics (Spitzer, 1970, is an important early contribution). Some authors also refer to the cases discussed in this section as *probabilistic cellular automata*. Useful (and relatively gentle) introductions to interacting particle systems are found in papers by Durrett and Levin (1994b) and Liggett (2010). More challenging and complete treatments of key concepts and results are presented in specialised monographs by Liggett (1985, 1999).

An interacting particle system (IPS) consists of a lattice of locations– generally, for the cases we are interested in, a two-dimensional grid, with cells indexed by integer values. The lattice is often denoted $\mathbb{Z}^2$. Any particular site at time t, at location (x, y) has an integer-valued state z, which we denote $z_t(x, y)$ or simply z_t, where the meaning is clear. As in a cellular automaton, a transition rule governs how the state of each site changes, dependent on its current state and on the state of neighbouring sites $N_{x,y}$, with a variety of neighbourhood definitions possible. Neighbourhoods at longer ranges are also possible (see Section 6.3.2).

State updating in interacting particle systems may proceed either *synchronously* or *asynchronously*. While synchronous updating is closer to the cellular automata case that we have just considered, we focus on the asynchronous update case because most of the literature presents models in this way. Mathematical results are more easily proven for asynchronous updating because the system state changes one site at a time, compared to the synchronous case when many sites change state simultaneously. In practical terms, this means that each model time step involves randomly selecting a single site in the grid and updating it in accordance with the model transition rule. Section 6.1 considers the distinctions between synchronous and asynchronous update in more detail.

3.4.1   The contact process

The key difference between interacting particle systems and cellular automata is that state transitions are probabilistic. The most basic interacting particle system is the *contact process*, where each site is in some integer state z. Most simply, $z \in \{0, 1\}$ and the state indicates presence or absence of a 'particle', indicating some entity or condition. Although the contact process is conceptually simple, many subtly different definitions appear in the literature (see, for example, Harris, 1974, Buttell et al., 1993, Durrett and Levin, 1994b, Liggett, 1999). The differences only become significant when precise mathematical analysis is required, so we focus on the two-dimensional version described by Harris (1974) with an asynchronous update rule as follows:

 (i) If the site is occupied, then with some probability δ the particle dies.
 (ii) If the site is vacant, then a new particle is born with a probability given by the proportion of neighbouring sites that are currently occupied.

Practically, the second rule is easily implemented by assigning to the vacant site the state of a randomly selected neighbouring site. The *basic contact process* in two dimensions is the application of this transition rule to $\mathbb{Z}^2$ and von Neumann neighbourhoods are used. This set-up produces surprisingly interesting behaviour over a range of values of δ. In particular, there is a critical value δ_c of the death-rate above which the process always dies out, but below which the process will settle to an equilibrium density of occupied

3.4

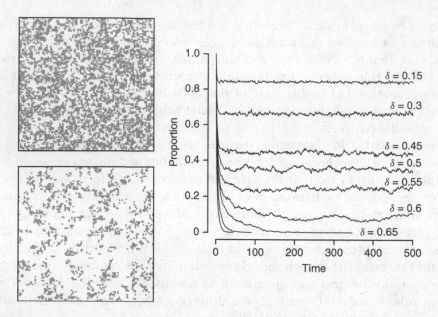

Figure 3.12 The basic two-dimensional contact process on a 100 × 100 lattice. Snapshots of the system are shown at $t = 250$ for $\delta = 0.45$ (upper) and $\delta = 0.6$ (lower). The latter case is close to estimated values of δ_c (the critical value for the process to persist). The graph shows time evolution of the proportion of occupied sites starting from 100% occupancy for different death rates δ. Here, a unit of time is a single 'generation', meaning a number of individual site state changes equal to the total number of sites in the lattice. The two unlabelled rapidly decaying time series are for $\delta = 0.9$ and $\delta = 0.75$.

sites, with the density continuing to change over time. At $\delta = \delta_c$ the process will eventually die out, but it survives for a very long time (Figure 3.12).

According to Buttell et al. (1993, page 459) δ_c for this system is around $\delta \approx 0.607$ and for $\delta = 0.6$ in Figure 3.12 it is clear that the system is long-lived in spite of the low overall site occupancy rate over time. For higher death rates the system dies away quickly, and for lower death rates it quickly settles around a long-run equilibrium occupancy rate. As we might expect, the equilibrium occupancy rate of the system falls as the death rate increases. Precise values for the equilibrium occupancy level are not known, although Durrett and Levin (1994b) provide empirical estimates. Snapshots of the spatial configuration of the process are shown in Figure 3.12 at $\delta = 0.45$ and $\delta = 0.6$, and it is apparent that they are clustered relative to a pure random distribution of occupied sites, although the aggregation effect is more apparent in the sparsely populated case. The contact process may be used as a simple model for growth processes and is closely related to standard models of epidemic spread (see also Chapter 5).

3.4.2 Multiple contact processes

A variation on the basic contact process is to introduce multiple non-vacant states each with different birth and death rates. In such cases a site's state is $z_i \in \{0, 1 \ldots, n\}$, meaning that sites may be vacant ($z_i = 0$) or occupied by one of n particle types. If there is no direct interaction between the states, so that the presence of one has no effect on the other, this set-up is the *competing contact process* (see Durrett, 2009). Snapshots of the competing contact process make its aggregated nature clear. For example, Figure 3.13(a) shows competing contact processes (implemented as for the basic process described above) both with $\delta = 0.5$. The overall proportion of live sites is around 0.35, as for the basic process in Figure 3.12, and the two different non-vacant states are aggregated into regions dominated by their own type.

Of interest here is the fact that competition for space between the two live states leads to surprising outcomes even when both states have relatively low death rates. Thus, in case (c), although δ_2 is only 0.4, which would normally be expected to result in widespread occurrence of that state (see Figure 3.12), competition for space with a state that has slightly lower mortality ($\delta_1 = 0.38$) sees it reduced to occupying only a small proportion of the space. In fact, it turns out that in this case, *where the neighbourhood for each state's contact process is identical* indefinite coexistence of the two states is impossible unless $\delta_1 = \delta_2$, although where the two rates are similar, the states can coexist for extremely long periods. More generally, and again *when the two processes have the same neighbourhood*, if rates for the birth process are introduced,

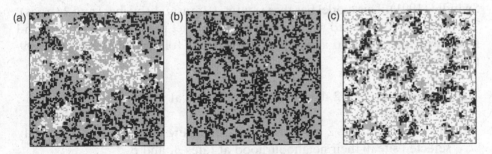

Figure 3.13 Snapshots of the (two-state) competing contact process. (a) Both live states have the same death rate $\delta_1 = \delta_2 = 0.5$. Distinct regions of state 1 (white) and state 2 (black) against the vacant state 0 (grey) are apparent. In (b) $\delta_1 = 0.58$ and $\delta_2 = 0.5$, and although state 1 might be expected to survive alone, the space denied it by sites occupied by state 2 will ultimately lead to its extinction. In (c) $\delta_1 = 0.38$ and $\delta_2 = 0.4$, and although both states have low death rates and are expected to do well, competition for space sees state 2 struggling to survive, even though the difference in death rates is small. Note that the appearance of vacant sites clustering around state 2 sites is an optical illusion: vacant states are more likely between regions of high concentration of *either* non-vacant state.)

so that replacement of the vacant state $z = 0$ with the state of a randomly chosen neighbour is not guaranteed but occurs with a probability β_i for each state $i > 0$, then the critical parameter is β_i/δ_i, and the eventual winner will be the state with the largest value of this ratio.

3.5 Where competing contact processes operate on different neighbourhoods, then a process with lower β_i/δ_i may be able to coexist with an otherwise dominant process. This result was proven by Durrett and Schinazi (1993) in a reworking of an earlier model of competition between a perennial species spreading by local contact and an annual spreading by seeding (Crawley and May, 1987). Unfortunately, the result is difficult to demonstrate in a simulation, as it requires δ_c (or β_c/δ_c where appropriate) of the local contact process to be close to the critical value. Coexistence is possible because the local process leaves vacant areas that can be exploited by the non-local process, allowing it to survive. A second local contact process would be unable to exploit these vacancies because many of them would already be occupied by the process itself. Bolker et al. (2003) discuss these effects in the context of plant community dynamics in some detail, and real-world systems show many features similar to these simple models.

The long-term dynamics of interacting particle systems with many states are difficult to predict, especially when direct interactions are introduced between different states, rather than having them merely compete for vacant space. In an important paper Durrett and Levin (1994a) provide guidelines for what to expect in such cases. They recommend inspection of expected rates of change in the number of sites in each state at an imaginary equilibrium. To see

3.6 how this works, we consider the succession model described by Durrett and Swindle (1991). This model has three states: 0 (grass), 1 (bushes) and 2 (trees). The model is formulated as two competing contact processes for bushes and trees, with grass treated as the 'vacant' state, as follows (see Durrett, 2009, page 481):

(i) Sites in state 1 or 2 die, reverting to state 0 at rates given by δ_1 and δ_2, respectively.

(ii) Sites in state 1 or 2 give birth and send new particles to a randomly selected site in their neighbourhood at rates β_1 and β_2.

(iii) If a new particle in state i arrives at a site where a particle in state j is already present, then it replaces that particle only if $i > j$.

The last rule is a *succession* relationship, since it means that trees can replace (succeed) bushes or grass, and bushes can succeed grass (but not trees).

Adopting the approach suggested by Durrett and Levin (1994a) we write approximate ordinary differential equations for the proportion p_i of the landscape in each of the states $i \in \{1, 2\}$ under *mean field* assumptions. The mean field approach assumes that every location in the model is identical and subject to rates of change that are a function of global relative occupancy rates or other conditions, rather than of local conditions. A mean field assumption ignores spatial aspects of the model such as neighbourhood and aggregation effects. This makes any mean field analysis approximate, but provides a more tractable starting point for analysis. We can see how this works in this simple case if, we form equations for the expected rates of change of each state:

$$\frac{dp_1}{dt} = \overbrace{\beta_1 p_1(1 - p_1 - p_2)}^{\text{birth}} - \overbrace{\delta_1 p_1}^{\text{death}} - \overbrace{\beta_2 p_1 p_2}^{\text{succession}}$$

$$\frac{dp_2}{dt} = \beta_2 p_2(1 - p_2) \qquad - \delta_2 p_2 \tag{3.5}$$

The first term in each equation estimates the birth rate for each state, based on the state's birth rate and the space available for it to colonise, which for state 1 is the space not occupied by states 1 or 2, that is, $1 - p_1 - p_2$, while for state 2 it is the space not occupied by itself, $1 - p_2$. The second terms similarly estimate the death rates, which are a simple function of the death rate and occupancy rate for each state, $\delta_i p_i$. Finally, state 1 faces further losses due to succession to state 2 as represented by the last term in the first equation. We estimate final stable proportions of each state in the system, p_i^*, by setting these rates of change to zero and solving for the right-hand sides. For state 2, this gives us:

$$\beta_2(1 - p_2^*)p_2^* - \delta_2 p_2^* = 0$$

$$p_2^* = (\beta_2 - \delta_2)/\beta_2 \tag{3.6}$$

as an estimate of the equilibrium site occupancy by state 2. Substituting this result back into Equation 3.5, after some algebraic manipulation, we get:

$$p_1^* = 1 - \frac{\delta_1}{\beta_1} - \frac{(\beta_1 + \beta_2)(\beta_2 - \delta_2)}{\beta_1 \beta_2} \tag{3.7}$$

According to Durrett and Levin's (1994a) conjecture, if there are positive solutions for the equilibrium occupancy rates of both states, then there are model parameter combinations that will produce long-term coexistence. The expression for p_2^* is positive provided $\beta_2 > \delta_2$ because state 2 is a simple contact process that can invade sites occupied by either of the other two states. For state 1 we can set the right-hand side of Equation 3.7 to be greater than zero and some more algebra gives us:

$$\beta_1 \frac{\delta_2}{\beta_2} > \delta_1 + \beta_2 - \delta_2 \tag{3.8}$$

for $p_1^* > 0$. Following Durrett (2009) if, to make things simpler, we consider only cases where $\delta_1 = \delta_2 = 1$ this reduces to

$$\beta_1 > \beta_2^2 > 1 \tag{3.9}$$

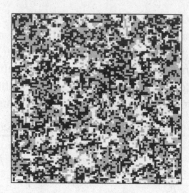

Figure 3.14　Snapshot of the grass–bush–trees succession model with $\delta_1 = \delta_2 = 1$, $\beta_1 = 6$ and $\beta_2 = 2$ so that long-term coexistence is possible. Grass is shown in grey, bushes in white and trees in black.

so the birth rate of state 1 must be greater than the square of that of state 2 for coexistence. Because this analysis makes the mean field assumption it is not precise, but tells us where in the parameter space to look for interesting outcomes. A snapshot for this model consistent with the condition allowing long-term coexistence is shown in Figure 3.14.

3.4.3　Cyclic relationships between states: rock–scissors–paper

3.7
The mean field analysis method is applicable to many models of this kind. Durrett and Levin (1994a) suggest that a number of interesting outcomes are possible, and these are reviewed by Durrett (2009). An example with a particularly interesting spatial behaviour involves a slight adjustment of the relationships between states in the succession model, so that each state has one state that it can invade, and one which can invade it. This produces a cyclic relationship between states so that $1 > 2 > \ldots > n > 1$ for the n states in the system, and we can say that state 1 dominates state 2, or that state 1 defeats state 2. The essence of this situation is found in the familiar game of rock–scissors–paper where rock beats scissors, scissors beats paper and paper beats rock. In ecology such systems are called competitively intransitive networks and there is some (disputed) evidence of real-world examples (Buss and Jackson, 1979).

An interacting particle system based on this interaction for a three-state system is easily formulated (see Frean and Abraham, 2001):

(i) At each time step, select a particle i at random and also one of its neighbours, j.

(ii) If the states $z_i = z_j$, then nothing happens.

(iii) If the neighbouring site dominates the central site, that is $z_j > z_i$, then the state of the central site changes to that of the neighbour, $z_i \rightarrow z_j$.

(iv) Otherwise, if $z_i > z_j$ then the state at j changes to that of i, $z_j \rightarrow z_i$.

The mean field differential equations governing this system each have the form

$$\frac{dp_i}{dt} \propto p_i p_{i+1} \tag{3.10}$$

because the rate at which the state can grow is proportional to its prevalence in the system and to the prevalence of the state that it dominates. For a three-state system it is clear that there is a possible equilibrium when $p_1 = p_2 = p_3$, provided the rates at which states invade one another are the same. The interesting feature of this system is that this equilibrium is not static when the system plays out in space. This is because at the interface between regions occupied by a state and the state it dominates the dominant state will tend to advance into the region held by the weaker state. But each state has another state that it dominates, and so they all have somewhere to go. The net effect is constant motion as contiguous regions of each state advance into regions held by the state they dominate and retreat from contiguous regions of the state that dominates them.

In Figure 3.15 this is clearly seen in the snapshots and time series display. The overall occupancy rates of each state in the system cyclically follow one another. When the presence of the black state peaks at $t \approx 9, 17$ and 27, this leads to an increase in its invader state (grey) over subsequent time steps. When the presence of the grey state peaks, the white state benefits, leading to a subsequent peak in the presence of the white state from which the black

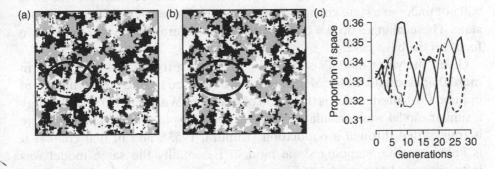

Figure 3.15 The rock–scissors–paper model. Successive snaphots show how the system is in constant motion. The colours shown are grey > white > black > grey. The indicated region in the first panel sees the grey regions invade neighbouring white regions. The overall system proportions in each state are shown in the time series: black regions in the heavy line, grey regions in the dashed line and the white regions by the lighter line.

state benefits in its turn. This pattern repeats endlessly, with the whole system in constant motion and cycling at or close to equilibrium occupancy rates determined by the relative invasion rates. In the illustrated example, and in the rules presented above, we have assumed that the invasion rates of each state of others are equal. It is easy to adjust the invasion rates, which will alter the equilibrium around which the system cycles, but not the qualitative nature of the dynamics. Durrett and Levin (1998) provide a good overview of many models of the general type we have considered here.

3.4.4 Voter models

The driver of the contact process is localised birth and death processes, combined with spatial exclusion, that is, each available site may only be occupied by a single state. This framework lends itself to many elaborations. In essence, all that is required, having selected a site, is a probabilistic rule for updating that site's state.

 3.8
A rule closely related to both the majority rule cellular automaton and to IPS models of successional dynamics is the voter rule, which can be formulated in one of two ways. An approach similar to the majority rule CA is for the site to deterministically change to the modal state in the neighbourhood, with ties broken by random selection between the tied states. The behaviour of this rule is the same as that of majority rule cellular automata models, so we do not consider it further here (see Section 3.3.1). The other approach is that having selected a site, it changes state to match that of a randomly selected neighbour. This is equivalent to making the probability of transition to a different state equal to the fraction of the neighbourhood that is occupied by that state. The model behaviour can be adjusted slightly by making state transitions occur at a specified rate, so there is a chance that selected sites will not undergo a state change. Transition rates can also be differentiated by state. These changes do not qualitatively affect overall behaviour, and so we focus on the basic case.

Given the voter model's simplicity, it is no surprise that it has origins in more than one field. In social science it has been used to study the dynamics of opinion formation, which is the origin of its name (Weidlich, 1971). In biology a similar model was formulated even earlier to explore how genetic change is transmitted through a population (Kimura, 1953), and in that context it is known as the 'stepping stone model'. Essentially the same model was later presented as a model of invasion or conflict processes by Clifford and Sudbury (1973), who also had the analytically valuable insight that changes in the patterns produced by the model happen only at the interfaces between regions in different states. Liggett (1999) discusses the mathematical details, but for our purposes a qualitative understanding is sufficient.

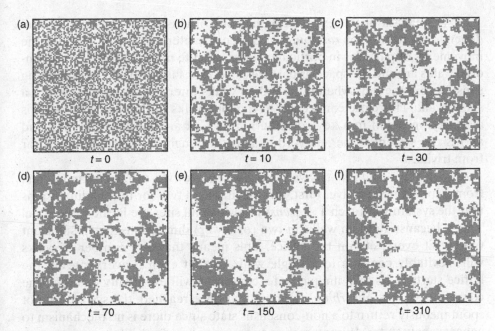

Figure 3.16 Successive snaphots of a two-state voter model on a 100 × 100 toroidally wrapped grid using a von Neumann neighbourhood. Note that the number of generations between snapshots doubles from each panel to the next. The emergence of clusters of one or the other state is apparent, as is the relative stability of large clusters over time.

Most obviously, relative to the majority rule automaton, which freezes to a stable configuration (see Figure 3.6), the ongoing random selection of neighbouring states introduces persistent dynamism into the voter model. However, the constant shifting of states is not without some regularities. Extended homogeneous regions of the same state, usually termed *clusters*, emerge and are persistent. The sequence of system states in Figure 3.16 makes it clear that although the system is constantly changing, larger clusters are relatively stable over extended periods. This effect is also apparent in Plates 2, 3 and 4. Large clusters are more stable than smaller ones because changes in the state of a cluster can only happen starting from sites on the boundary with a cluster of sites in a different state. At the cluster boundary the chance of a particular site-switching state is likely to be around 0.5, but at a distance k sites 'in' from the boundary, a flip in state will only happen with a probability of the order of 0.5^k, since at least k intervening sites must all change state before the interior site can change. As a result, the rate at which a large cluster can be invaded and completely overtaken by a neighbouring cluster is very low. It is notable that the stability of large clusters is an effect also observed in experimental systems (see, for example, Stoll and Prati, 2001).

Exact analysis of the expected behaviour of clusters in the voter model depends on results for *random walks* (see Chapter 4). This is because the movements of particles in particular states execute multiple intersecting random walks around the space as particles invade or fail to invade neighbouring sites. In this analogy, when two random walks intersect they *coalesce* into a single walk. This close connection to random walks in various dimensions is discussed in detail in a key contribution by Cox and Griffeath (1986), and allows the system to be analysed exactly, although the mathematics is far from trivial.

Among the results that have been proven are two of particular interest from a spatial perspective. First, it is inevitable in two (or fewer) dimensions that the system will reach a *consensus* in which all sites are in the same state. This is because random walks in two (or fewer) dimensions, given sufficient time, visit every site on the lattice. This means that any number of walks will inevitably coalesce to a single walk, so that even if every site on the lattice starts in its own unique state, consensus will eventually emerge. The consensus state is *absorbing*, meaning that once reached, the system cannot spontaneously return to a non-consensus state since there is no mechanism to generate new states. It is interesting to consider how long it takes a system to reach (global) consensus. Cox and Griffeath (1986) showed that the expected number of steps s_N for N lattice sites to reach consensus each starting in a unique state is given by $s_N \approx (1/\pi)N^2 \ln N$ for large N (see also Scheucher and Spohn, 1988, Durrett and Levin, 1994b, Castellano et al., 2009). This can be a very long time indeed, for example a 100×100 lattice of $10\,000$ sites is expected to take around 3×10^8 state changes (or 3×10^4 generations) to arrive at consensus!

Another useful result is that the distribution of cluster sizes after s state changes will include clusters at all length scales from $s^0 = 1$ up to $\sqrt{s/N}$ (Sawyer, 1977, Scheucher and Spohn, 1988, Castellano et al., 2009). These results are only properly true for an infinite lattice, since the maximum cluster size is limited on any finite lattice, but it indicates that we can expect clusters to grow with the square root of the elapsed time, meaning that their rate of growth slows quite quickly and that if clusters take s time steps to reach a particular size it will take $4s$ steps for them to double in size. In a finite system the slowing in cluster growth rates also reflects the limit placed on the maximum cluster size by the overall system size N.

3.4.5   Voter models with noise mutation

An important variation on the basic voter model is to introduce random switching of states with non-adjacent sites in the system and random mutations of states. A simple example might, with low probability p_n, update the selected site not by selecting a neighbouring site, but by randomly selecting any other site in the system. At the same time, with low probability p_m, update might be by *mutation* to an entirely new state not present elsewhere in the system.

The implementation of these options will vary from case to case, but the general effects are qualitatively similar regardless of the details. Non-local updating disrupts the formation of very large contiguous clusters of a single state, although ultimately, with no way for the initial number of states to change, consensus remains inevitable. More relevant from our perspective is the more fragmented pattern that is produced, which might be of interest if this type of model were being used as a phenomenological model of landscape structure. Examples of a two-state voter model with non-local update are shown in Figure 3.17(a) and (b). These should be compared with Figure 3.16(e), which shows the same elapsed model time starting from a random initial state as for these cases. Even the low probability of non-local updates in Figure 3.17(a) strongly disrupts formation of large clusters. With $p_n = 0.2$, as in Figure 3.17(b), the probability of a site updating to the state of a site anywhere in the lattice becomes equal to that of each of its neighbours, and the effect on the pattern is marked. With $p_n > 0.2$ the model rapidly transitions to effectively random noise.

Mutation has a dramatic effect on the model, even at very low rates. Most fundamentally the system can now *never* reach consensus because the disappearance of some states is counterbalanced by the appearance of new states. The snapshot shown in Figure 3.17 (c) shows that in a system which starts with only two states it is hard for new states arising from mutations to 'invade' the system. New states are rare when they first appear, and so tend not to persist for very long, and many states appear for only a brief period

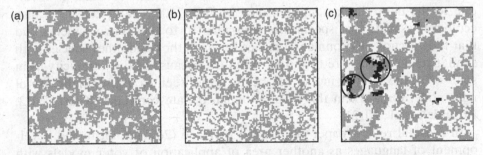

Figure 3.17 The effects of non-local update and mutation on a two-state voter model: (a) has $p_n = 0.01$, (b) has $p_n = 0.2$. Case (c) shows the effect of even a low mutation rate and has $p_m = 1.5 \times 10^{-4}$. Although the initial two states remain dominant in the system, ten other states are present in this snapshot (two are circled) and 238 different states in total have existed in the model over time. All snapshots after 150 generations on a 100 × 100 lattice.

of time. However, over time, mutation injects enduring dynamic activity into the system.

One way to think about this model is as a hierarchical set of nested spatial levels of interaction. State updating from neighbouring sites on the lattice is a local process. Updates from non-local sites anywhere in the model space are a global (whole-model) process. This spatial element is crude and it is easy to imagine altering it so that replacement states could be drawn from within some distance range in the space, rather than from any site. Finally, the introduction by mutation of new states into the model can be thought of as representing immigration from a wider world outside the model space.

As the terminology 'mutation' might suggest, such models, albeit aspatial ones, have been developed with essentially this structure to explore the dynamics of genetic evolution in biology (Kimura, 1968, 1983) and more recently to examine the dynamics of ecological communities (Hubbell, 2001), where the overall structure is closely related to earlier models of island biogeography (MacArthur and Wilson, 1967). The ecological analogy is perhaps easier to appreciate. The idea is that the sites are spaces available for occupation by plants of various species in a densely packed forest. All individuals have the same mortality rate, so one at a time we randomly select one and remove it. The space opened up is most likely to be filled by an individual of the same species as a neighbouring individual (the local process), but, with low probability, it may be replaced by a more remote individual, and even less probably an individual of a species new to the region may arrive from outside. This neutral model—neutral because all species at a given trophic level are assumed equivalent (i.e. no species has any competitive advantage)—produces species–abundance distributions that match empirical distributions remarkably well. The model has been controversial, however, due to the neutrality assumption (see McGill, 2003, Clark, 2009) because ecologists assume that species are well-adapted to their environments and that they occupy environmental niches in which they and other species with similar requirements are favoured. As much because of this simplicity as in spite of it, the neutral model has begun to be seen as a potentially useful null model for work in this area, although debate continues (Clark, 2012, Rosindell et al., 2012).

In a useful review paper Blythe and McKane (2007) point to the development of languages as another area of application of voter models with mutation. It is also easy to see how the idea of mutation and influence at a distance can be imported back to the context of opinion formation and influence from which the voter model gets it name. That such a simple model provides food for thought in so many diverse fields is impressive in itself. We anticipate that understanding spatial aspects of such models will be an important area of research in future (see Etienne and Rosindell, 2011).

3.5 Schelling models

Among the most influential of all spatial models is Thomas Schelling's model of segregation (Schelling, 1969, 1971). This can also be considered an example **3.10** of movement and search (see Chapter 4), but the aspect emphasised by Schelling himself, and also in the literature, is the segregated outcomes that the model produces. The set-up is as follows. Two types of individual are located in cells on a two-dimensional grid. A proportion of cells, p_v, must remain vacant to allow individuals to reorganise themselves. Individuals tolerate individuals of opposite type in their neighbourhood (a Moore neighbourhood is often used), but desire to be in locations with some minimum proportion of neighbouring individuals p_{like} of the same type as themselves. Any individuals dissatisfied with their current location move to the nearest available location at which their requirements are satisfied. Rounds of relocation of individuals are repeated until all individuals are satisfied or until no more can be successfully relocated. It is possible to formulate this model (or a very close relative of it) as an interacting particle system, when it is an *exclusion process*, where sites change state only by swapping with sites in other states so that the total number of particles in each state remains fixed (see Liggett, 1999). However, an agent- or individual-based formulation as presented here is favoured and **3.11** dominates the literature: it was how Schelling originally presented the model, and this perspective focuses attention on the intentions of the individuals, which are central.

A typical random starting arrangement and two final configurations are shown in Figure 3.18. Schelling's interest in the model lay in an apparent disparity between the preferences of the individual actors in the model—their *micromotives*—and the aggregate outcome—the *macrobehaviour* (see Schelling, 1978). The disparity is evident in Figure 3.18, where although individuals only desire that $p_{like} = 37\%$ of their neighbours be like themselves, the final stable state has a 'like neighbour' proportion as high as 75%.

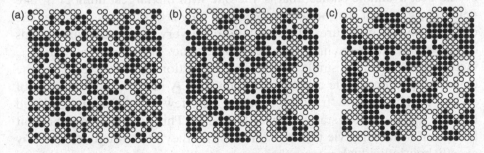

Figure 3.18 Typical start (a) and final configurations (b and c) for the Schelling model, here run on a 25 × 25 grid, with $p_v = 0.15$ vacant cells, even numbers of black and white individuals and a desired minimum proportion of like neighbours $p_{like} = 0.37$. The actual mean proportions of like neighbours arrived at in the final configurations are 0.718 in (b) and 0.749 in (c).

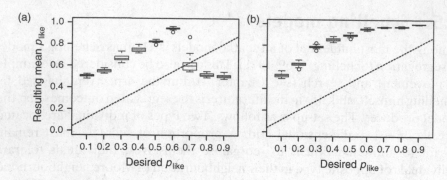

Figure 3.19   The disparity between desired and final stable proportion of like neighbours in the Schelling model (a) when individuals move only if the destination meets their requirement for like neighbours and (b) when individuals move to the best available location, even if it does *not* meet their requirements. The dashed line in both figures is where the resulting proportion of like neighbours is equal to that desired. Boxplots show results for 30 simulations at each setting of p_{like}.

We can explore this further by running an experiment where we change p_{like} and record the proportion of like neighbours that result (see Figure 3.19 (a)). Two points are noteworthy. First, up to $0.6 < p_{like} < 0.7$ the final stable proportion of like neighbours that results is considerably higher than the proportion desired by individuals, peaking when $p_{like} = 0.6$ at close to one, meaning that almost all individuals live among neighbours exclusively of their own type. Second, while we might expect similar outcomes as we further increase p_{like}, the segregation behaviour actually fails. This is because it becomes impossible for individuals to improve their situation by moving because p_{like} is so high that although they are unsatisfied, they are, in the initially random state, unable to find any alternative satisfactory location.

On the other hand, we can make a small change to the behaviour of individuals so that when they move they demand only that the new location be the best available choice, that is, the one with the largest number of like neighbours, even if that number is less than they require. This produces the outcome shown in Figure 3.19(b), where even very demanding individuals can be satisfied and the final arrangement becomes more segregated as p_{like} increases. This effect is similar to that seen in Section 3.3.1 with regard to the contrast between simple and twisted majority CA rules. Greedy pursuit of an end goal can create 'frustration' where very few individuals are satisfied because a non-optimal state becomes locked in. This points to an important feature of Schelling-style models, which is that they are strongly affected by exactly how individuals pursue their goal of finding a satisfactory location.

Another seemingly minor change to the model that may produce dramatic changes in outcomes is to introduce a background relocation rate regardless of whether or not individuals are dissatisfied. Such random relocations can free up space and so allow dissatisfied individuals to find suitable locations.

Schelling models are also sensitive to the order in which the requirements of dissatisfied individuals are resolved. For example, significantly different outcomes may result if all the individuals dissatisfied at the beginning of a 'round' of relocations are given an opportunity to relocate before any newly dissatisfied individuals. Typically, this leads to more 'frozen' configurations than when one dissatisfied individual at a time is selected for relocation. This is because the latter approach favours avalanches of relocations triggered by a single relocation.

Many variations on the Schelling model have been explored and many of the effects discussed above have been investigated in an extensive literature. The most important paper in this regard is Fossett's (2006) exhaustive exploration of the interactions between the demography of the model, that is, the relative proportions of individuals of each type and the preferences of each type. Aspects such as how the size of the local neighbourhoods considered by individuals when relocating affect outcomes (Laurie and Jaggi, 2003) have been studied, as have some of the effects of individual choice discussed above (see, for example, Bruch and Mare, 2006, Zhang, 2009).

It remains controversial how relevant such a simple model is to the real world (see Clark, 2006, Goering, 2006, for contrasting opinions). For example, a key finding from Fossett's (2006) work is that even if a minority group (one that forms, say, 10% of the population) is very tolerant of the majority, desiring to have (say) as few as 20% of neighbours like themselves, the disparity between their global presence (10%) and desired local presence (20%) can still drive the system to segregation, irrespective of majority group preferences. This result is important, but it is also important not to over-interpret it. The 'individuals' in this model are still just particles, not households. Fossett's finding does not imply that minority groups are somehow to blame for finding themselves in ghettos. The language of 'preference' and 'tolerance' surrounding the Schelling model can give the appearance that such claims are being made. It is important to keep in mind that we could just as easily interpret the movement of minority households into friendly neighbourhoods as arising from an inability to access neighbourhoods with a high presence of the majority, a reading that would make the driving mechanism not preference but discrimination. Such debates are not about the model's outcomes but about its interpretation and what can be inferred from it.

Recently, the Schelling model has come to the attention of more mathematical disciplines and it seems likely that there will be rapid progress in understanding the dynamics of Schelling-class models in the abstract (see, for example, Vinkovic and Kirman, 2006, Rogers and McKane, 2011). It is not clear, however, how much this will advance understanding of social segregation *per se*, given the difficulties of measuring the actual residential preferences of households (Clark and Fossett, 2008). From a spatial perspective, arguably the most urgent need is to understand better the interaction between individual preferences and the spatial arrangement of suitable opportunities: is it the requirements of individuals which drive the segregation behaviour of the Schelling model or the limited choices available, and how do these two effects interact spatially? This is an area ripe for further research, which could profitably revisit related earlier work (Portugali, 2000).

3.6 Spatial partitioning

All the models considered so far in this chapter produce aggregation and/or segregation as a result of local processes scaling up across a whole model space. Top-down models are an alternative way to generate such patterns. Top-down approaches tend to be more static in nature, although there is nothing in principle preventing them from being applied in an iterative fashion, much as an iterated Thomas process (see Section 2.2.5) involves iteration of an otherwise static model. We consider two approaches, to give a feel for the possibilities. Interested readers should consult the references provided for a more thorough coverage.

3.6.1 Iterative subdivision

3.12 It is common in urban land development for large parcels to be subdivided into smaller ones. We can model this process by starting with a square region and dividing it in two either east–west or north–south. Each subdivided area can be further subdivided in the same way. It keeps the shape of the resulting parcels 'nicer' if we take care at each subdivision to split the region in half perpendicular to its longer dimension, but this is not required. We can continue this process indefinitely so that the original region is subdivided down to arbitrarily small areas at whatever grain is desired. When subdivision occurs we assign the new parcels to one of two (or more) available states according to some rule, most simply at random.

Now, if instead of subdividing every parcel every time we randomly select just one, or make the subdivision process probabilistic so that it is not guaranteed to happen to all units, then interesting spatial patterns develop (see Figure 3.20). Other variations can make the state of each parcel the same as its parent parcel, but subject to chance mutations, or make subdivisions at each step uneven (that is the subdivision may lead to parcels of unequal size), or create more than two parcels, when patterns that conform more closely with organic patterns may develop. Any version of the process effectively grows a tree structure. The original whole region is the root of the tree and each subdivision produces two branches. At any stage, the 'leaves' are parcels that have not themselves been subdivided.

We are not aware of any published models based on this approach. However, it is closely related to various spatial data structures, particularly binary trees and quadtrees (see Samet, 1990), and also to the 'broken

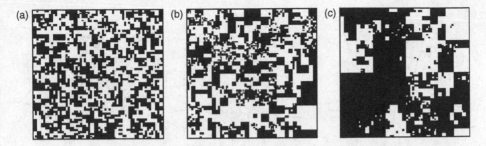

Figure 3.20 Three examples of maps produced by repeated subdivision. The space is a 128 × 128 lattice, with each parcel in one of two states (black or white) and subdivision carried out until 2000 parcels have been generated. (a) The result when parcels are subdivided with probability proportional to area and states are assigned with equal probability at the time of subdivision. (b) The outcome when probability of selection for subdivision is equal for all parcels regardless of size. (c) An outcome where every parcel is equally likely to be selected, but states are inherited from the parent, only changing with probability 0.05.

stick' model of species abundance (MacArthur, 1957). In unpublished work Morgan (2011) uses this method to initialise urban landscapes for subdivision by simulated property developers. Related approaches generate subdivision patterns by growing road networks (Barthélemy and Flammini, 2008, Courtat et al., 2011), drawing on ideas for simulating fracture networks (see Adler and Thovert, 1999, and Chapter 5) as well as leaf venation patterns (Runions et al., 2005).

3.6.2 Voronoi tessellations

The approach just described is based on repeatedly subdividing a space as items are added to the system. A Voronoi tessellation on the other hand is an 'all at once' partitioning where the space is divided with respect to a set of generating points P such that the polygon associated with each point is the region of the space closer to the generating point than to any other point in P. The Voronoi tessellation of a point pattern produced by a homogenous Poisson process is shown in Figure 3.21(a). This construction has been discovered and rediscovered many times, and goes by various names: Voronoi, Thiessen, Dirichlet and proximity polygons. The standard reference is *Spatial Tessellations* by Okabe et al. (2000), which provides comprehensive coverage of the history and of the numerous variations on the basic idea. The distances used in the procedure can be generalised to any reasonable metric (Okabe et al., 1994) and recently the idea of determining distances over an intervening network has also been explored (Okabe et al., 2008).

Many possible iterations that could be applied to the tessellation procedure almost immediately suggest themselves. If each original point is moved to the

3.13

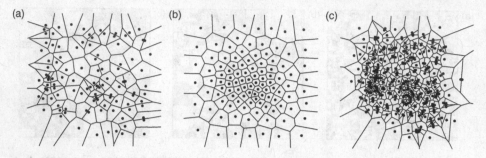

Figure 3.21   (a) A Voronoi tessellation from a point pattern generated by a homogeneous Poisson process, (b) after 10 iterations of the centroidal Voronoi tesselation procedure and (c) after two iterations of the vertex Voronoi tessellation. Note that the three figures are not at the same scale. The area covered by the centroidal tessellation grows with each iteration, while the vertex iteration patterns tend to shrink.

centroid of its associated polygon, and the tessellation recalculated, then an evenly spaced arrangement of points, with polygons of similar sizes, develops (see Figure 3.21(b)). This is a *centroidal Voronoi tessellation*—a process that has found application in a range of areas (Du et al., 1999, provide a useful review). Alternatively, if the vertices of the polygons in a tessellation are used as a new set of points for further iteration, a progressively denser and more clustered set of points develops, as shown in Figure 3.21(c). Some of the properties of the patterns produced by this procedure are explored by Boots and Shiode (2003), and similar ideas have been used to model the spatial distribution of galaxies (Martínez et al., 1990).

Both these top-down approaches generate a set of regions, and an assignment of states to the regions is required to fully define any resulting pattern. The extent to which the generated patterns are aggregated or not will depend on any autocorrelation in the assignment of states.

3.7   Applying these ideas: more complicated models

Before we consider examples of more complicated models that ultimately rest on the simple models presented in this chapter, it is useful to take stock. We have covered many different models that produce characteristic aggregated patterns and it is helpful to step back to appreciate what these examples have in common.

The driving mechanism of all the bottom-up models considered is a localised process that tends to make near things (most often grid cells) more similar, and which scales up from local similarity to a broader system-wide pattern. We have seen this idea expressed in a variety of ways. A local averaging process makes neighbouring cells more similar as it is iteratively applied. A majority rule CA produces a similar effect in binary: the local majority can be seen as an integer approximation to a local mean. Other CA rules do not lead to such obviously aggregated or segregated landscapes, but they do generate localised structure or pattern that persists over time. Interacting particle systems demonstrate that determinism is not required for a process to reliably produce aggregated patterns. In these cases, the patterns are more subtle. They are not necessarily clearly divided into distinct patches, but rather fragmented into regions more or less dominated by subsets of the available cell states. Even in the presence of action at a distance, such as occurs in variants of the voter model with mutation, distinctly clustered patterns are maintained even as they change over time. Finally, the Schelling model shows that relatively small pressures to segregate, combined with an aggregation mechanism, rapidly produce aggregated patterns, and that segregated and aggregated patterns are effectively each other's inverses. Most interestingly in this case, the dynamic of the process accelerates the overall trend: initially small differences between locations are amplified by a positive feedback effect inherent in the mechanism, and local effects quickly scale up to system-wide pattern. This is a good example of process driving pattern driving process . . . and so on.

Of course none of these models are detailed representations of anything in particular. They are generalised processes that can be used as building blocks for more complicated and potentially more realistic models of specific systems of interest. In the remainder of this chapter we consider some more specific examples informed by these simple models.

3.7.1   Pattern formation on animals' coats: reaction–diffusion models

A good illustration of the value of even the very simple models we have considered is provided by work on the mechanisms by which the rich variety of patterns of animals' coats—or, more generally, their actual body structures—might develop. Turing (1952) addressed the deep and difficult question of how embryonic development in animals can occur in the apparent absence of a detailed set of instructions for the assembly of such complicated entities. Turing suggested that a simple generic mechanism might account for such structures or patterns. He presented a model of competing activation and inhibition processes using a traditional partial differential equation-based diffusion model, and showed how such systems operating in a two-dimensional space can create highly varied patterns, following only localised rules. Turing readily acknowledged the limitations of his approach, noting the formidable mathematics involved and that, '[i]t might be possible [. . .] to treat a few particular cases [. . .] with the aid of a digital computer' (1952, page 72).

Turing's theoretical *reaction–diffusion* model has inspired a large body of work across various fields (see Koch and Meinhardt, 1994, Painter et al., 2012). His suggestion concerning computers proved especially prescient, and a simplified cellular grid version of his model was presented by Young (1984, see also Bard, 1981, Murray, 1981). Young's model strips the original down **3.14** to its bare essentials. A grid of cells may be in either of two states, activated or not. Each cell has an activation region of all other cells inside an ellipse centred on the cell and aligned to the x and y coordinate axes. An elliptical 'annular' inhibition region is defined by a larger ellipse but does not include cells inside the activation region. Cells update their state by determining the number of active cells n_A in the activation region and in their inhibition region, n_I. Then, if

$$n_A - w n_I > 0 \tag{3.11}$$

cells become (or remain) active, or if the condition is not satisfied they become (or remain) inactive, where w is the strength of the inhibition effect. Cell updating can occur synchronously or asynchronously and this may make a small difference to the outcomes (see Section 6.1). Patterns resulting after a few generations of this model are shown in Figure 3.22 (see also Plate 5) for a range of values of w. Recent experimental evidence suggests that elements of Turing's original model may indeed explain aspects of development in vertebrates (Economou et al., 2012) and may eventually prove to be a 'textbook' vindication of science advancing by model-based thought experiment!

It is notable that this model shares some of the characteristics of an early model of the *brousse tigrée* systems pictured in Figure 3.1(b) presented by Thiéry et al. (1995), insofar as both can be viewed as variants of a majority rule automaton with long-range neighbourhoods and both positive (activation) and negative (inhibition) effects. On a deeper level, this is unsurprising.

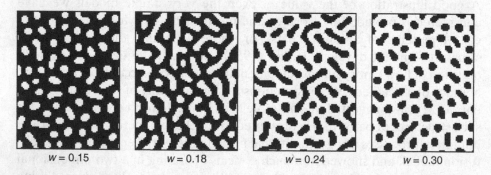

$w = 0.15$ $w = 0.18$ $w = 0.24$ $w = 0.30$

Figure 3.22 Typical outcomes from Young's (1984) cellular automaton version of Turing's (1952) morphogenetic reaction–diffusion activation–inhibition system for varying inhibition levels w as shown, and with circular activation and inhibition regions of radius 2.3 and 6.1, respectively.

Turing's reaction–diffusion framework has proven applicable to models of pattern formation in chemical oscillators such as the Belousov–Zhabotinsky reaction (Greenberg and Hastings, 1978) and the Liesegang phenomenon (Chopard et al., 1994), as well as vegetation in arid environments (Klausmeier, 1999) and patterns in insect colonies (Theraulaz et al., 2002), among others. This work is closely related to particle-based models of sediment transport, which provide insight into the formation and dynamics of dune fields (Werner, 1995, Baas, 2002) and beach cusps (Werner and Fink, 1993, Coco et al., 2000). These models are, in turn, related to interacting particle systems (Section 3.4), to the Schelling model (Section 3.5), since they involve mobile entities that stick to one another when they touch, to models of mobile entities that alter their environment (see Section 4.4), and to more general models of spread (see Chapter 5). A detailed treatment of reaction–diffusion systems would take us well beyond the scope of the current work. Ball (2009) provides a highly readable overview, while a more mathematical treatment with a particular biological focus is offered by Murray (2002).

3.7.2 More complicated processes: spatial evolutionary game theory

Game theory is a long-established field, originating in economics and concerned with decision making in social situations, where the outcomes for each party depend not only on their own decision, but on the decisions of others (Poundstone, 1992, Sigmund, 2009). The theory revolves around the analysis of stylised *games*, where two or more players have a number of *choices* and a *pay-off matrix* tells us for each combination of choices by players what reward (or punishment) will accrue to each. A *strategy* is a means for each player to make a choice when the game is played. The key idea is that players must consider what choice they think the other player might make in weighing up their own choice.

The canonical game is the *prisoners' dilemma* (PD). Here two players who are unable to communicate must choose to cooperate (C) or to defect (D). If both cooperate, they each receive a pay-off R (the *reward* for cooperation), if both defect they each receive P (the *punishment* for defection) and if one player cooperates while the other defects they receive respectively S (the *sucker's pay-off*) and T (the *temptation* to defect). If $T > R > P > S$, then regardless of belief about the other player's intention, it makes sense to defect. If I believe that you will cooperate, because $T > R$, I should defect, while if I believe you will defect, since $P > S$, it is also better to defect. Thus a rational player will choose D. The dilemma lies in the fact that the social outcome—the total pay-off to both players taken together—is one of $2P$, $2R$ or $T + S$, and since $R > P$ in cases where $2R > T + S$, it would be mutually more beneficial to cooperate. Both players know this, but cannot communicate to arrange the mutually better outcome.

What is the relevance of this abstract model to real-world decision making? In fact, the PD set-up captures the structure of many real-world choices rather well. Should I drive to work or take the bus? In most cases, the selfish (D) choice is to drive because it is individually better (temptation pay-off T), but the collective outcome is congested roads, which is bad for everyone (pay-off P). Anyone behaving altruistically (C) and taking the bus when others are driving is a sucker and suffers pay-off S. The mutually best outcome, where most people choose the bus, the roads are quiet and everyone benefits (pay-off R), is collectively hard to achieve because at an individual level the cooperative choice is irrational unless a mechanism can be put in place to enforce cooperation. In this context, road-pricing schemes, congestion charges and public transport subsidies are intended to provide the necessary incentives.

Game theory is rich in mathematical analysis (see, for example, Sigmund, 2009, or for a less formal account Poundstone, 1992), which confirms that in a one-off PD game, the rational choice is always D. If the game is repeated in an *iterated prisoners' dilemma* (IPD), then this may change things. Knowing that we will play the same player again, and that if they remember previous encounters they may punish us, changes the nature of the choice. However, in a finite series of games, if we know when it will end, it still makes sense always to choose D since the last game is equivalent to a single game and reasoning back from this point, so is the second to last, and so on. However, if we do not know that a game will be our last with a particular opponent, then it may be rational (as shown by Kreps et al., 1982) for players to cooperate for long periods of time.

The IPD can be transferred to a spatial context. Sites on a lattice each choose a strategy and play rounds of the game with their neighbours, accruing a total reward each generation depending on the outcomes. Each generation, lattice sites then update their strategy, usually based on observing which other sites in their neighbourhood have done well and switching to the most successful strategies locally. This framework is termed *spatial evolutionary game theory*, since players' strategies are changing over time in a spatial setting. Spatial games display many of the dynamics of contact processes (competitive exclusion), rock–scissors–paper (dependence on the density of dominated strategies) and voter models (clustering and persistence), and so often yield interesting dynamic patterns over long periods. Relative to interacting particle systems, the added complexity of the interactions in spatial

3.15 games makes rich dynamics possible even in deterministic cases (Figure 3.23). Nowak and May (1992) describe a simple deterministic evolutionary PD game that displays this type of rich behaviour.

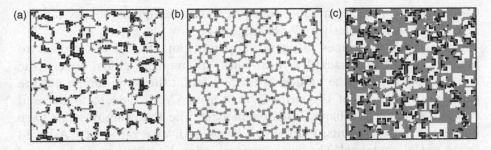

Figure 3.23 Typical configurations of the spatial prisoners' dilemma described by Nowak and May (1992) for varying levels of the defection temptation T. White regions are current and previous turn cooperators, grey regions are current and previous round defectors, while black areas have changed strategy in the current round. Cases (a) and (b) are relatively fixed, while case (c) is in constant dynamic motion, even though the model is entirely deterministic.

It is generally thought that in spatial IPD strategies favouring cooperation are more successful (Grim et al., 1998). This is because strategies tend to aggregate, and aggregation benefits cooperative strategies because they enter virtuous circles of cooperation and outperform defectors caught in vicious circles of defection. Whether spatial effects favour particular strategies in other games is an interesting research question (see Doebeli and Hauert, 2005). It is important to note that while simulation is an attractive approach, there is a substantial body of analytical results in game theory and that some influential results derived from simulation (Axelrod, 1984) have not always impressed the more analytically inclined in the field (Binmore, 1994). This clearly points to a need for theoretically well-informed model design and analysis (see Wu and Sun, 2005).

3.16

3.7.3 More realistic models: cellular urban models

It is useful to consider how the simple lattice- or grid-based models discussed in this chapter fare when we use them as a basis for more realistic models of real-world systems. One field that has grappled with this challenge is urban modelling. Among the earliest examples of a cellular model was Tobler's (1970) model of Detroit, an example of what he later termed 'cellular geography' (Tobler, 1979), which is entirely in keeping with CA-style models. An even earlier antecedent of this work was Hägerstrand's model of innovation diffusion in rural communities (Hägerstrand, 1968). However, bottom-up cellular or site-based models did not come to the fore in urban modelling until the 1990s. Important early articles pointing the way are by Couclelis (1985, 1988), who picks up on the by then considerable excitement around CA models, much of it associated with the discovery of

Conway's life and its remarkable properties, and subsequent more systematic investigations, particularly by Wolfram (1984).

Couclelis (1985) homes in on the apparent potential of CA models to link micro-scale processes (the local rules) to emergent global effects (the macroscopic patterns the models produce) but at the same time notes the probable limitations of the strict formalism of CA if it were to be used to develop realistic urban simulation models. The following decades have borne out her comments. During the 1990s, a diverse range of CA-based urban models appeared. The basic structure is for states to represent types of development (residential, commercial, industrial and so on) and for transition rules to describe how land uses interact to promote or hinder change to new uses. Yichun Xie's (1996) general framework sticks more closely to a strict CA than many later examples, but still succeeds in demonstrating a great variety of interesting behaviour (Batty and Xie, 1996, 1997, 1999). More typical are models that have eventually been used as planning support tools, such as those developed by Roger White and others (White and Engelen, 1994, 1997, White et al., 1997), and Clarke's SLEUTH model (Clarke et al., 1997, Clarke and Gaydos, 1998). The best overview of the field by the latter half of the 1990s is found in a special issue of *Environment and Planning B*, a journal which published many of the most important developments in this area (see Batty et al., 1997).

After a decade of innovation and excitement about the approach, a number of themes are apparent. Few models aiming at realism stuck closely to the formalisms of CA or interacting particle systems (IPS). Many early models employ a localised cellular process to determine relative suitabilities for land conversion (to new land uses) of all cells in the system and then probabilistically apply the associated state changes. Thus, they combine elements of the CA and IPS models of this chapter with the percolation and related models that we consider in Chapter 5. Many models rely on synchronous update, and it is interesting to speculate about how the outcomes might vary if this were not the case. The generic suitability calculation mechanism set the stage for a switch in the 2000s to agent-based approaches, in which individual land users (farmers, developers, planners and so on) make varying evaluations of the relative suitabilities of a (usually) cellular landscape and act accordingly (Parker et al., 2003, Matthews et al., 2007). Another common feature of many models is action at a distance. The incorporation of transport systems with their non-uniform effects on distance presents clear difficulties for lattice-based models. Some models cope by incorporating snapshot images of regional transport networks into the model mechanisms with snapshots updated outside the model, making longer timeline historical outcomes to some degree 'forced' (Clarke et al., 1997), but allowing exploration of alternative transport infrastructure scenarios. Where there was an interest in looking at fine-grain urban change, the idea that lattices are too restrictive a spatial structure arose (see Couclelis, 1997, and earlier Takeyama, 1996, Takeyama and Couclelis, 1997). Difficulties in evaluating the detailed forecasts produced by models were also apparent, with few well-developed methods for comparing models snapshots with real-world maps (an issue we consider in Chapter 7).

Nevertheless, progress was rapid and continues apace, with regular innovation in the structural properties of models and their level of detail. For example, *irregular CA* models running not on a regular lattice but on a graph (O'Sullivan, 2001), while initially limited in scope (O'Sullivan, 2002, Semboloni, 2000, Shi and Pang, 2000) because of their greater computational demands, have now become relatively commonplace (see, for example, Norte Pinto and Pais Antunes, 2010, Stevens and Dragićević, 2007). Another approach to more detailed representation of spatial structure is a multiscale grid (van Vliet et al., 2009). On the evaluation front, snapshot comparison approaches have evolved but remain problematic (Pontius et al., 2008, Pontius and Millones, 2011), while newer ideas also consider the path-dependency effects inherent in models (Brown et al., 2005), which may mean that two very different maps actually point to a similar general effect, for example suburbanisation to the north-west rather than to the east of the urban core.

The use of cellular urban models has become 'normal science', so that work in this field now focuses on the urban questions at hand rather than on the models. The complicated nature of current models has seen them blend into agent-based models because it is difficult to separate changes in land use from the decision-making of human actors bringing about those changes. For this reason, the simulation method research focus has moved on to agent-based models (Parker et al., 2003, Matthews et al., 2007). This evolution has seen the abandonment of simple CA/IPS models in favour of more detailed, (hopefully) realistic representations. This is certainly desirable where models are to be used to inform planning and policy decisions in real-world settings. These developments have to some extent obscured the relevance of this chapter's simple models to their complicated and increasingly distant relatives.

Nevertheless, insights from the simple models remain valuable, as Batty (2005) makes clear. Local processes or rules that favour similarity in the states of near neighbours will, over time, produce patchy aggregated landscapes. Where states compete with one another (in the urban context, land uses compete for land) then the relative rates at which they invade one another and the vacant spaces in the model will be critical to the long-term dynamics. And while probabilistic rules will lead to different outcomes each time a model runs, they are unlikely to disrupt the general tendencies and patterns produced. Understanding the simple model building blocks we have discussed is an essential precursor to approaching the simulation literature in this and other fields, even when the state of the art lies in detailed and complicated multi-component models.

4
Random Walks and Mobile Entities

In this chapter we consider a fundamental stochastic model, the *random walk*, which is the basis for a wide range of simulation models. Random walks, whether actually spatial or more metaphorically so—such as when stock price fluctuations are represented as random walks—are central to modelling change of any kind. The stochastic models considered in Section 1.4 focus on the expected distribution of aggregate outcomes of simple additive or multiplicative processes. Random walk models focus not on the aggregate outcomes of such processes, but on how individual members of the population arrive at the locations that combine to give their aggregate distribution. The key idea behind random walk models is that if we know something about the increments (the 'steps') in the walk, then we can say something useful about how the sequence of locations visited by the walk will develop over time. While none of this presumes that a walk plays out in space, random walk models are directly and naturally applicable to spatial cases, when they become a null model for processes involving the movement of entities in space. The interest of ecologists, geographers, anthropologists and social scientists more generally in spatial random walk models derives from this wide-ranging applicability.

4.1 Background and motivating examples

Examples of the sorts of phenomena where the concepts explored in this chapter may be relevant are illustrated in Figure 4.1. The movement of pedestrians is a critical driver of social systems, even in the contemporary heavily networked city. Pedestrian flows affect the success of retail businesses

Spatial Simulation: Exploring Pattern and Process, First Edition.
David O'Sullivan and George L.W. Perry.
© 2013 John Wiley & Sons, Ltd. Published 2013 by John Wiley & Sons, Ltd.

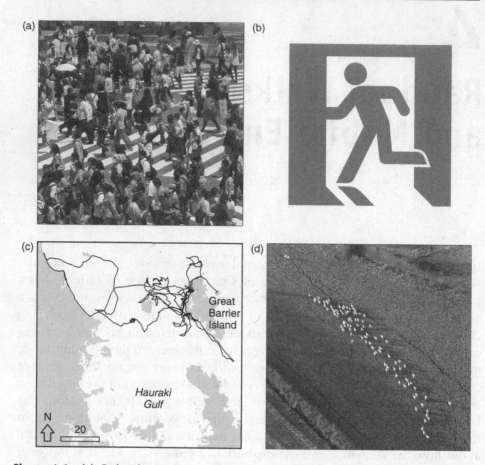

Figure 4.1 (a) Pedestrians on a crowded crossing in Shibuya, Tokyo. (b) Understanding movement in emergency situation is critical. (c) GPS tracks of nine Australasian gannets (*Morus serrator*) in the Hauraki Gulf, Auckland, New Zealand. (d) A flock of sheep in Warwickshire, England. *Sources*: (a) authors' collection, (b) © Foundation for Promoting Personal Mobility and Ecological Transportation, (c) image courtesy of Dr Todd Dennis, School of Biological Sciences, University of Auckland, and (d) photograph by Paul Englefield, 27 August 2006, http://commons.wikimedia.org/wiki/File:Sheep herd.jpg.

and have an immediate effect on how we experience cities as places to live and work, as in Shibuya, Tokyo, in Figure 4.1(a). Understanding how such flows behave in emergency situations is particularly critical, when such behaviour can be a matter of life and death (Figure 4.1(b)). In ecology and the social sciences, movement data are now routinely collected and used to aid in understanding how species and individuals interact with one another and their environment. Recent developments in the global positioning system (GPS) and other tracking technologies allied with geographical information systems (GIS) have seen an explosion in the availability of highly detailed empirical data describing the movement of people, animals and other entities. For example, the full extent of the long-range migrations of many animals, particularly birds, are now being explored using tracking, as shown in Figure 4.1(c). The complexities of herding and flocking behaviour are slowly being unravelled (Figure 4.1(d)) using both models and tracking technologies.

Random walks are a natural starting point for understanding both how to analyse such data and for simulating the processes that produce them. We develop a little of the mathematics behind simple random walks where individual steps in the walk are strictly independent of one another (Sections 4.2.1 and 4.2.2) before considering elaborations of this model where later steps in the walk are affected by earlier ones (Section 4.2.3) or by the environment (Section 4.2.4). An important class of walks undergoes super-diffusion as a result of non-finite variance of its step lengths, making their behaviour quite different, and we consider these cases in Section 4.2.5 (see also Chapter 5). In all these cases, the walk's trajectory is governed entirely by random effects and by its own previous evolution. In most real applications this is not a plausible representation, and so in later sections we consider how the evolution of a walk might be affected by the environment in which it unfolds (Section 4.3). We also look at simple models where the walk alters the environment in which it is evolving (Section 4.4), and at models of flocking where the most salient feature of the environment is the presence of other moving individuals (Section 4.5). The chapter closes with an overview of some of the ways in which random walk models have been applied in a range of disciplines (Section 4.6).

4.2 The random walk

4.2.1 Simple random walks

Although we can be fairly certain that few real mobile entities move entirely randomly, it is useful to start with a null model that incorporates no assumptions about underlying processes and to build from there. In physics, where particles often do move randomly or at least without conscious intention, and in related fields, there is an extensive literature on various types of random walk (see, for example, Berg, 1993, Spitzer, 2001, Rudnick and Gaspari, 2004, Révész, 2005, Lawler and Limic, 2010). Rudnick and Gaspari's book is recommended and Codling et al. (2008) review applications in theoretical ecology and biology, two fields which have contributed significantly to and drawn from the theory of random walks. These literatures deal with highly simplified and abstract models of random walk processes and form a natural starting point for our own exploration of these ideas.

Consider the two-dimensional integer coordinate plane $\mathbb{Z}^2$ and a walk beginning at the origin $(0, 0)$. Now, assume that a walk consists of a series of 'steps' of unit length in one of the four cardinal directions, north, south, east or west. Such a walk can only land at integer coordinate locations (x, y), so that it is moving on a *lattice*. The choice of movement direction each time step is completely random, with no bias in any particular direction and no relationship between the direction of the previous step and the current one.

4.1 What sort of outcomes do we expect from such a model? How far will a walk(er) have moved from its starting point after a given number of steps? Will it ever return to the origin? Before considering some analytical results, it is helpful to show an example. Building a model to produce this type of walk is simple, requiring only the ability to randomly select which of the four neighbouring lattice locations to move to at each step. An example walk of 1000 steps is shown in Figure 4.2. Some features are immediately apparent, particularly the fact that the walk retraces its steps often. At any moment there is a one in four chance that the next step will reverse the previous one, so this is unsurprising. The overall effect is that the walk does not get as far as we might expect. This walk of 1000 unit-length steps only gets to $(19, -25)$, a journey that could have been made more directly in a mere 44 steps.

Such tortuous progress is typical. Figure 4.3 shows the points reached after 1000 time steps of 500 such random walks. It is clear from the density plot in (b) that many walks end near the origin, even after many steps. The mean distance of these 500 walks from the origin is 27.9 units, although the average *displacement* is 0 because there is no directional bias in step direction.

To derive an expectation of the distance from the origin of a lattice walk of n steps is relatively straightforward and is a result with many connections to fundamental results in statistics and mathematics (recall Galton's board on

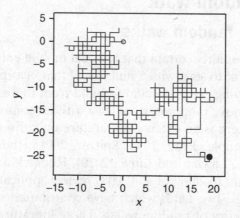

Figure 4.2 A simple 1000-step random walk on a lattice. The open circle indicates the start location at the origin (0, 0), while the black circle shows where the walk has reached after 1000 steps.

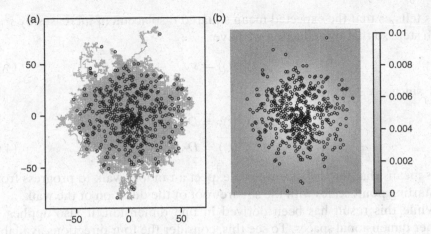

Figure 4.3   The end points of 500 random walks of 1000 steps (a) and their density (b).

pages 23ff.). The simplest approach is to consider a one-dimensional walk, where each step displaces the walker either to the right or the left along the number line. If the displacement at each step is a random variable S such that $s(t) = \pm D$, so that the walk moves either D units to the right or the left, then we can construct a *difference equation* for the position $X(t)$ after t steps, relative to its position at the previous time step:

$$X(t) = X(t-1) + s(t) \tag{4.1}$$

We can immediately deduce from this that the expected displacement at any time t is $X(t) = 0$, since the initial displacement is zero and each step is equally likely to take us in either direction. A more useful result avoids negative quantities by considering the *square root of the mean squared distance*, $\sqrt{\langle X^2(t) \rangle}$. Squaring each side of Equation 4.1 we obtain:

$$X^2(t) = X^2(t-1) + 2s(t)X(t-1) + s^2(t) \tag{4.2}$$

If we have N such walks, then the mean squared displacement is:

$$\langle X^2(t) \rangle = \frac{1}{N} \sum_{i=1}^{N} [X_i^2(t-1) + 2s_i(t)X_i(t-1) + s_i^2(t)] \tag{4.3}$$

The second term in this summation drops out because positive and negative values are equally probable, giving a difference equation for the mean squared displacement after t steps in terms of the mean squared displacement after $t-1$ steps:

$$\langle X^2(t) \rangle = \langle X^2(t-1) \rangle + s^2(t) \tag{4.4}$$

This tells us that the expected mean squared displacement increases by s^2 at each step, so that by deduction we have:

$$\langle X^2(t)\rangle = s^2 t \tag{4.5}$$

and, since $s = \pm D$, we have:

$$\langle X^2(t)\rangle = D^2 t$$
$$\sqrt{\langle X^2(t)\rangle} = D\sqrt{t} \tag{4.6}$$

This means that the distance we can expect a random walk to progress from its starting point scales with the square root of the duration of the walk.

　While this result has been derived in one dimension, it also applies in higher dimensional spaces. To see this, consider the four directions available to a lattice walk in two dimensions. Since each direction is independent of the other, the overall progress parallel to each coordinate axis is modelled by a one-dimensional random walk of half the duration because each step makes progress parallel to only one of the axes and no progress in the other direction. Thus we have:

$$\langle X^2(t)\rangle = \langle Y^2(t)\rangle = D^2 \left(\frac{t}{2}\right) \tag{4.7}$$

where $X(t)$ and $Y(t)$ denote the expected displacements parallel to the x and y axes. The root mean square distance covered $\sqrt{\langle R^2\rangle}$ is given by Pythagoras's theorem and

$$\langle R^2\rangle = \langle X^2\rangle + \langle Y^2\rangle$$
$$= D^2 \left(\frac{t}{2} + \frac{t}{2}\right)$$
$$\sqrt{\langle R^2\rangle} = D\sqrt{t} \tag{4.8}$$

Similar reasoning means that Equation 4.8 holds for a lattice walk in any number of dimensions.

　An obvious generalisation of the lattice random walk is a *simple random walk*. Here, each unit-length step may occur in any direction, randomly distributed on the interval $[0, 2\pi)$ radians. The result of Equation 4.8 again holds. In fact, this was the original model of a random or *drunkard's walk*, discussed by Pearson (1905a,b) and Lord Rayleigh (1905) in a quick-fire correspondence in *Nature*.

4.2.2　Random walks with variable step lengths

So far we have considered walks with a fixed step size. An obvious generalisation is to allow steps of varying lengths. Surprisingly, although Equation 4.8

no longer holds exactly, it turns out that the more general proportionality result:

$$\sqrt{\langle R^2 \rangle} \propto \sqrt{t} \qquad (4.9)$$

applies for many types of walk. This result is related to the classical *central limit theorem* of statistics. The very general proportionality result of Equation 4.9 holds if we allow steps of any length drawn from a probability distribution *provided it has finite variance*. Thus if we draw walk step lengths from (say) an exponential distribution with mean $\lambda = 1$, although $\sqrt{\langle R^2 \rangle}$ is greater than before it still grows proportionate with the square root of the walk duration t. Examples of a simple random walk and an exponential step-length random walk are shown in Figure 4.4. The rate of increase in the root mean square distance covered by 100 exponential random walkers is shown in Figure 4.5, where the square root shape is evident, although the absolute value of the distance is greater than that of a simple walk with the same mean step size.

4.2.3 Correlated walks

Ecologists and biologists have used simple random walks such as those described in the previous section as models of animal movement and search strategies over several decades (see, for example, Pearson, 1906, Wilkinson, 1952, Morrison, 1978, Hoffmann, 1983). It is apparent, however, that animals, or for that matter people, rarely wander completely aimlessly in the manner of such walks—not even Pearson's drunkard! We can make our random walks more 'purposeful' or *persistent* by biasing the walk direction from step to step so that the direction of the next step is chosen by selecting a random *turn angle* relative to the previous step (Marsh and Jones, 1988). If we permit only

4.1

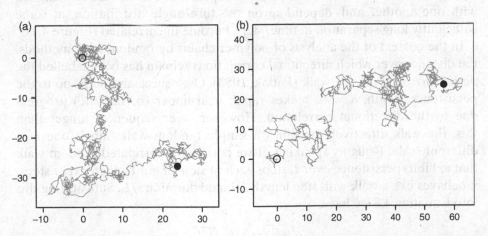

Figure 4.4 (a) A simple random walk and (b) a random walk with exponentially distributed ($\lambda = 1$) step lengths.

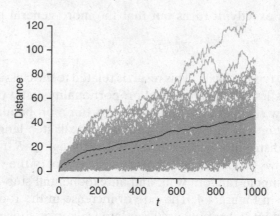

Figure 4.5 The root-mean-square distance traversed by 100 exponential step length random walks. The distances of all walks are shown in grey, and the square root of the number of steps (dashed) and root mean square distance of all walks (solid) are also shown.

small turn angles between steps, then walks move in near straight lines, and as we increase the probable turn angle walks become more tortuous. This effect is shown in Figure 4.6, where turn angles are chosen from a wrapped normal distribution with a standard deviation of, respectively, 10°, 30° and 50°. The more rapid progress of the first walk with its smaller turn angles (10°) is clear, keeping in mind the different scales on the axes of each plot.

Even so, with finite step lengths the square root proportionality relation *still* applies. The reason for this is that the directional correlation between steps is purely local so that over sufficiently long sequences of consecutive steps the correlation disappears. Thus successive steps are highly likely to be in the same or similar directions, but steps ten apart will be less strongly aligned with one another and, depending on the turn angle distribution, at some sufficiently long separation in time, steps become uncorrelated (Figure 4.7).

4.2 In the context of the analysis of polymer chains by random walk methods, the distance over which directional correlations remain has been labelled the *persistence length* of the walk (Patlak, 1953). Over successive steps up to the persistence length, a walk makes rapid, near-linear (or *ballistic*) progress due to the directional correlation. However, over sequences longer than this, the walk effectively becomes a simple random walk, albeit one at a different scale. Roughly speaking, if we consider a correlated random walk that exhibits persistence over L steps each of step length D, then over t steps, it behaves like a walk with step length DL and duration t/L. Substituting this into Equation 4.8 we have

$$\sqrt{\langle R^2 \rangle} = DL\sqrt{t/L}$$

$$= D\sqrt{tL} \qquad\qquad (4.10)$$

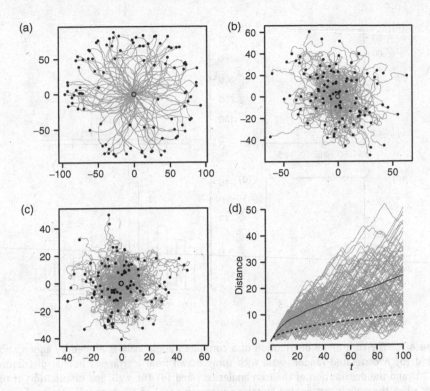

Figure 4.6   (a)–(c) 100-step random walks with turn angles chosen from a normal distribution, $\mu = 0°$ and $\sigma = 10°$, $30°$ and $50°$, respectively. Note the different scale of each plot. (d) The distance from the origin of such walks still has the characteristic $\sqrt{t}$ growth rate, shown here for the $\sigma = 50°$ case. The dashed line shows the expected (slower) rate of progress for a simple random walk.

The net result is that the expected distance associated with the walk still scales with the square root of its duration t. Since the $\sqrt{L}$ factor is a constant dependent on the degree of directional correlation, highly correlated walks will have longer persistence lengths and correspondingly larger scaling factors relative to the simple case. This point is clearly visible in Figure 4.6(d), and is even clearer in Figure 4.8, where the upper and lower straight lines show linear and square root proportionality, respectively. In each case, correlated walks show an initial linear trend but after some time this 'rolls off' to the slower square root relationship, with the persistence length decreasing progressively as the variability in the turn angle distribution increases (see Bartumeus et al., 2005).

A complete analysis of the expected distances for such walks is more complex. As Bovet and Benhamou (1988) explain, drawing on previous work by Tchen (1952) and Mardia (1972) (see also Batschelet, 1981, Mardia and

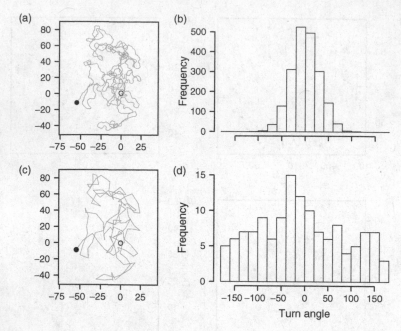

Figure 4.7 The turn angle distribution of a correlated angle walk is altered by aggregation. (a) and (b) A correlated random walk with turns drawn from a wrapped normal distribution $\sigma = 30°$ and the distribution of the turn angles. (c) and (d) The walk and distribution of turn angles when the walk is aggregated by taking successive sets of 15 steps (effectively changing the grain of the walk). The distribution of turn angles is much more uniform and similar to what we would expect of a simple random walk. Aggregation over more steps will increase this effect.

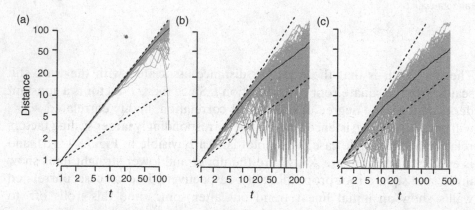

Figure 4.8 Expected distance of correlated random walks with turn angles chosen from a normal distribution, $\mu = 0°$, and (a) $\sigma = 10°$, (b) 30° and (c) 50°. Note the logarithmic time and distance scales.

Table 4.1 Scale factors for correlated random walks

Turn angle, σ (°)	r	Scale factor $\sqrt{\frac{1+r}{1-r}}$
10	0.985	11.5
20	0.941	5.7
30	0.872	3.8
40	0.784	2.9
50	0.683	2.3
60	0.578	1.9
70	0.474	1.7
80	0.377	1.5
90	0.291	1.4

Jupp, 1999), the scaling factor for the expected distance relative to a simple random walk can be approximated by:

$$\sqrt{\langle R^2 \rangle} = D\sqrt{t(1+r)/(1-r)} \qquad (4.11)$$

where r is a measure of the correlation between the directions of successive steps of the walk. For a wrapped normal distribution, as in the cases shown, $r = \exp -\sigma^2/2$ where σ is the standard deviation of the normal distribution measured in radians. Table 4.1 shows the resulting scale factors for a range of increasingly wide turn-angle distributions.

The rapid acceleration that results from a very strongly directed walk is apparent and, of course, in the limiting case with $\sigma = 0°$, the root mean squared distance varies linearly with the duration of the walk, and it becomes ballistic point-to-point movement. Although subsequent work has provided more accurate estimates of the increased rate of progress expected due to directional correlation in successive steps (see Wu et al., 2000), the values in Table 4.1 demonstrate the qualitative impact of this refinement of the model and are reliable except at higher values of r, say when $r \gtrsim 0.9$ (Benhamou, 2004).

Directionally correlated walks are an example of a broad class of walks where each step is dependent on one or more of the previous steps. For example, steps might be similar to previous steps in length *and* direction. Even something as simple as a rule preventing an immediate return to the previous location makes steps non-independent of one another. Mathematicians have explored a variety of this broader class of correlated random walks.

Lazy walks are a modification of the simple random walk where with some probability in each time period, rather than take another step the walk remains in its current location. The most likely application of this sort of walk in a spatial context would be where the walker was engaged in some other activity, such as shopping or feeding, although the mechanism driving movement in such a model would most likely relate to the requirements of that other activity (see Section 4.3). Lazy random walks are closely related to the ant in the labyrinth model discussed in Section 5.2.3, where pauses are enforced because not all movement directions are available at each step.

Reinforced walks are walks where the probability of making a particular step is greater if the link between two lattice sites has been crossed previously (edge-reinforced, see Davis, 1990a) or if the site has been visited previously (vertex-reinforced, see Pemantle, 1992). Pemantle (2007) provides an overview of both types of walk in the context of stochastic processes with 'memory' effects more generally, which are closely related to sampling processes without replacement. Such walks can exhibit surprisingly complicated behaviour, and they are relevant to understanding how terrain is explored and a territory, activity space or a path network is established by humans or other animals (see Boyer et al., 2012).

Self-avoiding walks are forbidden from returning to any previously visited location, or in more complex cases are excluded from approaching within some specified distance of any previously visited location (Madras and Slade, 1996). Such walks are commonly used as models of long-chain polymer molecules. They might also be considered a possible model for animal or human exploration in a new environment, since they force the walker to visit previously undiscovered sites.

In general, it is difficult to derive estimates of the effect of such varied inter-step dependencies on the expected distances traversed by a walk, although provided the step length distribution has finite variance, a reasonable null expectation is that the root mean squared distance traversed by a walk will be proportional to the square root of the walk duration.

4.2.4   Bias and drift in random walks

While correlation between the steps can produce the appearance of purposeful movement in a walk, another possibility is a persistent bias in the preferred movement direction. Such bias is easily incorporated into the models we have been considering by preferentially choosing the direction of each step with respect to some fixed direction. This approach can be applied equally well to lattice walks or walks in continuous space. The overall effect is to shift the peak of the probability surface associated with a walk away from the origin in the direction of the movement bias. Patlak (1953) provides a detailed, but mathematically demanding, account of this type of walk, while Turchin (1991) provides a more accessible introduction. Our interest in such walks is more practical than theoretical, and arises in situations where mobile entities or 'walkers' interact with an environment, in particular when they respond to environmental gradients, with movement biased in the direction of increasing or decreasing availability of some resource. This kind of behaviour might be considered as foraging or search, and we examine the models that arise when walks interact with an environment in more detail in Section 4.3. For now we merely note that in practice it can be difficult to distinguish the effects of between-step directional correlation and directional bias in movement paths (Cai et al., 2006), especially over short time periods, yet another example of equifinality. Over longer time periods, if bias persists it will usually be obvious, since walks with correlated step directions ultimately show no directional bias.

4.2.5 Lévy flights: walks with non-finite step length variance

As has been noted, the additive central limit theorem governed distance–time relationship of Equation 4.8 holds for walks with step-length distributions of finite variance. When the step-length distribution does not meet this criterion, the outcome is quite different. Specifically, when step lengths are distributed according to a heavy-tailed distribution (e.g. a Pareto or a Cauchy distribution), then the walk is referred to as a *Lévy flight* (Viswanathan et al., 2011). An example is shown in Figure 4.9, with step lengths drawn from a Cauchy distribution. This walk is characterised by periods of short localised steps interspersed with long-range 'jumps'. A distinction is sometimes (not always consistently) made between Lévy flights when steps in the walk are 'instantaneous' and Lévy walks where successive steps occur at constant velocity over time periods whose durations have a probability distribution with

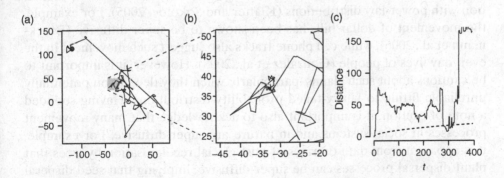

Figure 4.9 A walk of 281 steps, where each step is in a direction drawn from a uniform random distribution and step lengths are drawn from the Cauchy distribution with location parameter $x_0 = 0$ and scale $\lambda = 1.0$. (a) The full walk, (b) a zoomed-in region (grey area in (a)) and (c) the distance from the origin of the walk compared to a $\sqrt{t}$ expectation.

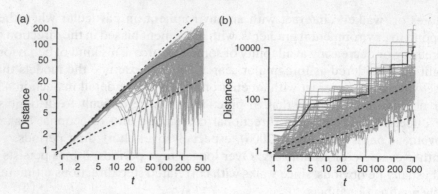

Figure 4.10 Comparison of the progress made by (a) 100 correlated random walks with exponential step lengths and (b) 100 Lévy flights with Cauchy distributed step lengths. Note the change in scales on the vertical axis and that both axes are logarithmic. The roll-off to diffusive behaviour already noted (see Figure 4.8) is clear for the correlated walks but does not occur for Lévy flights.

non-finite variance or alternatively where step length is constant but velocity varies (ben-Avraham and Havlin, 2000, Shlesinger and Klafter, 1986). For present purposes, the difference between flights and walks is unimportant, relative to a feature they share, which is that the overall progress made by the walk is dominated by the long-range jumps. As a result, such walks cover more ground more quickly than the other kinds of walk considered so far, as is clear in Figure 4.10. This characteristic of Lévy flights is termed *super-diffusive* behaviour (see also Section 5.2.3).

Super-diffusive walks, particularly Lévy flights with step lengths D, distributed according to a power law $P(D) \propto D^{-\mu}$ where $1 < \mu \leq 3$, have attracted considerable attention, at least in part because of the recent fascination with power-law distributions (Klafter and Sokolov, 2005). For example, the movement of dollar bills has been shown to be super-diffusive (Brockmann et al., 2006), while cell phone tracks also suggest such movement in the everyday lives of people (González et al., 2008). However, it is important to be cautious about such claims, particularly when they depend on potentially unreliable fitting to heavy-tailed probability distributions. Having sounded a note of caution, it is important also to acknowledge that many movement processes in social systems and in nature are super-diffusive. For example, it is apparent from data on rates of post-glacial recolonisation by trees that plant dispersal processes can be super-diffusive, implying that seed disperal kernels are heavy-tailed (Clark, 1988, Clark et al., 1998).

However, the existence of super-diffusive movement in populations does not imply that the underlying processes are Lévy flights. Take, for example, human movement. Even cursory consideration of how and why people move suggests that no single process accounts for the combination of habitual everyday movement (the daily commute), movement arising from regular but less frequent activities (such as weekend sports), movement over longer distance (such as annual vacations) and infrequent global-scale travel (perhaps due to migration or 'once in a lifetime' vacations). Thus an obvious alternative movement model to Lévy flights might incorporate multiple processes at a variety of scales, with each process modelled as a different kind of random walk. It is well understood that heavy-tailed probability distributions can result from composites of exponential distributions (see Petrovskii and Morozov, 2009), which further suggests that we might expect to find step lengths in empirical movement data consistent with Lévy flights if the generating mechanism were a combination of ordinary diffusive movement processes at a variety of scales (see, for example, Fritz et al., 2003).

In any case, all of the random walks that we have considered so far are limited as representations of any kind of intentional movement, in that they lack context or purpose, even when the resulting traces *look like* they might represent purposive movement. The mechanisms that drive these walks are entirely probabilistic and internal to the walks themselves, so that as models they are purely phenomenological: they produce effects that appear similar to observed patterns but are clearly not driven by the sorts of mechanisms that drive real entities in the real world, unless those entities are unthinking particles in random motion. In the next section we consider more purposeful models of movement.

4.3   Walking for a reason: foraging and search

The Nobel prize-winning economist and early originator of artificial intelligence Herbert Simon, considering the path of an ant, presents a simple parable:

> We watch an ant make his laborious way across a wind- and wave-moulded beach. [. . .] I sketch the path on a piece of paper. It is a sequence of irregular, angular segments—not quite a random walk, for it has an underlying sense of direction, of aiming toward a goal. [. . .] Viewed as a geometric figure, the ant's path is irregular, complex, hard to describe. But its complexity is really a complexity on the surface of the beach, not a complexity in the ant. (Simon, 1996, page 51)

Simon's analogy suggests that the explanation for observed movement patterns, whether in terms of step lengths, turning angles, diffusivity or otherwise, may lie as much in the structure of the environment in which movement occurs as in the internal logic of the moving entities.

In earlier work, Simon (1956) explores the question of how simple the search behaviour of an organism can be, yet still allow it to survive in an environment with randomly distributed resources. The context for this paper is the debate around the limits to (human) rationality in decision-making (see also Simon, 1955, Conlisk, 1996). In economics, many models posit perfectly rational, *optimising* decision-makers able to make best choices among options whose relative merits are difficult to assess (the so-called *Homo economicus*). Simon suggests that limits on the availability and accuracy of information, and on the calculative abilities of organisms, argue for *bounded rationality* rather than perfect rationality. Under bounded rationality decisions are made using 'good enough' reasoning, as in, 'this may not be the best choice, but it is good enough', an approach Simon dubbed *satisficing*. Satisficing behaviour is a feature of many individual- and agent-based models of decision-making, and Simon's (1956) paper provides a good starting point for exploring how such decision-making in search processes can account for observed movement patterns.

4.3   Simon's (1956) model is not explicitly spatial, but is easily implemented in a spatial model. Consider a 'walker' searching for a target location on a two-dimensional lattice. At each step it may move in one of four available directions. The walker has a detection radius, or 'vision', v on the lattice, so that when it comes within v lattice sites of the target it switches 'mode' to move directly to the target by the shortest available path. Once at the target, the walker collects the resource, a new target location is randomly created and the walker begins searching again. Each movement step costs a single unit of energy and the walker starts with an energy level E. When the walker arrives at a target, its energy level is recharged to E, but if at any time the energy level falls to zero the walker dies.

Simon sets out a simple analysis of this model (see Simon, 1956, page 131ff), which we rehearse here, adjusted for our two-dimensional lattice setting. Critical variables are the proportion of lattice sites that provide resources, p_R, the maximum energy level, E, and the range of vision v. At each step, if the walker is moving forward into a previously unvisited part of the lattice, then $2v + 1$ new lattice locations come into view inside the range of vision. If the walker is turning to left or right, however, only $v + 1$ new sites come into view (see Figure 4.11). If the walker returns to a previously occupied site then no new sites come into view. As a result, the number of newly visible sites each step depends on the nature of the walk and the range of vision. For now, in order to develop the argument, we will assume that the mean number of newly visible sites, $\langle v \rangle \approx v$, meaning that v new lattice sites become visible each step, although we revisit this assumption below. The probability ρ that none of the newly visible sites will be a target location is:

$$\rho = (1 - p_R)^{\langle v \rangle} \tag{4.12}$$

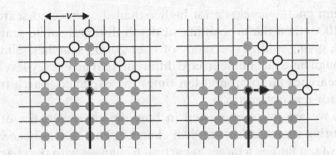

Figure 4.11 The number of newly visible sites on a lattice where the range of vision v is $2v + 1$ when the lattice walk moves forward into an unexplored region, but only $v + 1$ when the walk takes a turn to right or left. The latest movement step is shown as a dashed line. All grey spots are sites that have already been seen; black-outlined white circles are those newly visible once the step is taken.

The probability $P(k)$ that no target will be found in $k - 1$ moves, only for a target to be found on the kth move, is given by

$$P(k) = (1 - \rho)\rho^{k-1} \tag{4.13}$$

and the expected number of moves to find a target, M, is given by

$$M = \sum_{k=1}^{\infty} k(1 - \rho)\rho^{(k-1)}$$

$$= \frac{1 - \rho}{(1 - \rho)^2} = \frac{1}{1 - \rho} \tag{4.14}$$

The probability of death P_{death} is equal to the probability that the number of moves to find a target will exceed $E - v$ and is given by

$$P_{\text{death}} = P(M \geq E - v) = \rho^{E-v} \tag{4.15}$$

Equations 4.13, 4.14 and 4.15 are standard results for the *geometric distribution*. The expected number of searches for which a walker will survive is similarly given by $1/P_{\text{death}}$, which we may consider the 'life expectancy' of the walker.

To get a feel for the numbers, if we consider a case where $p_R = 0.001$ and the range of vision is ten, then ρ from Equation 4.12 is 0.990 and the expected number of moves to spot a target, from Equation 4.14, is around 100. If the maximum energy level E is also 100 then the probability of death during a single search from Equation 4.15 is given by $0.990^{90} \approx 0.405$, giving a short life expectancy of only $2.46 = 1/0.405$ searches. If, on the other hand, the maximum energy level is a lot higher than the expected search time at (say) $E = 1000$, then the probability of death during a single search is only 4.99×10^{-5}, giving a life expectancy of over 20 000 searches. Simon (1956) notes, although his exact numbers are different, that an energy level sufficient to sustain searches up to around ten times the expected search time seems reasonable for real organisms.

For now, we are more interested in how the characteristics of the walk executed during search affect survival chances. The critical issue is how the probability of not finding a target at each step changes with the characteristics of the walk. As described above, Simon's forager executes a simple random walk, and so is as likely to return to a lattice site just vacated as it is to move to one of the three other adjacent sites. This fact dramatically affects the exponent in Equation 4.12 . We can demonstrate this experimentally by measuring the total area searched for random walks with differing degrees of directional persistence and fixed vision.

Results are shown in Figure 4.12 based on running a number of random walks with varying probability of turning, from zero (ballistic movement in a straight line) to one (a simple random walk). As the probability of turning increases, the walk's directional persistence decreases, and it becomes more tortuous and more likely to revisit previously visited sites. The number of sites seen within a range v, fixed at ten lattice units, falls so that a non-turning

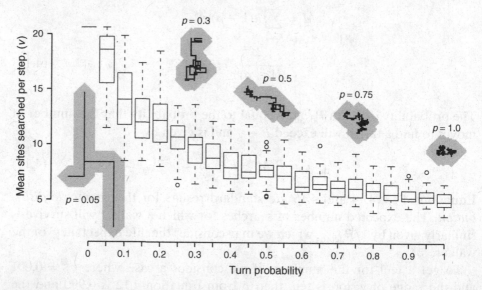

Figure 4.12 Variation in the region searched on the lattice with vision $v = 10$ for random walks as the probability of turning (p) at each step changes from zero to one. Examples of some walks and the associated 'sausage' or region explored are shown. Note that the lack of variation in the $p = 0$ case is because such walks are ballistic in a straight line with no change from run to run.

walk sees around 20 new lattice sites each step (i.e. $2v$) while the random walk sees only around five. These experimental findings are supported by work on *Wiener sausages*, the name given to the region within a specified distance of a random walk in various dimensions (see, for example, van den Berg et al., 2001). From this literature, the growth in area in two dimensions is not expected to increase linearly with time t, but to slow gradually, proportional to $t/\log t$. For now, we can treat the deceleration as a second-order effect and focus on the large differences in $\langle v \rangle$ that different walk geometries produce. Here, the change from ballistic motion to random walk behaviour sees the effective vision in terms of the exponential in Equation 4.12 fall from approximately $2v$ to $v/2$. For the numerical example given above, this changes the probability of not finding a target each time step from around 0.98 to 0.995. This may not seem like much, but the overall effect is to dramatically reduce the life expectancy when E is (say) 500 from 18 122 to just 11.6 searches. To get comparable life expectancy with the less efficient random walk movement pattern requires an approximately four-fold increase in the organism's storage capacity E. Thus, as we might expect, there are substantial benefits to be gained by searching more efficiently.

Although it is rarely recognised, probably due to the separation of the relevant literatures across economics, ecology and anthropology, much of the subsequent work on models of spatial search or foraging behaviour is effectively an exploration of elaborations of Simon's model. Below we briefly consider some of the many possible refinements of the model.

4.3.1 Using clues: localised search

If resources are not distributed purely at random, but instead tend to aggregate or cluster, then it may be profitable for a forager to adopt search behaviour that takes advantage of this. The most obvious change in behaviour is to assume that where resources have already been found, more resources are likely to be found (following Tobler's first law from Chapter 3). This entails behaviour where the region around a known resource is thoroughly explored, and also that between finds any available clues to the probable presence of resources are followed. A simple example of a clue to resource presence is to associate a resource density with each lattice location, using a kernel density method (see O'Sullivan and Unwin, 2010). This might represent how the environment provides cues suggesting that a particularly area is likely to be rich in the resource. Presented with a resource density signal an obvious search strategy is *hill-climbing*, where the forager moves to the adjacent lattice site with the highest resource density signal.

While this strategy is effective when it produces a clear preference for the next step in the walk, it can be problematic if, given no cues, the search reverts to a simple random walk. If instead of simple random behaviour between resource concentrations the forager pursues less convoluted walk behaviour, then more ground is covered and the chance of encountering a new region of higher density is increased. Two other adjustments to search behaviour can also help with more effective searching.

Path memory involves the forager avoiding locations visited previously and found to have no resources. This requires the forager to have a memory of where it has previously been so that it can avoid those locations, when the walk is a form of self-avoiding walk, or revisit them, if they have been good sources in the past, when it is a reinforced walk. This approach can lead to some problems, particularly if self-avoidance behaviour is rigidly applied, when a walk may easily 'trap' itself because all surrounding available lattice sites have already been visited. Nevertheless, early work by Benhamou (1994) suggests that memory, as we might expect, improves the likelihood of successful foraging.

Mental (cognitive) maps build on memory-based approaches and involve the foraging organism constructing a mental map of the search region. A simple mental map might record locations which have been rich in resources on the previous visit, along with the time of that visit. A more detailed map might also include pathways to known resources and some concept of any time constraints on resource availability, such as seasonality of fruiting or opening hours of facilities.

The model we present in Chapter 8 uses mechanisms similar to these, and clearly departs some way from a simple random walk. Given how complicated the memory and mental maps of foraging organisms are (whether real or in models!) it is difficult to generalise about their effects on the resulting movement patterns.

4.3.2   The effect of the distribution of resources

As Simon's parable of the ant suggests, the explanation of empirical movement data may lie in the landscapes where the movement occurs and the behaviour which the landscape causes. A more realistic example than those based on Simon's model is provided by a model of spider monkeys foraging fruit-bearing trees (Boyer et al., 2006), which produces movement patterns similar to those observed empirically (Ramos-Fernández et al., 2004). It is worth nothing that very similar approaches to that described by Ramos-Fernández et al. (2004) have been used to explore the frugivorous foraging behaviours of birds by Morales and Carlo (2008) and, albeit in a more empirically and mechanistically grounded way, by Uriarte et al. (2011).

This example provides us with the basis for a plausible model of foraging or, more generally, resource search behaviour that is easily implemented.

Consider a landscape consisting of fruit-bearing trees distributed according to a homogeneous Poisson point process (see Section 2.2.2). Each tree is assigned a total quantity of resources (say fruit or seeds) z_i drawn at random from some distribution, say the exponential. Now, a single individual, starting from a tree near the centre of the study area, first exploits the resource, reducing its remaining resource to zero. In subsequent time steps the individual moves to the tree j that maximizes the quantity z_j/d_{ij}, where z_j is the resource available and d_{ij} is the distance to it. Having reached a new resource, the resource available is again reduced to zero. A typical walk produced by this process is shown in Figure 4.13, along with the distributions of the site resources and walk step lengths. Notably, the two distributions appear very similar. In Boyer et al.'s paper (2006) power-law distributed resource availabilities produce similarly distributed step lengths in the resulting walks. As a further example, Figure 4.14 shows how the walk and walk step-length distribution are changed when the resource distribution is log normal. Although little detailed work has been done on this question, it is plausible that different resource distributions will produce different walk step-length distributions.

More elaborate versions of this simple strategy produce surprisingly little variation in the resulting walk step-length distributions; for example, the search strategy may be unreliable in identifying the best next site and instead pick one of the n best sites at random. Another variation is to prevent the searcher from detecting available resources beyond some maximum distance from the current location. Except in cases where such a restriction severely affects the number of possible next sites, resulting movement patterns are

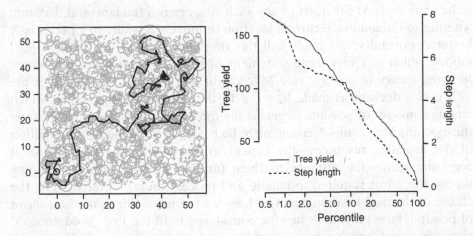

Figure 4.13 A typical walk produced by the process described in the text, when each step of the walk maximises the yield per distance moved z/d and the walk does not return to a previously visited resource. The right-hand panel shows the distribution of resources available at sites and the step lengths in the walk.

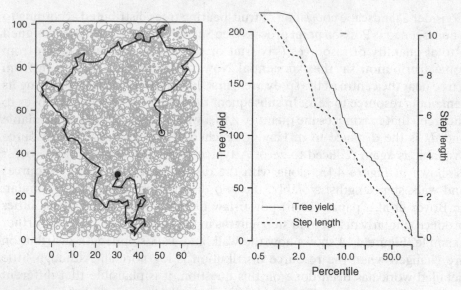

Figure 4.14 A typical walk produced when the movement dynamic is as in Figure 4.13, but resources are lognormally distributed.

surprisingly stable for a given resource distribution. This remains an active research area, and while it is likely that it will remain difficult to disentangle the effects of landscape pattern and search strategies, models will be an important research tool in this field (Mueller et al., 2010).

This example provides a simple framework for movement arising from search strategies in a landscape that provides some distribution of resources to be exploited. At the heart of any such strategy is a fundamental decision: whether to remain in the current location or move elsewhere to a potentially better opportunity. We might call this the 'Should I stay or should I go?' model, which is a very loose statement of the central problem of *optimal foraging theory* in ecology (see MacArthur and Pianka, 1966, Pyke et al., 1977). If a decision is made to go, a further decision, determining which among some set of possible targets is the most promising, is required. As in the example above, this decision might be based on an immediate evaluation of the available resources—the benefit available—relative to the cost of accessing them—the distance to them (and in classical optimal foraging theory the effort required to handle and process them). Alternatively, the choice of the next location might be less deterministic. For example, a series of possible target locations may be considered until the first 'good enough' choice is found, another example of Simon's satisficing concept. The search process might also incorporate memory effects, so that an individual has prior knowledge of the available choices, based on previous experience, and newly discovered options are considered relative to one another and to locations in memory. As an additional complication, memory may not be completely reliable, either in retaining details of the resources available at remembered locations or even in retaining them at all.

4.3.3 Foraging and random walks revisited

Recent research on animal foraging behaviour includes an extensive literature emphasising the importance of Lévy flights as search strategies (see Viswanathan et al., 2011, and the references therein). Much of this work followed assertions of Lévy flights in albatross movement (Viswanathan et al., 1996), although this claim was subsequently found to be erroneous (Edwards et al., 2007). Similar claims made for the movement of other species have also been found to be questionable (Edwards, 2011), the problem being both the technical difficulty of fitting empirical data to heavy-tailed distributions and over-optimistic claims based on relatively small variations in the scale of movements—say two or three orders of magnitude, when more are required to properly sustain claims for heavy-tailed distributions. This debate has been further energised by work showing that under conditions where resources are sparse and randomly distributed, Lévy flight search patterns are an efficient search strategy, perhaps even an optimal one (Viswanathan et al., 1999). However, it remains unclear how relevant such conditions are in practice (Benhamou, 2007) and recent work suggests that the specific model used by Viswanathan et al. (1999) has rather narrow applicability (James et al., 2011).

From the foregoing discussion, it should be clear that it is unsurprising that Lévy flight-like movement would be more efficient than simple random walks or even correlated walks as a search pattern *in some environments*. Super-diffusive Lévy walks cover ground much more quickly, and consequently have much larger associated search regions (see Figure 4.12, page 114). However, as is also clear from the foregoing, an organism responding to its environment may exhibit movement patterns that depend strongly on the structure of the environment, which may even mimic the heavy-tailed distributions of walk step lengths characteristic of Lévy flights. This is a good example where disentangling the mutually reinforcing effects of process and pattern is extremely difficult—a difficulty made even more challenging when movement processes also alter the environment in which they occur.

4.4 Moving entities and landscape interaction

The model we will examine here is similar to the previous one, but unfolds on a 60×60 (nontoroidal) lattice, which for convenience we say has N sites—obviously N is 3600. Sites are assigned a sequence number, representing the order in which they become productive, that is, when resources will be

4.6

available at the site. A patchy pattern is enforced by running 30 iterations of a 12-state voter model (see Section 3.4.4) and then assigning a sequence number between 0 and $N - 1$ to each site in the lattice ordered by the end states of the voter model. This produces a roughly clustered sequence in which sites yield resources. We will refer to the sites as s_i, where i is their sequence number. All sites from s_{N-n} to s_{N-1}, where n is a number of recently visited sites that we want to observe, have their initial yield z set to one, and all others are set to zero. As the model runs, at time step t, the site s_i, where $i = t \bmod N$, whose turn it is to become available, has its yield set to one, and the clock t is advanced by one. Thus we have a patchy environment in which resources become available in a 'seasonal' manner. An individual forager is then placed in this environment, at a randomly selected site with nonzero yield.

Each model time step, the forager first takes the resources at its current location by setting z to zero. It then moves by picking the closest site which currently has resources available. Once the forager has moved to the chosen site, the model proceeds to the next step. While one new site comes available each model time step and one is removed by foraging, sites will persist on average for n time steps. Although it is unsurprising, the interesting feature of this model is that it produces repeatable 'seasonal' behaviour over time, as shown in Figure 4.15, where at the same point in the model's N step cycle, in successive 'years', the forager is exploring similar regions of the map and following similar paths. This example is deliberately designed to reinforce the point that the external drivers of movement in terms of the resources (or activities or, generally, the attractions) available across the landscape are as important to movement patterns as any internal logic of the movement itself.

This model was suggested in part by Boyer and Walsh (2010), who present a much more complicated model, with seasonally predictable fruiting of trees, distributed across a landscape. Monkeys foraging in this landscape with memory capability and a cognitive map of the environment end up following regularly repeating paths for a wide range of search behaviour parameter settings. Given the addition to this model of considerably more complexity in terms of both the resource landscape and the foraging behaviour, the mutual interaction of landscape and movement behaviour is intriguing. One finding is that it is important for the overall foraging success of the monkeys that they do not adhere too rigidly to a deterministic decision-making method for identifying their next target tree. Rather, they do better if they retain some randomness in their search behaviour. This prevents them from getting stuck in a 'loop' where they repeatedly re-exploit the same already known resources, rather than exploring and identifying potential new sites. This paper suggests that while, as Benhamou's (1994) early work demonstrated, memory supports successful foraging, it is also important that an exploratory urge be maintained for long-term success. These issues are relevant to the model

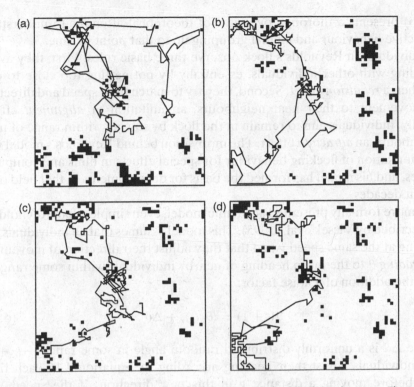

Figure 4.15 Seasonal patterns in the foraging model described in the text. These four snapshots show the currently live sites (black squares) in the model space and the most recent 200 steps of the forager's movement for four snapshots, 3600 model steps apart, that is, separated by a full model 'year'. Broadly the same regions in the space are active each time, and in all but one of the snapshots (a) the forager has recently been exploring similar areas of the map.

we present in Chapter 8. It is also noteworthy that the general framework of entities moving in and changing an environment as they move is closely related to the particle-based reaction–diffusion models briefly discussed in Section 3.7.1.

4.5   Flocking: entity–entity interaction

Models of *flocking* behaviour explore the conditions of imitation, attraction and repulsion between individual entities, under which spontaneous concerted movement of large groups—in other words, flocking—can emerge and be sustained. The most frequently cited original source for simulation models of this type is a conference paper by Craig Reynolds (1987), which presents the basic features of a model that allows the graphical simulation of flocking behaviour. Reynolds's model was not the first to consider flocking; Okubo

(1986) presents a thorough overview of theoretical attempts to understand collective behaviour and animal grouping up to that point in time.

Individuals in Reynolds's flock observe three basic rules. First, they avoid colliding with other individuals, essentially by not getting too close to one another, a *repulsion* effect. Second, they try to match their speed and direction of movement to their near neighbours, an imitation or *alignment* effect. Finally, individuals aim to remain in the flock by staying within range of near neighbours, an *attraction* effect. The motivation behind Reynolds's model was the simulation of flocking behaviour for special effects in films and computer games, and his model has formed the basis for developments in that field over recent decades.

 4.7
A more formally presented academic model, even simpler than Reynolds's, is described by Vicsek et al. (1995). This model assumes that all individuals are moving at the same speed v and that they adjust their direction of movement or *heading* θ to the mean heading of nearby individuals within some range r, with the addition of a noise factor:

$$\theta(t+1) = \langle \theta(t) \rangle_r + \Delta\theta \tag{4.16}$$

where $\Delta\theta$ is a uniformly distributed random angle in some range $[-\eta, +\eta]$. All individuals adjust their heading according to Equation 4.16 each time step before moving a distance v in this new direction. If the speed v is considered a fraction of the neighbourhood range r and we set $r = 1$, then this model has only three free parameters: v, η and the density ρ of individuals per unit area. Figure 4.16 shows a typical sequence, starting from randomly located individuals with random headings. The rapid development of 'flocks' is apparent. Of course, this is unrealistic for many applications, where we might expect individuals to be moving at a range of different speeds and also to actively avoid one another, rather than stay apart solely due to the random term. Even so, this simple model demonstrates that flocking behaviour is not hard to generate and that the relationship between the search radius for alignment and the speed of movement is a critical parameter.

Elaborating this model rapidly becomes rather complex and demands some care in how the neighbouring individuals (or flock-mates), which affect the alignment and relative movement of individuals are determined. Figure 4.17(a) shows that the simplest approach using a square grid in which all individuals sharing the current grid cell are considered flock-mates may result in many more flock-mates than a more natural range-limited circular neighbourhood. Such a circular neighbourhood can be further elaborated by restricting it to an angular range, usually centred facing forward with respect to the current heading. One individual is excluded from the set of neighbours by this method in Figure 4.17(b). This diagram also shows how a preferred separation distance criterion (see the next paragraph for details) can be used to

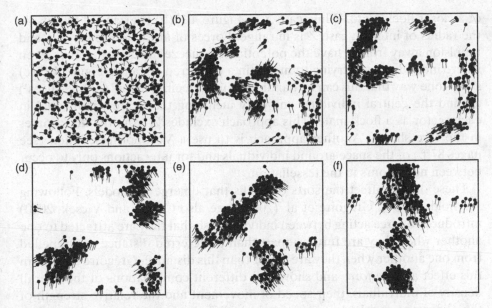

Figure 4.16 Development of flocks in Vicsek et al.'s (1995) model. Panels from (a) to (f) show system state at 30 time-step intervals, with $v = 0.3$, $\rho = 1$ and $\eta = 10°$ in a 25×25 toroidally wrapped space. This example uses a simple grid-based method to select flock-mates. Tails on the individuals show movement over the previous five time steps.

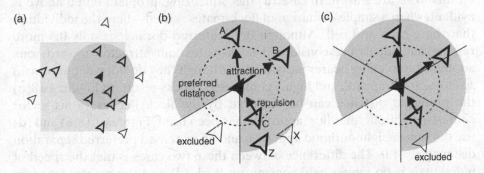

Figure 4.17 Details of the selection of flock-mates and the body-force. (a) Alternative definitions of flock-mates, one based on square cells in a lattice, which may include many more mates (heavy outlines) than those included by a circular radius. (b) Exclusion of some flock-mates by a cone angle restriction, and the attraction and repulsion effects associated with a preferred distance setting. (c) Exclusion of a second order mate by a 'pie slice' criterion. See text for details.

introduce forces of attraction and repulsion between individuals. For coherent behaviour to emerge, individuals accelerate towards, that is, are attracted by, nearby individuals more remote than the preferred distance (cases A and B) and are repelled by closer individuals (case C). This introduces a difficulty of how to handle individuals within the range of interest, but on the far side of a closer flock-mate. While case X in Figure 4.17(b) is excluded as outside the radius of interest, case Z is in range. Forces of attraction towards Z and repulsion away from C have the potential to squeeze the flock together, even where individuals are trying to maintain a preferred distance. Figure 4.17(c) shows one way that this can be handled by introducing sectors (or 'pie-slices') around the central individual and only including the nearest individual in each sector as a flock-mate. This approach excludes Z from the flock-mates in the case shown. Another approach is to use a Voronoi tessellation (see pages 87ff.) of the space around individuals and for interactions only to occur between neighbours in the tessellation.

These details affect the sorts of flocks that emerge in models. Following Reynolds (1987), Grégoire et al. (2003) (see also Czirók and Vicsek, 2000) introduced a force acting between individuals so that they are attracted to one another when they are further apart than a preferred distance and repelled from one another when they are closer than this distance. Grégoire et al. term this effect a *body-force* and show that different combinations of the overall density of individuals, their speed of movement and the relative strength of the alignment and body forces produce different movement regimes, such that the behaviour resembles that of molecules in gaseous, liquid or solid states.

4.7   In Figure 4.18, a number of flocks produced by a model loosely based on this work are shown. In case (a) the 'squeezing' problem noted above is evident, when a simple definition of flock-mates is used—here the individuals that share each grid cell. Although the preferred distance is only 0.5 more remote individuals are considered flock-mates and attracted towards one another. Intervening nearer individuals, although they do not want to remain so close, are 'trapped', and highly compressed flocks result. In Figure 4.18(b) the preferred distance can be achieved by the flock because a six-sector nearest-neighbour 'pie slice' approach has been used. Figures 4.18(c) and (d) use the same neighbourhood approach and an increased preferred separation distance $r = 1.0$. The difference between these two cases is that the speed of individuals is no longer held constant in flock (d) and the greater freedom of manoeuvre available to individuals results in a more evenly spaced flock without the bunching evident in (c). Flocks (e) and (f) increase the preferred separation distance to $r = 2$. This distance is greater than is possible given the overall density of individuals. Even so, in case (e) a stable well-spaced flock still forms, albeit with some zig-zagging in the movement, as individuals attempt to space themselves more widely than the available space permits. The difference in Figure 4.18(f) arises from the relative strength of the body-force being increased four-fold. This increases how quickly individuals attempt to adjust their speed and direction of movement to remain well-spaced, and makes the system unstable because individuals overcompensate for the presence of others nearby and overshoot. The result is that a coherent flock is unable to form.

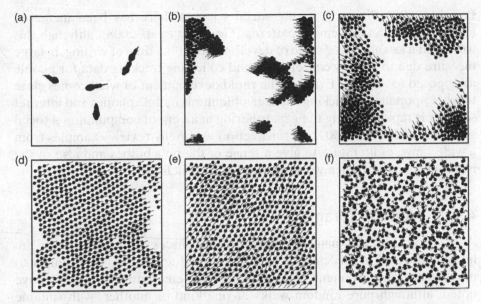

Figure 4.18 Differing outcomes for a more complex model loosely based on Grégoire et al.'s (2003). Each panel shows the model state after 150 iterations starting from a random initial state. In all cases the model space is a 25 × 25 toroidally wrapped grid and the density of individuals is $\rho = 1$. Other parameters as follows: (a) body-force with preferred distance $r = 0.5$, fixed speed $v = 0.3$, flock-mates based on shared grid cells; (b) body-force $r = 0.5$, $v = 0.3$, $k =$ six-sector nearest neighbours; (c) $r = 1$, $v = 0.3$, $k = 6$; (d) $r = 1$, $k = 6$, speed allowed to vary in the range $0.2 < v < 0.6$; (e) $r = 2$, $k = 6, 0.2 < v < 0.6$; (f) $r = 2, k = 6, 0.2 < v < 0.6$, but with the body-force increased four-fold.

Clearly, a wide variety of collective movement behaviours can be generated using an overall attraction–repulsion–alignment framework. As a result models with broadly this structure have been widely used to explore the behaviour of flocking animals. Schellinck and White (2011) provide a useful overview and cite many examples, particularly models of schooling fish (see, for example, Hemelrijk and Hildenbrandt, 2008, Huth and Wissel, 1992), but also of herds of land animals (Gueron and Levin, 1993, Turner et al., 1993). Perhaps more surprising is that similar models can be used to investigate the behaviour of human crowds (see Torrens, 2012, for an extensive review), particularly in emergency evacuation situations when the options available to people are limited (Helbing et al., 2000).

4.6 Applying the framework

Movement is a fundamental driver (and outcome) of many processes in social and environmental systems. Movement ecology has become an important focus of that discipline in recent years (Nathan et al., 2008, Sugden and

Pennisi, 2006). Similarly, in the social sciences there has been increased interest in human movement patterns (González et al., 2008), although this work is at a less advanced stage of development at the time of writing, in large measure due to privacy concerns around collecting tracking data for people as opposed to animals. Even so, the rapid development of what comes close to whole-population tracking via near-ubiquitous mobile phones and internet services is rapidly closing the gap, ushering in an era of computational social science (Lazer et al., 2009). In this section we briefly review examples from a wide range of literature to give a sense of the possibilities and also of the issues encountered when working with these models.

4.6.1   Animal foraging

As is clear from the many examples already discussed, models along the lines considered in this chapter find widespread application in attempts to understand the movement of animals in their environments. As we have noted, although pure random walks of one kind or another, with suitable tuning of turn-angle distributions and step-length distributions, can be used as phenomenological models, it is widely recognised that these are not satisfactory representations of most real movement behaviour, although they have a role as null models in the study of empirical data. Thus, for example, the considerable body of work in geographical information science on the analysis of moving point objects can utilise random walks as a basis for assessing real movement data (see, for example, Laube and Purves, 2006).

We have already seen work focused on the movement of animals driven by an environment in which resources are non-randomly distributed. Because they provide a plausible reason (or motivation) for the movement behaviour at different scales, such models are more convincing than purely internally driven random walks. Generally, movement related to exploitation of a resource is highly localised, while search behaviour is more wide ranging. Where resources are plentiful, localised movement from one site to the next is a viable strategy, but as resources are depleted more long-range and rapid search movement is called for (Sims et al., 2012). An interesting idea in the foraging theory literature is that most search behaviour is *saltatory*, meaning that it is 'stop–go' movement to a greater or lesser extent. A key consideration is then the length of moves during which no search happens and the duration of stops when search takes place, and the considerations discussed with respect to Simon's model and Figure 4.11 (page 113) concerning the overlap between previously searched space and the newly accessible space become relevant (O'Brien et al., 1990).

More complicated mechanisms such as memory and path reinforcement of movement have also been developed. The paper by Boyer and Walsh (2010) is a case in point, deserving more detailed consideration. Their model features a high-resolution landscape (a 200×200 lattice at $1\,\mathrm{m}$ resolution, with model ticks every $0.5\,\mathrm{min}$) populated by fruiting trees, which bear power-law distributed resource loads, so that resources are concentrated in a small proportion of all trees. Trees fruit for a 30-day period starting at uniform randomly distributed times. Monkey foragers in this model can remember the location, resource yield and fruiting state of a tree when they last visited it, although this memory is lost if the tree fruits again when they are not present. Based on this information, the monkeys can predict when a tree will be in fruit and how much fruit it is likely to have available, at least for a limited period after they last encountered it. Because the model is intended to explore the relative benefits of memory and exploration, the monkeys' movement involves either moving to a known tree in memory or with probability p taking a random walk step. If the decision is to move to a known tree, then the choice of which tree is based on a cost-benefit comparison similar to that in an earlier model (Boyer et al., 2006) but with the expected resource availability based on information in memory (see Boyer and Walsh, 2010, page 5649 for details). Additionally, the probability p of a random step is linked to the expected yields of known trees, so that when no 'good' trees are known the probability of random exploratory movement increases.

Although this sounds complicated, it is a highly abstract representation of the reasoning processes of an intelligent animal and does not require that we assume an implausible level of knowledge and reasoning skill—in this sense it meets Simon's criteria for bounded rationality. Findings from this model point to an optimal balance between memory and random search. Memory improves the success of foraging but only up to a point because over-reliance on memory leads to non-discovery of important resources. In an interesting spatially-informed analysis, the authors show that habitual movement paths emerge, that is, sequences of visited trees that are repeated through a model run. Paths in the model also have heavy-tailed step-length distributions. Overall, the model confirms earlier findings that the distribution of resources is an important control on likely movement patterns. It also points to the importance of moving on from sterile debates about the optimal efficiency of single mechanism null models of movement, which are unlikely to apply to many real foragers. A general framework for the effects of memory on models of movement is suggested by Gautestad and Mysterud (2010), who coin the term 'memory dependent kinetics' and provide pointers to much of the relevant literature.

Another interesting variant in the foraging model literature looks at how flocks or herds forage. In these cases a tension between safety in numbers (staying close to the group) and the competition for food resources and finding new resources (pushing individuals away from the group) is central, and has been reported in models (Getz and Saltz, 2008, Bonnell et al., 2010) that directly consider the urge to find new resources relative to the need to stay within a safe distance of the group. More general questions as to how animal groups make collective decisions revolve around resolving a similar tension between the interests of individuals and groups, and are reviewed by Conradt (2012). Work on very large herds has adopted a continuous space differential equation-based approach (Okubo, 1986, Gueron and Levin, 1993, Illius and O'Connor, 2000) rather than the flocking model methods discussed here, perhaps reflecting the difficulties of scale in such models (see the concluding remarks of this chapter).

4.6.2   Human 'hunter-gatherers'

There is no difference in principle between animal foraging behaviour and human hunter-gatherer behaviour, although it is likely that the decision-making behaviours of human hunter-gatherers and of animals differ, and will be represented differently in models. The cognitive capacities of human hunter-gatherers will most likely be more complicated to represent than animals in models, particularly when cultural aspects are included, but even this will depend on the model scale and the goals behind building it. Boyer et al. (2012) are careful to note that their conclusions regarding the effects of memory on foraging success and on spatial outcomes are equally applicable to primates and humans. As they note, 'How scaling laws emerge from the interplay between memory and landscape features remains elusive' (page 846). It is important to recognise that they make this point not with respect to human hunter-gatherers but in relation to models of contemporary human movement patterns, which typically occur in dense urban networks with numerous 'resources' of interest, such as places of employment, schools, shops, leisure activities and so on.

With regard specifically to early human hunter-gatherer activity, humans appear to have exhibited complex and varied relationships with environments in terms of the choices they made about how to survive in a particular landscape, and many of these choices related quite specifically to movement (Kelly, 1983). This is a context where the use of simulation models appears attractive because data are often limited and because of the impossibility of experiments. Exploration of how field data on artifact distributions might be explained by different movement strategies is an idea that has been explored in models of early human hunter-gatherer societies (Lake, 2000, 2001). Interestingly, the issues associated with how memory affects outcomes were identified early in the development of this field (Costopoulos, 2001). Movement at migration scales has also been a focus to a greater extent than in animal movement models (Mithen and Reed, 2002, Nikitas and Nikita, 2005), these models being at large scale (the whole Eurasian–African land mass) and more closely related to random walk models or even the contact process discussed in Section 3.4.1.

Models of human movement in the contemporary period are most prevalent in the transport literature, where origins and destinations distributed over networks are the dominant mode of representation and the methods discussed in this chapter are less relevant. More recent interest in random walk based models has derived from attempts to understand general patterns of human movement (Brockmann et al., 2006, González et al., 2008) and here the debate about Lévy movement has again appeared. While humans tend to range more widely than most animals, the argument that different scales of movement arise from different processes appears even stronger. Pedestrian models of human movement (Section 4.6.4) are more closely related to the models discussed in this chapter.

4.6.3 The development of home ranges and path networks

The repeated revisiting of pathways in Boyer and Walsh's (2010) model is immediately suggestive of a link between movement and the development of home ranges or activity spaces. Generally speaking, home range and movement approaches have tended to be treated separately (Börger et al., 2008), with the former analysed statically (perhaps using spatial partitioning models, see Section 3.6) and the latter more amenable to dynamic approaches. When repeatedly reused spaces emerge from a movement model, the potential arises to link these important perspectives on how space is used. Various types of reinforced random walks (Davis, 1990a, Pemantle, 1992) are one possible basis for such links, although as we have just seen, there is no necessity for purely phenomenological approaches to be adopted. Instead, memory effects, either in searching individuals or in the landscape itself (or both), might produce the outcomes. The latter case, where the movement has effects on the landscape and these subsequently strengthen the likelihood of repeated traversals of the same paths, has been explored in the context of the formation of informal pathways in an urban park (Helbing et al., 1997). Essentially the same 'active walker' model can be applied to the development of networks connecting settlement systems (see Batty, 2005) and to paths formed by animals or human foragers.

4.6.4 Constrained environments: pedestrians and evacuations

In the previous section, moving individuals might contribute to the formation over time of a network of paths across a landscape, which they are more likely to use than not. Pedestrian movement, on the other hand, almost always occurs in confined environments where, in addition to other pedestrians, individuals in motion must steer a path through many and varied obstacles. This is not in principle a difficult additional feature for movement models to deal with. Some locations can be excluded from consideration for movement. Alternatively, the repulsion effects applied in flocking models can be applied to street furniture and other obstacles to enable pedestrians to avoid them. Early models of pedestrian movement used some combination of these features (see, for example, Haklay et al., 2001).

Importantly, it is immediately apparent in representing this kind of activity that several levels of movement behaviour are involved. Pedestrians are generally in a particular urban location for a reason, and are likely to be going somewhere, with some kind of plan. In pursuit of that plan, they may have a more immediate or intermediate goal in mind, such as the next street intersection. More immediately, they will be steering a path through any surrounding crowd and avoiding obstacles in the form of other pedestrians, street furniture and so on. While models based on the simple attraction–repulsion mechanisms of the flocking models in Section 4.5 are reasonably useful for the lowest level of behaviour in this scheme, they perform poorly at the higher levels. Torrens (2012) reviews models in this general area, although he is critical of most, and goes on to describe an intricate model that addresses the perceived failings by incorporating multiple levels of spatial decision-making and behaviour. This multi-level structure is not so different from that of some earlier models, although the end result is considerably more sophisticated.

Where such highly detailed models are most likely to prove their worth is in the critical context of planning for evacuation from complex buildings in emergency situations. It is here, where individual options are most limited, that the details of body size and shape, and of specific cognitive reactions are most likely to make a difference to overall outcomes. Additionally, there are rarely many data on which to base an assessment of the accuracy of models, and experiments are not an option. In this data-poor environment, a conservative approach using complex, detailed models may be appropriate. More to the point, highly simplified physically-based models, while they capture general patterns, are a risky choice on which to make potentially life and death decisions about maximum rates of egress from a building in an emergency. Notwithstanding the likely need for richer models, many simple models of emergency evacuation have been developed (see, for example, Helbing et al., 2000, Klüpfel et al., 2005, Pan et al., 2007, Moussaïd et al., 2011). These models all use mechanisms similar to those in the flocking models discussed in this chapter, but the constrained situation leads to emergent dynamics with wave effects, delays and other phenomena, the safe management of which is important in emergency situations. Insights from models can inform the detailed design of spaces to enhance flows, but testing such models remains a problem—Shiwakoti et al. (2009) even suggest using panicking ants as a 'hardware model' (see Section 1.1.2) with which to evaluate simulation models, or perhaps even as a model for panicking humans!

4.6.5 Concluding remarks

A key challenge in all movement models is to retain a clear idea of the model scale and (especially) grain, and of the model scope, both spatially and temporally, in the context of the questions that are being addressed. Combining minute-by-minute routine foraging behaviour with seasonal and continental-scale migrations is never going to be easy, and it is important to consider if it is even necessary. If the focus of attention is on detailed movement at small scales over relatively short time periods, then longer range reorientations of movement may not even occur inside the model's scope, but may instead be better represented by the disappearance of individuals that have finished feeding and the arrival into the model of new individuals. This allows the model to focus not on search, but on foraging itself (see, for example, Mercader et al., 2011). On the other hand, where longer range, seasonal or migratory movements are of interest, then localised, search-related movement may become irrelevant to the model structure given the disparate scales involved, and feeding need only be represented (if at all) by resource depletion effects 'in place' at a particular location. Indeed it may not even be necessary to represent individual entities and their behaviours, and in some cases the frameworks presented in Chapters 3 and 5 may be more appropriate. We revisit some of these issues of model scale and representation in a more general sense in Chapter 6.

5
Percolation and Growth: Spread in Heterogeneous Spaces

'Spread' encompasses a wide range of environmental and social processes and ensuing patterns, occurring across a wide range of spatial and temporal scales. For example, wildfire spread, spread of forest into grassland habitats, encroachment of urban environments into the wildland interface, the movement of invasive species, the spread of epidemics such as the black death and the movement of gossip through social networks are all processes that can be represented using simple spatial models. In this chapter we will look at two broad types of spatial model that have been used to represent these types of dynamics: percolation and growth (or aggregation) models. These two approaches provide a broad framework for discussing issues relating to the dynamics of spread processes and the patterns formed by them. We will conclude by considering how we might apply these general approaches to more targeted questions about specific environmental and social systems.

5.1   Motivating examples

Soetaert and Herman (2009) recognise two fundamental spread processes: advection and diffusion. *Advection* is the *directed* movement of a substance in a fluid (for example transport of pollutants in groundwater), whereas *diffusion* is the *random* movement of materials from regions of high concentration to regions of low concentration (for example the movement of oil from a ruptured tank on a ship to the surrounding environment). Soetaert and Herman (2009, pages 82ff.) consider the formulation of models of advective and

Spatial Simulation: Exploring Pattern and Process, First Edition.
David O'Sullivan and George L.W. Perry.
© 2013 John Wiley & Sons, Ltd. Published 2013 by John Wiley & Sons, Ltd.

diffusive fluxes in one, two and three dimensions in continuous and discrete space in detail. While there are fundamental differences in the physico-chemical mechanisms that underpin advection and diffusion, pragmatically, in environmental and social settings it may be difficult to distinguish them (similarly, it is hard to distinguish bias and localised correlation in random walks, see Section 4.2.4). Thus, it is probably more useful to think about the processes that drive spread *per se* and those that influence the environment through which the spread is occurring.

In some of the examples in Figure 5.1 (see also Plate 6) the process is inherently one of active spread or *contagion* (wildfire or epidemics), while in others

Figure 5.1 Examples of spread processes: (a) spread of fire through shrubland on the Eneabba sandplain (north of Perth, Western Australia), (b) the northward expansion of the city of Auckland (northern New Zealand), (c) a beaked plague doctor, the 17th century response to the spread of the black death through Europe, and (d) altitudinal treelines in Rocky Mountain National Park (USA). *Sources*: (a) photograph kindly provided by Dr Ben Miller, (b) and (d) authors' collection and (c) Paul Fürst (1656), copyright expired, http://commons.wikimedia.org/wiki/File:Paul_F%C3%BCrst,_Der_Doctor_Schnabel_von_Rom_%28Holl%C3%A4nder_version%29.png.

the spread is a response to a change in underlying conditions (expansion of an altitudinal or latitudinal treeline in response to warming). Another way to look at this is to consider some spread processes to be predominantly controlled endogenously and others exogenously. Contagious spread processes exemplify how a process can both generate and be constrained by a pattern. For example, wildfires create a 'mosaic' of vegetation of different ages, each with different vulnerabilities to burning during subsequent fire events, and so constrain future fires. The processes that drive wildfires and fire mosaics occur at many different scales (Whitlock et al., 2010). Fire mosaics arise from multiple fire events, driven by processes ranging from the very fine-scale physico-chemical processes of combustion, to the broad-scale accumulation of biomass as a function of climatic conditions and successional dynamics. That many processes across many spatio-temporal scales influence fire as a process suggests that the wildfire regime will show pattern at multiple scales too.

In a social context, contagious disease spread works in an analogous way, with the disease spreading from individual to individual, with pockets of susceptible individuals remaining unaffected, but perhaps not acquiring immunity. We also have the added complexity that the space through which diseases spread is social, not just geographic. For example, the people you have most contact with are not necessarily those you live next door to, and spread may occur via schoolmates or colleagues in the workplace (Bian, 2004). Recent research (Stein, 2011) has highlighted other complexities in the spread of diseases by demonstrating that not all affected individuals (both in humans and other animal species) have the same chance of spreading the disease to unaffected individuals. So-called *super-spreaders* may be important in explaining the progress of global pandemics such as SARS, smallpox and measles (Lloyd-Smith et al., 2005, Stein, 2011) as they may be responsible for the bulk of transmissions (this is an example of the '20–80 rule'). While cases such as Typhoid Mary—the first healthy transmitter of typhoid in the USA, who is believed to have infected more than 50 other people—have long been anecdotally recognised, classical models of disease spread have tended to ignore such individual-level variability (Lloyd-Smith et al., 2005). In a similar way to both fire and disease spread, the growth of the urban margin is constrained by previous decisions as well as by the nature of the physical landscape, which is effectively static over the time scales of most urban development. Batty (2005) shows how models of contagious spread can be applied to an urban development context. That contagious spread processes are affected by spatio-temporal heterogeneity across many different scales is one reason why they are so challenging to model mechanistically.

The movement of tree species range margins at the end of the Last Glacial Maximum provides an example of a spread dynamic that, at least on the face of it, appears to be controlled by an exogenous factor (in this case climate). In 1899 Clement Reid, in his *Origin of the British Flora*, pondered the apparently anomalously rapid rate of migration by tree species in response to climate shifts in the early Holocene. Based on their typically very short dispersal distances he could not see how species could have migrated as quickly as the pollen record suggests that they did—this became known as 'Reid's Paradox' (Clark et al., 1998). Recent research shows that the rapid rate of spread can be understood if the distribution of seed dispersal distances (the dispersal kernel) is closely examined. Typically such distributions are heavy-tailed (Clark et al., 1998), meaning that long-distance dispersal events occur more frequently than is the case for a distribution such as the Gaussian, so that the front's expansion *accelerates* over time. Occasional long-distance dispersal events result in complicated spatial dynamics, with outlying populations forming ahead of the main front before eventually coalescing back into the main population. This dispersal dynamic can be seen as a process similar to the Lévy flights introduced in Section 4.2.5, but with walkers leaving a 'mark' (an individual tree or a population of trees) at each site they visit. The migration of tree species is not, however, simply top-down controlled by climate. Processes at many different scales, from the individual to the landscape, exert an influence on tree species range shifts (see Harsch and Bader, 2011, in the case of altitudinal treelines). Landscape structure will also affect the spread process, with different parts of the landscape acting as barriers or conduits for dispersal (With, 2002). As with fire and disease the multi-scalar nature of the controls on these spread processes makes modelling them challenging (and also interesting!).

Many models of 'spread' have been developed, ranging from simple models of spread in homogeneous spaces, grounded in the theory of random walks introduced in the previous chapter, to more complicated mechanistic models of specific processes. In this chapter we will consider two broad frameworks: *percolation theory* in Section 5.2 and aggregate *growth models* in Section 5.3. Percolation considers the nature of the media through which spread is occurring, while aggregate growth models focus on the dynamics of the spread itself. The primary literature on both of these approaches is technical, but an appreciation of it helps place the more specific and idiosyncratic models presented in Section 5.4 into a wider context. It is also important to recognise the fundamental and important links between the spread processes considered in this chapter and the reaction–diffusion systems introduced in Section 3.7.1. Reaction–diffusion systems, for example, have frequently been used to model the spread of populations (Hastings et al., 2005, Holmes et al., 1994), building on the much earlier seminal work of Fisher (1937) and Skellam (1951). Okubo (1980), Cantrell and Cosner (2003) and Murray (2003) provide substantial coverage of reaction–diffusion models (with an ecological flavour), while Wilson (2006) points out the close relationship to spatial interaction models of urban systems. Likewise, the random walks and

associated models introduced in Chapter 4 are closely intertwined with the
issues considered here.

5.2 Percolation models

5.2.1 What is percolation?

One way to think about spread is to focus on the structure of the system
(or landscape) through which it is occurring. Start by imagining a grid or
lattice of sites (cells), some 'susceptible' and some 'immune' to the spread of a
disease during an epidemic. Alternatively we can imagine a lattice where every
site is susceptible but only some fraction are connected. If the epidemic begins
in a susceptible site and then spreads to neighbouring (or linked) susceptible
sites, and from there to susceptible sites neighbouring susceptible sites and so
on, how will the epidemic progress or *percolate* through the landscape? How
will the final size of the epidemic be affected by the proportion of immune
sites? Instead of disease we could think of other contagious phenomena such
as fire, when susceptible sites are dry or flammable ones and immune sites are
non-flammable, or the spread of urbanisation into sites of varying suitability
for development, or the spread of gossip in schools, or any number of other
examples.

Percolation theory is concerned with such questions. More generally, percolation is concerned with the question that Broadbent and Hammersley (1957,
page 629) asked in their seminal paper: 'How (do) the random properties of a
"medium" influence the percolation of a "fluid" through it?'. The literature
on percolation is extensive. Grimmett (1999) attributes the wide interest in
percolation systems to three factors:

(i) Despite their simplicity they can produce macroscopically realistic
outcomes.
(ii) They are (relatively) tractable stepping-stones to more complicated
systems.
(iii) They are rich in mathematical interest.

Stauffer and Aharony (1994), Gouyet (1996) and Christensen and Moloney
(2005) all provide thorough, readable accounts of percolation systems and
their dynamics, starting with one-dimensional systems before considering
Bethe systems, two-dimensional systems and other more complex cases.
Percolation theory (and the related models considered here) has received
considerable attention from those seeking to develop models of environmental and social processes such as urban growth and the spread of invasive

species, for example Solé and Bascompte (2006) outline their use in ecology and Castellano et al. (2009) consider their application in the social sciences.

5.2.2   Ordinary percolation

To begin, consider a lattice of $L \times L$ sites, all 'empty'. If we independently and randomly select and occupy some proportion p of the cells on the lattice where $0 \leq p \leq 1$, then we add 'disorder' to the system (*sensu* Christensen and Moloney, 2005) and obtain structures like those in Figure 5.2. Before we consider spread through such lattices, we need to characterise the geometry of the *clusters* that are formed by neighbouring sites. Two types of percolation structure are recognised, as shown in Figure 5.3. First, we can consider a square lattice with only some proportion p of sites occupied and clusters defined by the state of neighbouring sites—this is termed *site percolation* (as in the examples in Figure 5.2). Second, there is the case where *all* sites on the lattice are occupied, with only some proportion p of them linked—this is termed *bond percolation*, and clusters are then defined as sets of sites linked together by bonds.

We can also consider percolation on lattices with different geometries, such as hexagonal (honeycomb) lattices or networks, and with more than two dimensions (Christensen and Moloney, 2005), as well as with different neighbourhood definitions (Malarz and Galam, 2005), but we will restrict ourselves here to square lattices in two dimensions. Likewise there are other percolation rules such as continuum percolation where the process occurs in continuous space with adjacency defined by objects overlapping (ben-Avraham and Havlin, 2000) and bootstrap (k-sat) percolation where occupied sites with fewer than some specified number (k) of occupied neighbours themselves become empty (Adler and Lev, 2003). Site and bond percolation processes are closely related, and bond percolation problems can be converted to site percolation problems, although there is no easy mapping from site to bond percolation (Berkowitz and Ewing, 1998, Grimmett, 1999). We will focus on site percolation.

Percolation theory is very much concerned with the geometric properties of clusters on lattices with different characteristics. What is the average size of a cluster? What proportion of clusters will comprise s sites? What is the their frequency-size distribution or *cluster number density*? At what value of p do clusters that extend to the full width or height of the lattice, known as *spanning clusters*, emerge? As Christensen and Moloney (2005, page 1) state, 'The challenge in percolation lies in describing its emergent structures rather than understanding its defining rules.'

A central concern and important finding of percolation theory is the behaviour of the system as p changes and, in particular, how the system

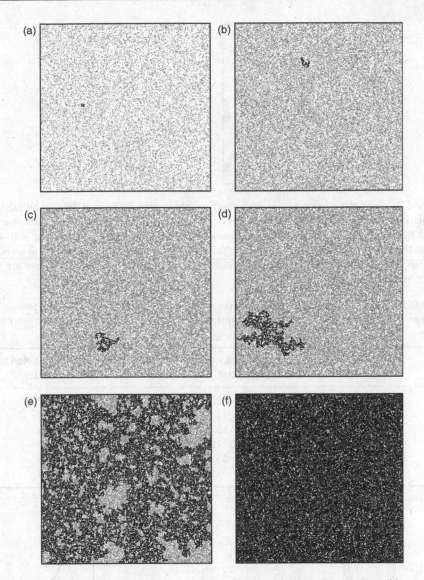

Figure 5.2 Examples of site percolation on a 256 × 256 lattice. (a) $p = 0.2$, (b) $p = 0.4$, (c) $p = 0.5$, (d) $p = 0.55$, (e) $p = 0.6$ and (f) $p = 0.8$. Note how the geometry of the clusters, particularly their size, changes as p increases. White cells are unoccupied, grey cells are occupied and black cells are occupied and belong to the largest cluster on the lattice, which for (e) and (f) is a spanning cluster.

changes around some critical threshold p_c. As Table 5.1 and Figure 5.2 suggest, and Figure 5.4 confirms, for site percolation on a square lattice, when p reaches a critical value close to 0.6 the nature of the clusters appears to change. At a critical value p_c the probability of a spanning cluster appearing increases very rapidly. The threshold p_c for a two-dimensional site percolation

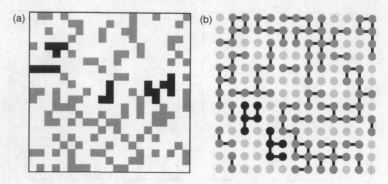

Figure 5.3   Examples of clusters (in dark grey) on percolation systems as formed by (a) site percolation, where contiguous grey cells form clusters, and (b) bond percolation where sites linked by bonds form a cluster. In (a) all five clusters coloured black are size four with the two on the right *not* connected, because neighbourhoods are defined by the sharing of an edge between cells, and in (b) both clusters coloured black are size five and pale grey sites are not part of any cluster.

Table 5.1   Statistics describing the six lattices in Figure 5.2. Note the abrupt changes between $p = 0.50$ and 0.60 and that mean cluster size is size-weighted and excludes spanning clusters

p	Number of clusters	Mean cluster size, $\chi_{(p)}$	Probability site belongs to spanning cluster, P_∞
0.20	7 998	2.604	0.000
0.40	7 107	11.957	0.000
0.50	4 363	56.541	0.000
0.55	2 940	327.329	0.000
0.60	1 723	63.231	0.474
0.80	117	1.743	0.796

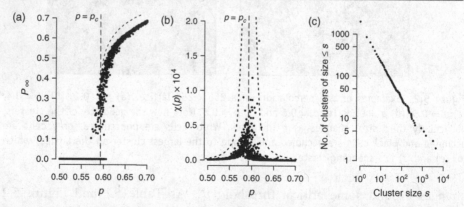

Figure 5.4   Aspects of the cluster geometry on a site percolation system with $L = 256$ as a function of p. (a) Probability of a spanning cluster occurring P_∞, (b) average cluster size $\chi(p)$ and (c) the power-law distribution of cluster sizes observed at p_c. Note that $\chi_{(p)}$ is the size-weighted average of the non-spanning clusters, and can be interpreted as the typical size of a cluster that an occupied site will belong to (Christensen and Moloney, 2005). Dashed lines are the theoretical expectations on an *infinite* lattice, see Equations 5.1 and 5.2.

lattice with a neighbourhood defined as the four adjacent cells sharing an edge (the von Nuemann neighborhood), estimated by computer simulation (see Malarz and Galam, 2005, Newman and Ziff, 2001), is $p \simeq 0.59274621\dots$. If we consider the Moore neighbourhood (the adjacent eight cells) then $p_c = 0.407\dots$. Malarz and Galam (2005) provide a list of site percolation threshold values for a range of neighbourhood structures in two dimensions, and Grassberger (2003) for site and bond percolation with von Neumann neighbourhoods in up to 13 dimensions. In fact, p_c for two-dimensional bond percolation has been analytically solved to be $1/2$ (Stauffer and Aharony, 1994)*. At $p = p_c$ the size-frequency distribution of clusters follows a power-law distribution, meaning that it is scale-invariant (see Section 2.1.3), so that clusters of all sizes are typical.

Points at which a system's properties change abruptly, such as $p = p_c$ in a percolation system, are called *phase transitions* and the parameter values at which phase transitions occur are known as *critical values* (Kesten, 2006, Solé, 2011). The existence of a phase transition in percolation systems is one feature that makes them interesting to mathematicians and statistical physicists. Understanding the dynamics of the system at or near this critical point is another topic of considerable interest in percolation theory.

Most formal analysis of percolation systems considers infinite grids and so-called infinite clusters. Strictly speaking, spanning clusters on a finite grid are only candidate infinite clusters. On an infinite grid the probability of an infinite cluster at p_c is zero—the very large clusters that appear at this point are termed *incipient infinite clusters* (Gouyet, 1996, Christensen and Moloney, 2005). As Gould and Tobochnik (2010) note there is, of course, a chance of spanning clusters forming at $p < p_c$, but the probability of this happening is so small that we can disregard it.

Various other important characteristics of the system change around p_c. The probability P_∞ that a randomly selected site will belong to the incipient cluster changes abruptly at p_c, and this threshold is what defines the value of p_c. As $p \to p_c$ (see Stauffer and Aharony, 1994, ben-Avraham and Havlin, 2000, Christensen and Moloney, 2005):

$$P_\infty(p) \propto (p - p_c)^\beta \tag{5.1}$$

where β can be theoretically derived for an (infinite) two-dimensional lattice, irrespective of the geometry, to be $5/36$ (Christensen and Moloney, 2005).

Two other properties of the cluster that change at this point are the average cluster size $\chi_{(p)}$, that is the size (or mass) of clusters *excluding* the spanning

---

*Note that it is conventional in percolation theory to report analytical results as fractions and those derived from simulation as decimals.

cluster, and the scaling of the clusters' *correlation length* (ξ)—that is, their typical length—around p_c (ben-Avraham and Havlin, 2000):

$$\chi_{(p)} \propto |p - p_c|^{-\gamma} \tag{5.2}$$

$$\xi_{(p)} \propto |p - p_c|^{-\nu} \tag{5.3}$$

γ has been theoretically derived for two-dimensional lattices to be $^{43}/_{18}$ and ν to be $^4/_3$, again irrespective of geometry (ben-Avraham and Havlin, 2000, Christensen and Moloney, 2005).

β, γ and ν are *universal critical exponents* (Stauffer and Aharony, 1994, ben-Avraham and Havlin, 2000, Christensen and Moloney, 2005) and they describe the dynamics of percolation systems around the percolation threshold in a way that depends on the percolation model's dimensionality (d), but not on other fine-scale detail of the system, hence their universality.

Finally, the spanning cluster around $p = p_c$ has fractal properties in that 'holes' of all sizes are typical of the object (Gouyet, 1996, ben-Avraham and Havlin, 2000), and as ben-Avraham and Havlin (2000) demonstrate, the fractal dimension (d_f) of the cluster (in dimension d) is itself universal:

$$d_f \propto d - \frac{\beta}{\nu} \tag{5.4}$$

While considerable attention has focused on the detail of these abrupt phase shifts, for our purposes such critical points are interesting because they point to some of the behaviours we might expect spatial models built around percolation systems to show.

5.2.3   The lost ant

While their underlying properties are intriguing, we are not interested in the geometry of these disordered systems *per se*, rather we want to understand how flow or spread occurs in and through them. A simple simulation that makes some of these issues a little more concrete is the 'ant in the labyrinth' introduced by Nobel laureate Pierre Gilles de Gennes (1976). Note that the term 'ant' is metaphorical: this model does not claim to represent the behaviour of real ants!

Imagine a lattice with sites occupied with probability p, in other words a site percolation model. We place an ant in a randomly selected occupied cell and let it explore the lattice subject to the constraint that it can move only in one of the four cardinal directions. Two types of ants are considered: 'blind' and 'myopic' (Havlin and ben-Avraham, 2002). Blind ants select one of the four neighbouring locations at random and move there only if it is occupied; myopic ants, on the other hand, select only from neighbouring cells that are occupied. The questions asked about this system include: what distance R

will the ant travel as a function of time t through this static, but disordered, landscape? And how is this distance affected by p?

As Figure 5.5 shows, p strongly influences the distance the ants travel, although this is hardly surprising. If $p = 1$ then the ants are free to roam across the entire landscape and their pattern of movement follows Brownian motion, as for simple random walks (see Section 4.2), and so is linear with time. On the other hand, if $p < p_c$ then the ants tend to remain trapped in small clusters and are unable to move very far across the landscape. However, the ants' behaviour changes at $p = p_c$, such that movement shows *anomalous diffusion*, the term given to the nature of spread through fractal entities such as percolation clusters around p_c (see Havlin and ben-Avraham 2002 for details).

5.2

Under normal (or Gaussian) diffusion, as, for example, in a simple random walk, $\langle R^2 \rangle \propto t$ (see Section 4.2). In a fractal structure, such as a spanning cluster in a percolation system, this is not the case. Instead $\langle R^2 \rangle \propto t^{2/d_w}$, where $d_w \neq 2$, meaning that movement is either slower than normal and is *sub-diffusive* ($d_w > 2$) or is quicker than normal and *super-diffusive* ($d_w < 2$). Under normal diffusion $d_w = 2$. Lévy flights (Section 4.2.5) are an example of super-diffusion, while the spread of contaminants through ground water systems and the movement of proteins across cell boundaries are potential examples of sub-diffusion (Klafter and Sokolov, 2005, provide other examples across biology, chemistry and physics). ben-Avraham and Havlin (2000), Metzler and Klafter (2000) and Havlin and ben-Avraham (2002) all provide thorough accounts of anomalous diffusion and related processes.

A process directly analogous to the ant in the labyrinth is the spread of a contagious phenomena such as wildfire or disease across the lattice under site percolation. In this case 'ignition' or 'infection' occurs in a randomly selected occupied cell, and the fire then spreads into the initial location's neighbouring cells. From each new location the fire or infection spreads further, moving from occupied neighbour to occupied neighbour. With each round of spread, new cells—termed a *chemical shell*—are added to the burning or infected area. In essence, this process is an algorithm that identifies the cluster to which the ignition location belongs. We can describe the behaviour of the 'fire' in terms of its duration (the number of shells it persists for) and, as Figure 5.6 shows, this value changes abruptly around p_c, as we might expect from our knowledge of percolation structures (see ben-Avraham and Havlin, 2000, for an analytical demonstration). In this case the number of chemical shells is a measure of how 'long' the fire lasts (the *transient time*), and provides a measure of the complexity of the percolation structure's geometry (Solé, 2011).

5.3

As it turns out, we can use this fire spread algorithm to find other useful geometric properties of the spanning cluster, such as the shortest path(s) through it and the occupied cells that would carry a current if it were applied to both edges of the cluster (the 'backbone'; Herrmann et al., 1984). As an

5.4

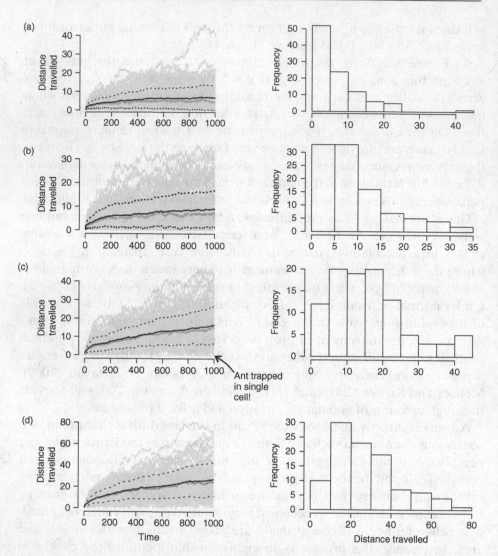

Figure 5.5   Distance travelled by 100 myopic ants in a labyrinth over 1000 time steps where (a) $p = 0.5$, (b) $p = p_c$, (c) $p = 0.7$ and (d) $p = 1.0$. At $p = 1.0$ the ants' paths follow Brownian motion with mean squared displacement proportional to the time travelled but at $p = p_c$ movement follows anomalous diffusion. Pale grey lines are distance travelled for each of the ants, the solid black line is the mean distance travelled, the dark grey lines the median distance travelled and dashed lines show the mean $\pm 1\,SD$ distance travelled. Histograms on the right show the distances travelled from the starting point after 1000 time steps. Note that x- and y-axes are not at the same scale across the graphs. Note also that for $p = 0.70$ one unfortunate ant was trapped in a single-cell cluster and was unable to move at all!

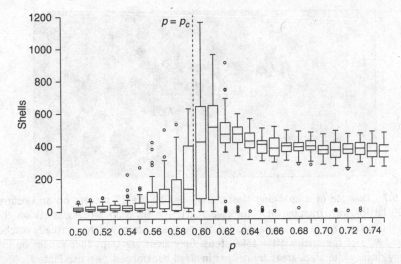

Figure 5.6 Number of chemical shells that fire events persist for as a function of p, based on 100 replicates per value, with $0.55 \leq p \leq 0.75$. The average fire size, as a function of p, can be estimated on the basis of the mean cluster size. Note the characteristic change in behaviour around the critical percolation threshold p_c. Around $p = p_c$ the complicated fractal shapes of incipient spanning clusters see high variability in the time for which fires may persist.

aside, this forest fire method of finding the shortest path is directly analogous to Dijkstra's (1959) algorithm for finding shortest paths in a graph.

5.2.4 Invasion percolation

Much of ordinary percolation (OP) theory focuses on the statistics of the system's cluster geometry (Stauffer and Aharony, 1994). Of more interest to us is how energy, matter and information flow through heterogeneous media. *Invasion percolation* (IP) models (Knackstedt and Paterson, 2009) are explicitly concerned with this issue, and were first studied in the 1980s and popularised by the work of Wilkinson and Willemsen (1983).

The basic IP model is easily described. On a large rectangular lattice each site is assigned a random number (usually $r \sim U[0, 1]$ although any range of values from any distribution will suffice), with r isotropic and uncorrelated. The r values do not change over time so the model is deterministic, with the rules and underlying field fixed after the initial assignment of site values. The invasion process starts on one side of the lattice, with spread occurring into the site with the lowest value of r that is adjacent to an invaded site. This process continues until the opposite side of the lattice is reached. The physical analogue to this process is that of one fluid displacing another, as can happen when a fluid moves into some porous media via the path of least resistance (Ebrahimi, 2010). During the IP process clusters of uninvaded

5.5

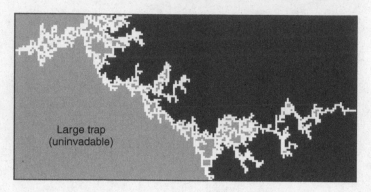

Figure 5.7 Example of a spanning cluster produced by the IP process on an uncorrelated 180 × 90 lattice with trapping in place (see text). The invasion process starts on the left edge of the lattice. White cells were invaded during the process, which eventually reaches the right-hand side of the lattice after 1596 steps. Grey areas are traps that cannot be invaded once they close, while black areas are neither invaded nor trapped. See also Plate 7.

sites may become surrounded by invaded ones. Such clusters are called *traps* (Wilkinson & Willemsen, 1983). In IP with trapping, trapped sites are immune to future invasion, whereas under IP without trapping such sites remain open to invasion (see Figure 5.7 and Plate 7.).

Knacksted and Paterson (2009) and Ebrahimi (2010) outline the key differences between OP and IP. For us the most crucial of these is that in IP *the sequence of events matters*. IP is a dynamic process, and over time more than one invasion front may take the lead in advancing the invasion across the lattice (see example in Plate 7). A spanning cluster always results from the IP process; there is no percolation threshold as there is in OP (Ebrahimi, 2010). IP has received much less attention and application than OP in the environmental and social sciences.

Inspection of the dynamics of a single invasion shows that it is a localised process. On initial invasion the 'easiest' sites surrounding the initial point of invasion (i.e. those with low r) are invaded but once these are exhausted, the focus of the invasion moves to a different location on the front. Thus, there is significant correlation in the r value of successively invaded sites; Roux and Guyon (1989) call this temporal dynamic *bursting*. Variations on the basic IP theme have been developed. Two that are of interest are gradient IP, where a gradient is superimposed on the random field, and temperature IP, which introduces stochasticity to r as the invasion process progresses (Knackstedt and Paterson, 2009, Ebrahimi, 2010). The introduction of a gradient causes a bias in the direction of the invasion process, and the introduction of even limited stochasticity makes the cluster formed by the invasion much more compact.

Stark (1991) used a subtly modified IP model to simulate drainage networks (Figure 5.8). Stark's model is essentially IP without trapping but with the

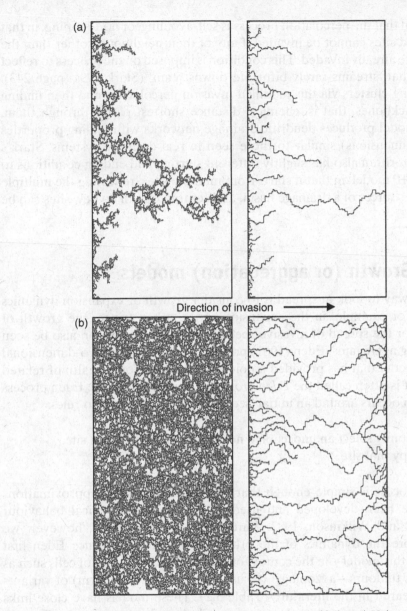

Figure 5.8 Two examples of drainage networks produced by Stark's (1991) IP-based model. In both cases the model runs on a lattice of 512×256 cells with the underlying field initialised with isotropic Gaussian white noise. In (a) the process was halted when one site on the right-hand edge of the lattice was invaded, and in (b) the process continued until there were no new sites available for invasion. The left-hand images are of the underlying invasion percolation process and the right-hand images show the drainage networks formed by isolating each cluster's elastic backbone (see Herrmann et al., 1984).

5.6

restriction that the percolation process is self-avoiding, or non-looping, in that uninvaded sites cannot be invaded if any of their neighbours, other than the source, are already invaded. This condition is imposed on the process to reflect the fact that 'streams rarely bifurcate downstream' (Stark 1991, page 243). By building clusters via this modified invasion percolation, and then finding elastic backbones, that is, chemical distance shortest paths, through them, Stark's model produces dendritic drainage networks with scaling properties (fractal dimensions) similar to those seen in real drainage systems. Stark's (1991) algorithm also has slightly different starting and ending conditions to the basic IP model, in that it starts from multiple sites, reflecting the multiple upstream sources of a drainage basin, and continues until no new sites can be invaded.

5.3   Growth (or aggregation) models

Another way to look at spread is to consider growth or expansion dynamics as seen, for example, in the expansion of the urban fringe, the growth of tumours or the spread of invasive species. These processes can also be seen as forming aggregates. Eden (1961) proposed a very simple two-dimensional growth model that has provided a point of departure for a wealth of related models. It is often called the *Eden growth process*. In the basic Eden process expansion occurs around an initial 'seed' site following just two rules:

(i) Randomly select an unoccupied neighbour of an occupied site.
(ii) Occupy that site.

This process is simple enough that analytically tractable approximations of it have been developed and used to investigate its critical behaviour (Edwards and Wilkinson, 1982, Kardar et al., 1986). Here, however, we will explore the dynamics of the process via simulation. Since Eden first proposed this model—in the context of the growth of colonies of cells such as cancerous tumours—a 'zoology' (to use Herrmann, 1986's term) of variants

5.7

have appeared in the literature (Table 5.2). These models have close links to the aggregation models described in Chapter 3, to random walks (see Chapter 4) and also to the percolation processes described in this chapter (Section 5.2.2). Another way to look at the Eden growth process is to imagine many random walkers starting from a central point on a lattice and stopping at and marking the first site they arrive at which has not previously been visited. An important difference between the Eden process (and related growth models) and the OP processes outlined above is that *history* plays a prominent role in the dynamics. Whereas OP processes are static in that the rules and geometry are fixed at the outset and do not change, growth processes,

Table 5.2 Some variants of the two-dimensional Eden growth process (Eden, 1961); see Gouyet (1996) and Herrmann (1986) for details

Model	Description	Reference
Classical model variant	1. Choose a randomly occupied site with at least one unoccupied neighbour 2. Occupy one of its neighbours	Herrmann (1986) and Gouyet (1996)
Inverse Eden	Start with a hole in the middle of the lattice and infill from edges using rules as above	Savakis and Maggelakis (1997)
Eden tree model	1. Select a perimeter site with *exactly* one occupied neighbour 2. Occupy that site	Dhar (1985), Dhar and Ramaswamy (1985)
Eden tip model	1. With $p = \frac{R}{1+R}$ select a site with only one occupied neighbour (a 'tip'), otherwise select any perimeter site 2. Occupy selected site	Sawada et al. (1982)
Williams–Bjerknes model	1. Select an occupied site 2. Grow into one of its unoccupied neighbours with probability α and render the occupied site 'healthy' (unoccupied) with probability β	Williams and Bjerknes (1972)
Diluted Eden (epidemics) model	1. Choose a random perimeter site 2. With probability p occupy it and with probability $1 - p$ render it unoccupiable	Alexandrowicz (1980), Bunde et al. (1985)

5.7 to 5.12

such as those represented by the Eden process, are more complicated because their dynamics at any point in time are strongly influenced and constrained by what has happened earlier, even when the rules are fixed.

5.3.1 Eden growth processes: theme and variations

The focus of analysis of the classical Eden growth process lies in understanding how the characteristic geometry of the cluster—a compact and dense interior with a diffuse and crenellated perimeter (the active zone)—emerges over time and, in particular, the scaling dynamics of the cluster's perimeter (Kardar et al., 1986, see also Figure 5.9 and Plate 8). In analysing the Eden process *noise reduced* forms, in which an unoccupied site must be visited a certain number of times (m) before it becomes occupied, are often used (see Figure 5.10 and Gouyet, 1996). This addition smooths the behaviour of the growth process (ultimately allowing more precise estimation of the process's critical exponents) and makes it more likely that sites adjacent to the oldest sites will be the next updated.

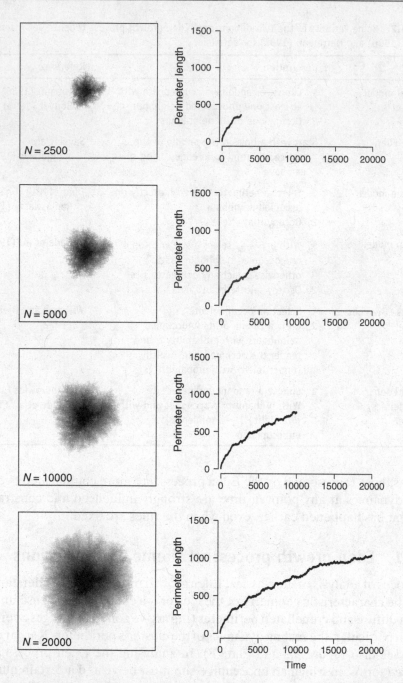

Figure 5.9 Examples of the basic Eden growth process at $N = 2500$, 5000, 10 000 and 20 000 steps. The grey scaling represents the relative time of the site being occupied (from light [young] to dark [old] within each time slice). The lattice comprises 512×512 cells.

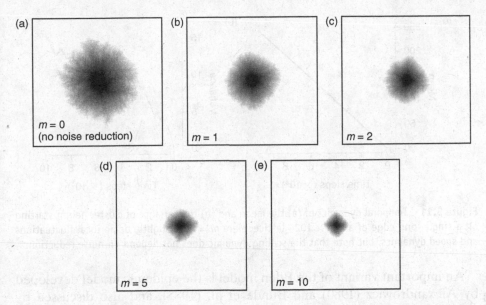

Figure 5.10 Eden growth at $N = 16\,000$ (a) without, and (b)–(e) with noise reduction; $m = 1, 2, 5, 10$ in (b)–(e), respectively. Note that as m increases the underlying geometry of the lattice becomes more and more pronounced, and the rate of spread drops markedly. Grey scaling is as in Figure 5.9.

The Eden process (and its relatives) is often explored by considering the dynamics of a front moving across a lattice, with the question of interest being how the cluster's 'height profile' (i.e. the distance it has advanced at each point on the front) changes over time. Edwards and Wilkinson (1982) consider the same problem analytically by asking how the height profile of a pile of sand being slowly and steadily dropped into a bucket will form. As Figure 5.11 shows, the increase in the mean height of the cluster is linear with time. The temporal dynamics of the variance in height (σ^2) are much more interesting, however. Initially σ^2 varies considerably but eventually it settles down. The value of σ^2 at which stabilisation occurs is a function of the width of the lattice L, and it occurs once a critical mean height $\langle h_c \rangle$ has been attained (Gouyet, 1996). The roughening of the front involves the development of a correlation length $\xi(t)$ over time (t), which increases as a power-law, $\xi(t) \simeq t^{1/z}$, until it reaches the width of the system L, at which point the system settles (Barabási and Stanley, 1995, O'Malley et al., 2009). Gouyet (1996) and Barabási and Stanley (1995) provide details of the scaling dynamics that underlie this transition, but again, as with percolation, a simple process is characterised by very different dynamics either side of a critical transition point. There has been considerable interest in the scaling of the roughening front as it is an example of a *self-affine interface*—phenomena that are common in a range of spread contexts (Barabási and Stanley, 1995).

 5.12

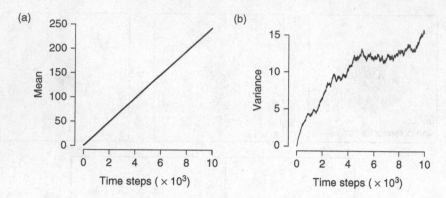

Figure 5.11 Temporal dynamics of (a) the mean and (b) the variance of cluster height starting on a single long edge of a 256×1024 lattice. Here $m = 1$ to slightly damp local fluctuations and speed dynamics, but note that the scaling dynamic does not depend on noise reduction.

5.13 An important variant of the Eden model is the epidemic model developed by Alexandrowicz (1980) and Bunde et al. (1985), and also discussed by Herrmann (1986). In this model, the Eden growth process is modified such that once a new cell is selected to become occupied then it either does become occupied (with probability p) or it becomes unoccupiable (with probability $1 - p$). In the context of disease spread, becoming occupied is analogous to becoming ill (with no chance of subsequent recovery) and becoming unoccupiable to becoming permanently immune. The questions of interest are how does the value of p affect the dynamics of the disease spread process and what sort of patterns form in this disease spread model? Clearly at $p = 1$ the epidemic grows as in the classical Eden model because there is no immunity. On the other hand, if p is very small then most sites become immune, and so growth will be limited and will stop when the perimeter is composed entirely of immune sites. The probability of indefinite spread at low p is effectively zero, which is similar to the critical value of the death rate in the basic contact process (see Section 3.4.1). However, above some critical value of p the growth process will spread indefinitely (Figure 5.12). It turns out that this critical point is exactly the same as that for OP considered above (Section 5.2.2), and that on completion the disease spread process will have generated a cluster very similar to a percolation cluster, albeit with some subtle geometrical differences (see Herrmann, 1986, Gouyet, 1996).

5.14 As detailed in Table 5.2, many variants of the basic Eden process have been developed, some of which have been applied to social and environmental systems. For example, Benguigui (1995, 1998) argued that a noise-reduced Eden growth model, with the variation that unoccupied sites adjacent to visited sites are also incremented, reproduces certain aspects of the growth of city margins (Figure 5.13), and O'Malley et al. (2009) use Eden growth fronts to represent the spread of invasive organisms.

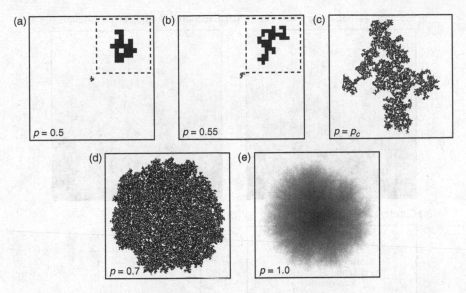

Figure 5.12 Growth of Eden 'epidemics' (Alexandrowicz, 1980, Bunde et al., 1985) with $p = 0.5$, 0.55, p_c ($0.59274621\ldots$), 0.7 and 1.0 on a lattice with $L = 256$. At $p = 1.0$ the epidemic model is the same as the classical Eden growth process. In the $p = 0.50$ and $p = 0.55$ examples the insets are blow-ups of the cluster and in the $p = 1.0$ example the grey scaling represents the time at which the site was invaded, showing the nature of the Eden epidemic growth over time (compare with Figure 5.9).

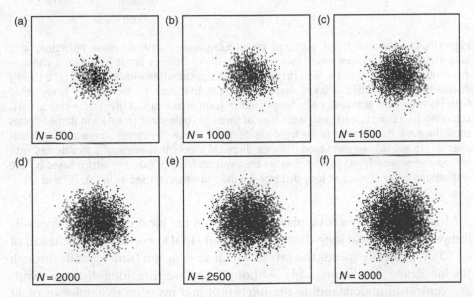

Figure 5.13 Growth of urban aggregation as simulated by Benguigui's modified Eden process on a lattice of $L = 128$ after $N = 500$, 1000, 1500, 2000, 2500 and 3000 settlement events. Note that this form of the model results in islands of urbanised land around the margins of a denser centre. The classical Eden model does not produce these islands.

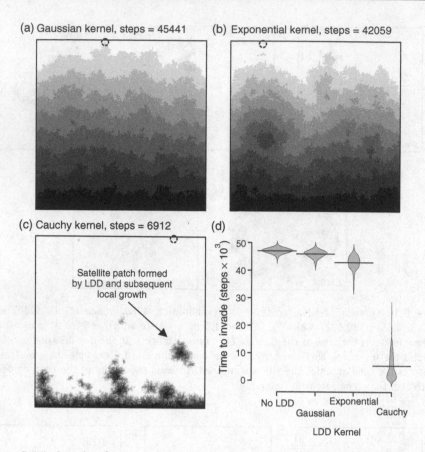

(a) Gaussian kernel, steps = 45441   (b) Exponential kernel, steps = 42059

(c) Cauchy kernel, steps = 6912   (d)

Satellite patch formed by LDD and subsequent local growth

Figure 5.14 Invasion fronts produced by an Eden model, with no noise reduction, with long-distance colonisations on a 256 × 256 lattice: long-distance jumps follow (a) a Gaussian distribution with $\mu = 0$ and $\sigma = 5$, (b) an exponential distribution with $\lambda = 5$ and (c) a Cauchy distribution with location = 0 and shape = 5. The distributions are increasingly heavy-tailed from (a) to (c). The probability of a long-distance jump occurring in a given time-step is 0.01. Colouring from light to dark represents time of invasion (light most recent) and dashed circles show the event that resulted in the front spanning the lattice. Dark areas surrounded by lighter ones in (a) and (b) represent long-distance dispersal events that eventually recoalesced with the main expansion front. (d) The time for the invasion front to span the lattice based on 100 realisations in the absence of long-distance jumps, and parameterised as in (a), (b) and (c).

5.15 Most invasions by exotic plant species combine local growth and sporadic jumps via occasional long-distance dispersal (LDD) events. The inclusion of LDD dramatically speeds the rate at which an invasion front spreads through the landscape (see Figure 5.14). Although they are rare, long-distance jumps are central to understanding the migration and invasion dynamics of plant species. This two-component dynamic is in part the outcome of the typically leptokurtic (heavy-tailed) nature of seed dispersal but is also amplified by human-assisted dispersal. We can modify the Eden growth process along a front (see Section 5.11) such that occasional LDD events occur and new sites away from the front are colonised. Long-distance dispersal events result in complex spatial patterns in which islands of invasion, sometimes called satellites, form beyond the main front and then as local growth occurs around the isolated sites they coalesce back into the main front. Kawasaki et al. (2006) explore a similar lattice-based model in more detail and compare it with analytical solutions for travelling wave fronts, and Clark et al. (1998) use a related model to explore the implications of heavy-tailed seed dispersal kernels for migration rates of tree species (see Section 5.1).

5.3.2 Diffusion-limited aggregation

Diffusion-limited aggregation (DLA), first described by Witten and Sander (1981) but see also Rosenstock and Marquardt (1980), is a growth process with close ties to the Eden model. It has been argued that it is also closely related to diffusion processes such as fire spread (Clarke and Olsen, 1996) and urban growth (Batty et al., 1989), and to many physico-chemical phenomena (see Sander, 2000). The DLA algorithm as originally described by Witten and Sander (1981) is simple:

(i) 'Seed' a single occupied site on the lattice (usually at the centre).
(ii) Release a random walker a long way from any occupied site and let it wander until it is in an unoccupied site neighbouring an occupied site, when it stops and occupies that site.
(iii) Repeat N times.

This process describes the aggregate structure built up by the collisions of many 'sticky' random walkers. Despite its simplicity the DLA algorithm results in objects that bear a close macroscopic similarity to many objects observed in the real world, and it has attracted considerable interest since being proposed. Sander (2000) attributes this interest to three things. First, its extreme simplicity; second, the apparent similarity of the objects produced by DLA to real objects; and third, the fact that despite its algorithmic simplicity its analysis is very difficult: 'this is a *devilishly* difficult model to solve, even approximately' (Sander, 2000, page 203). These attractions are remarkably similar to those which Grimmett (1999) identified as having made percolation theory so beguiling (Section 5.2.1).

The algorithm described above is called *on-lattice DLA* because it occurs within the geometric confines of a regular lattice. The alternative is *off-lattice DLA*, in which the walkers roam continuous space and stop when they are within some critical distance, usually the radius of the walker itself, to another walker that has come to rest. For very large objects ($N > 10^6$) on lattices, the shape of the DLA clusters starts to become distorted and off-lattice simulation is required to explore the geometry of such large objects (Sander, 2000). Here, we will solely consider on-lattice simulations for simplicity. DLA is a computationally expensive model and so we make a couple of refinements to the algorithm described by Witten and Sander (1981) to speed things up (following Meakin, 1983 and Sander, 2000):

(i) Release the walkers just outside the cluster radius—this relies on the fact that there is an equal probability of a walker intercepting any point on this circle if released sufficiently far from it, but saves a lot of the time involved in walkers 'finding' the cluster margin.

(ii) If the walker wanders more than a set distance from the original 'seed' cell, return it to a point on the circle from where it originated.

Both of these refinements address the problem that true random walkers are capable of wandering a long way 'off course', but do not change the probability that they will eventually encounter the edge of the DLA cluster. In other words, from the point of view of cluster growth, far-flung wandering is wasted (computational) effort, which these changes seek to minimise.

So, why does the DLA process result in the fractal shape that it does? Why don't the embayments obvious in Figure 5.15 in-fill over time? Sander (2000, page 204) provides the answer, 'Basically the reason that fjords do not fill up

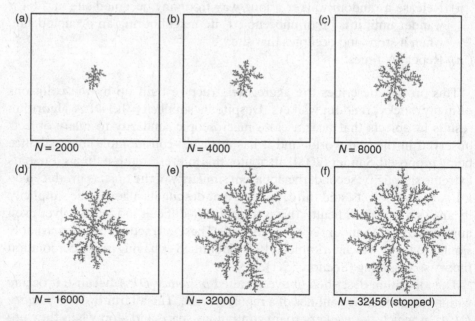

Figure 5.15 Growth of an 'on-lattice' diffusion limited aggregate at $N = 2000$, 4000, 8000, 16 000, 32 000 and 32 456 when the process reached the edge of the 640 × 640 simulation area. For such relatively small values of N the aggregate's geometry is unaffected by the underlying lattice.

in the cluster is that random walkers come from the outside and hit one of the branches before they can go very deep inside—the fjords are screened.' This is a nice example of *path-dependence*, where the state of the system in the future is strongly constrained by what has gone on in the past, an effect we have also seen in Eden growth processes.

Extensive interest has focused on the fractal properties of the clusters formed by DLA. On the two-dimensional plane D is approximately 1.7 (Meakin, 1983), although physicists have gone so far as to estimate fractal geometries of DLA clusters in up to eight dimensions (see Sander, 2000 for details)! Random walks on the clusters formed by the DLA process provide another example of the phenomenon of anomalous diffusion described in Section 5.2.3 (see ben-Avraham and Havlin, 2000, Havlin and ben-Avraham, 2002).

One area where a number of attempts have been made to use generic growth models based on DLA and related approaches is urban growth. This is certainly a context where the chronology of growth and subsequent path-dependence is important (see Frenken and Boschma, 2007, Arthur, 1994, especially Chapter 6). Cities are widely believed to show fractal shapes and this provides an apparent link to models such as DLA and the Eden process, many of which produce self-affine geometries (Batty and Longley, 1994).

Batty et al. (1989) argue that DLA provides an adequate representation of the spatial dynamics of urban expansion and compare various geometric measures of DLA clusters with those for the town of Taunton (Somerset, UK). Visually at least, however, the centre of the DLA aggregate (the 'central business district') is not compact enough to adequately represent real cities. As Makse et al. (1998) point out, this failing is likely because new growth is effectively restricted to the edges of the DLA cluster as the walkers are extremely unlikely to navigate their way through to the object's core. The closely related dielectric breakdown model (described by Gouyet, 1996) has also received attention from urban geographers as it too appears to capture some of the dynamics of the expansion of city margins, although the objects it produces are, again, too diffuse to be applied in their basic form (Batty et al., 1989). Makse et al. (1995, 1998) used a modified form of the percolation model motivated by two observations: that population density tends to decline following a power-law with distance from the urban centre, and that in cities the spatio-temporal sequence of development is correlated rather than purely random. The correlated percolation model described in detail by Makse et al. (1998) appears to provide an acceptable qualitative match to the observed geometries of cities both static and dynamic. While no single growth model seems likely to capture the myriad complexities of city growth—and it is, perhaps, naïve to imagine that any one model could—the growth and aggregation models presented in this chapter provide a range of potential starting points.

5.4   Applying the framework

Perhaps more so than those considered in Chapters 3 and 4 the models we have described in this chapter may appear rather divorced from 'real' social and environmental systems. In this section we consider some concrete examples of how models derived from percolation and aggregate growth approaches have been directly applied to a wide range of topics relevant to social and environmental sciences.

5.4.1   Landscape pattern: neutral models and percolation approaches

If we want to make inferences about how a particular spatial pattern influences a specific process then we need to know how that process operates in the absence of that pattern (Gardner et al., 1987). For example, if we are considering how a species may respond to habitat loss we might start by observing that habitat loss is usually spatially non-random. Thus, if we hope to understand how an organism may respond to habitat loss, one strategy is to simulate habitat loss in a way that is consistent with this spatial structure and to simulate it in a way that is not. This is a specific example of the more general Monte Carlo strategy introduced in Section 2.2.3. So, for example, if habitat loss is 40% we can produce landscapes with that amount of habitat removed, with removal either random or spatially structured. We can then simulate the process of interest and any difference in effect between the two types of landscape—one informed by the spatial structure of habitat loss, the other not—gives a sense of the importance of any spatial effects in habitat loss on the species we are concerned about. Such process-free models are called *neutral models* (Caswell, 1976) and many different approaches have been developed in the specific context of neutral models of landscape pattern (Gardner et al., 1987, O'Neill et al., 1992a), especially in terms of understanding fluxes of organisms, materials and energy through spatially complex systems (Wang and Malanson, 2008). The majority of such approaches draw heavily on the OP models introduced in Section 5.2.2.

5.17   The earliest and simplest neutral landscape models are directly derived from OP approaches (Gardner et al., 1987). Such landscape models are purely randomly constructed (hence they are devoid of process) and are usually binary in that they represent landscapes as habitat and non-habitat. In such neutral models, around the percolation threshold p_c connectivity will dramatically alter, which suggests that the response of ecological systems to habitat loss might also be nonlinear around this point. Unfortunately, simple percolation-based models produce landscapes (Figure 5.16(a)) with quite different characteristics—number and size of patches and boundary fractal dimensions—from real landscapes with a similar amount of habitat

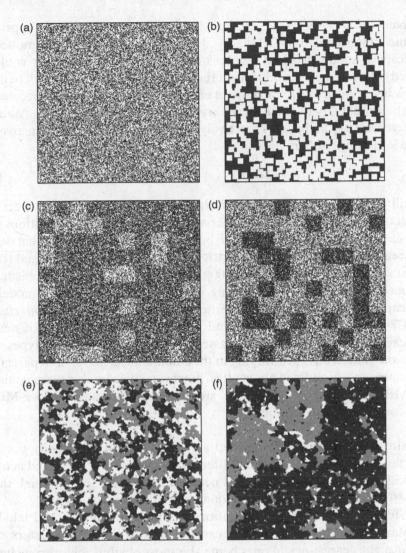

Figure 5.16 Examples of neutral models produced by (a) ordinary percolation $p = p_c$, (b) ordinary percolation $p = p_c$ but in blocks of 9×9 sites, (c) hierarchical percolation $p_1 = 0.20$, $p_{2A} = 0.65$, $p_{2B} = 0.40$, so that $p_m = 0.45$, (d) hierarchical percolation $p_1 = 0.30$, $p_{2A} = 0.20$, $p_{2B} = 0.60$, so that $p_m = 0.48$, (e) MRC $p = 0.5$, $n = 3$, $A_i = 1$ for all and (f) MRC $p = 0.58$, $n = 3$, $A_i = \{6, 3, 1\}$.

(Gardner et al., 1987, Saura and Martínez-Millán, 2000). Nevertheless, Boswell et al. (1998) used this approach to explore the implications of habitat loss on the army ant, *Eciton burchelli*, adding the variation that habitat loss occurred in blocks, so that small contiguous sets of cells were selected for habitat loss, rather than single cells (Figure 5.16(b)).

O'Neill et al. (1992a) describe a slightly more sophisticated approach, but one still grounded in OP theory. In their hierarchical binary model the landscape is subdivided into regular blocks (although this feature could be relaxed) of two types, A and B, with the probability of a given block being of type A being given by p_1. Each site in each block is then tested for occupancy, with the probability of occupancy varying between blocks of type A and B, and given by p_{2A} and p_{2B}, respectively. The expected occupancy across the entire lattice, p_m, is then:

$$p_m = p_1 p_{2A} + (1 - p_1)p_{2B} \tag{5.5}$$

O'Neill et al.'s (1992a) hierarchical algorithm allows spanning clusters (i.e. connected landscapes) to occur under a much wider range of conditions than is the case for the classical OP model, but still results in landscapes that depart from real landscapes in their geometric properties (Figure 5.16(c) and (d)).

5.18   Saura and Martínez-Millán (2000) developed a more sophisticated approach with the goal of producing more general and realistic models of landscape pattern than earlier methods. Their *modified random clusters* (MRC) method is based on OP and allows for the simulation of neutral landscapes containing multiple classes (for example land-use types, soil conditions and so on) each varying in their abundance, and so it departs from the binary models outlined above. It is, however, considerably more complex. The MRC model proceeds in four steps (see Saura and Martínez-Millán, 2000, for a detailed description):

(i)   Sites on the lattice are labelled as occupied with probability p as per standard OP. To ensure that different types can be represented across a wide range of abundances it is necessary to set $p < p_c$, although this is not required for landscapes strongly dominated by one type.

(ii)   Clusters in the simple percolation system are identified and labelled. Various neighbourhood rules can be used to build the clusters, four or eight nearest-neighbours being the most obvious, but also including asymmetric neighbourhoods, which will result in anisotropic landscape patterns.

(iii)   A type (or class) is assigned to each cluster, so that the large number of clusters produced in the previous steps will be reduced to a map of n types each with abundance A_i. At this stage only *occupied* cells are assigned so that $p \times L^2$ sites will have been allocated a type at the completion of this step.

(iv)   The landscape is in-filled by assigning a type to remaining unallocated sites. If a site already has one or more allocated neighbours, a majority rule is applied, and if none of its neighbours are allocated then it is assigned a type at random proportional to the required abundance of each type A_i.

These four steps are controlled by three parameters:

(i) The initial probability of occupancy, p.
(ii) The neighbourhood structure used to build the clusters.
(iii) The number of different types n, and their relative abundances $\{A_1, \ldots A_n\}$.

When p is close to p_c it becomes difficult for the algorithm to honour requested relative abundances because of the existence of incipient spanning clusters, but the synthetic landscapes produced by MRC are broadly controllable, as is evident from Figures 5.16(e) and (f).

All these neutral landscape models are *static*, but they show how a simple procedure—the random filling of sites on a lattice—can form the basis of methods that provide rich templates for dynamic and/or process-based models. Crucially these templates can be created with known statistical properties and so are useful in understanding interactions between pattern and process.

Using such methods it is possible to look at how processes operate along gradients in spatial pattern, and so perhaps isolate thresholds in spatial pattern at which a system's behaviour changes qualitatively. This is difficult to achieve using more empirical approaches where landscapes are not easily controllable. For example, drawing on the de Gennes (1976) ant in the labyrinth metaphor (Section 5.2.3), Malanson and Cramer (1999a, b) explore the implications of landscape structure for animal movement using neutral landscapes derived from simple percolation and more complicated hierarchical approaches. Malanson (2003) expanded on this earlier work to explore the implications of a continuous (0-1), as opposed to the (then) more commonly applied binary (0 or 1), representation of habitat structure. Malanson and Cramer's simulations again reinforced the importance of critical thresholds in landscape structure and connectivity around p_c, but showed that organisms following different patterns of movement, such as simple random walks, correlated random walks or edge-following strategies, are differently affected by fragmentation. They argue, perhaps optimistically, that such frameworks can be used to develop rules of thumb for identifying landscapes close to fragmentation thresholds and so priorities for habitat management. Irrespective, Malanson and Cramer's (1999a, b) approach shows the value of combining simple models, in this case percolation-based landscape models and models derived from random walks, to address questions about the interplay between spatial patterns and processes.

5.4.2   Fire spread: Per Bak's 'forest fire model' and derivatives

Models of spread are ubiquitous in the natural and social sciences, but one process that has received particularly close attention is fire spread, so much so that it is used as a metaphor for a raft of generic processes. In an example in Section 5.2.3 (see Figure 5.6) we considered the spread of a *single* fire as a contagious process through a site percolation system. In that model, a randomly selected grid cell is ignited and fire spreads from that cell to neighbouring cells and so on. This is obviously a gross simplification of the complicated physico-chemical process of combustion and the ecological and climatic processes that underpin flammability, but it again serves as a valuable point of departure. While the dynamics of fire spread are obviously of interest for emergency response, ecologists are more interested in repeated patterns of fire events in time and space—the fire regime. It is the fire regime, rather than individual events, that shapes ecological systems over long time scales. As a result there has been considerable interest in the dynamics of simulated landscapes where multiple fire events occur, and, perhaps surprisingly, models based on percolation processes *do* capture some of the macroscopic patterns seen in some ecological systems (Malamud et al., 1998, Zinck and Grimm, 2008, Moritz et al., 2011). A simple but much-studied model helps to show how this works.

Although the controls on fire are heterogeneous in time and space at multiple scales, as always, it is easiest to start by thinking about a much more abstract model, in this case the so-called 'forest fire model' (FFM, see **5.19** Bak et al., 1990 and Drossel and Schwabl, 1992). Bak et al. (1990) originally proposed the FFM to describe the dissipation of energy entering a non-equilibrial system at a constant rate. It is thus important to emphasise that the fire aspect in this model is really a *metaphor* rather than a mechanistic representation (see Millington et al., 2006, for discussion). Even so, the FFM has attracted much attention, partly because it does reproduce some of the macroscopic (aggregate) patterns seen in real fire regimes. The simplest form of the FFM can be summarised as follows:

(i) The simulation arena is a lattice of $L \times L$ sites, each of which can be in one of two states (empty or occupied), with only occupied cells able to burn. This is a simple site percolation system.

(ii) With frequency $1/f_s$ fire events occur. In time steps when a fire does not occur a randomly selected site is set to occupied (if it is currently empty).

(iii) If a fire occurs it proceeds as a contact process via spread into neighbouring unburned and occupied cells, with this process continuing until no further such site can be ignited. Thus, all of the percolation cluster in which a selected site is located will be 'burned'.

(iv) There is a clear separation in time scales between the fire and the rate
of revegetation such that cells change state empty → occupied at a rate
much slower than individual fires operate, since these spread to their full
extent in a single time step.

Thus, the FFM is very similar to the model of fire spread through a site percolation system introduced earlier, with the key differences that: the underlying
lattice changes over time as sites become occupied and are burned; and we are
interested in the patterns emerging across many individual fire events, rather
than in the extent of individual fires. The classical FFM can be reframed
in terms of the much-studied SIR (susceptible–infected–recovered) model
of disease transmission, with unburned occupied sites (healthy trees) being
equivalent to susceptible individuals, burning trees infectious individuals and
empty sites recovered individuals (Keeling and Rohani, 2008), and is analogous to a wide range of spatial ecological processes (Pascual and Guichard,
2005).

A fundamental characteristic of wildfire frequency-area statistics, the probability distribution describing how often events of different sizes occur, is that
they are heavy-tailed, as shown in Figure 5.17. A heavy-tailed distribution
means that large fires are more common relative to small fires than we would
expect if fire size-frequency statistics were adequately described by a thin-
tailed distribution such as a Gaussian. So how does the naïve FFM perform?
As Figure 5.17 shows, after a sufficient number of fires ($n > 10^4$) the FFM
produces heavy-tailed (approximating a power-law) frequency-size statistics.
Comparison with observational data, however, suggest that the power-law
exponent (the slope of the relationship between size and frequency) is too

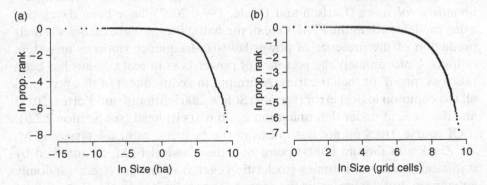

Figure 5.17 Fire size-frequency statistics arising from (a) $N_F \simeq 3.41 \times 10^6$ shrubland fires
in Portugal over the period 1980–2005 (data from the Portugese Rural Fire Database, Pereira
et al., 2011) and (b) from $N = 2 \times 10^4$ fires in the classical FFM. Here the data are plotted as a
rank-frequency (RF) plot with ln $\frac{j}{N}$ against ln x_j from largest to smallest where x_j is the size of
the jth largest fire event and N is the total number of fires (following James and Plank, 2007).

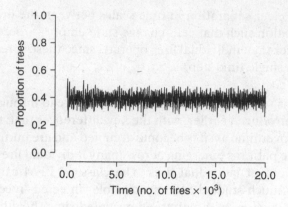

Figure 5.18   Self-tuning of the proportion of trees in the grid over $N = 2 \times 10^4$ fires for the same data as in Figure 5.17. The proportion slowly stabilises to settle on the same value irrespective of initial conditions.

low and that the FFM produces too many 'large' fires (Malamud et al., 1998, Zinck and Grimm, 2008).

The proportion of patches occupied by trees over time is also of interest. Irrespective of the initial proportional abundance of trees, the proportion of occupied cells approaches the same value (Figure 5.18)—in other words the system self-tunes. This dynamic has led to speculation that this model system is 'self-organising' and that analogous real-world systems may be as well. Because the fire size-frequency statistics approximate a power-law distribution this has led to the suggestion that the FFM shows the critical behaviour associated with a range of conceptual models describing the dynamics of nonlinear systems (see discussion in Moritz et al., 2011). Various theories, such as *self-organised criticality* (Turcotte, 1999) and *highly-optimised tolerance* (Carlson and Doyle, 1999, 2002), have been developed using models such as the FFM. One of the hallmarks of these theories is their prediction of the presence of power-law size-frequency statistics in system outputs. Unfortunately the presence of power-laws in real systems has been taken as 'proof' of such theories. Affirming the consequent in this way is an all too common logical error (see also Solow, 2005, Stumpf and Porter, 2012) and the issue of under-determination again rears its head (see Section 2.2.6).

Of course, trees do not just fall from the heavens, as in the classic FFM. As Zinck and Grimm (2008) point out, the classical FFM, as analysed by statistical physicists, assumes stochastic vegetation change (trees randomly appearing), but deterministic fire spread (as constrained by the formation of clusters). An ecologist, on the other hand, would more likely conceptualise such a system in reverse, with vegetation change being more deterministic, as a function of age, and fire spread more stochastic. We can modify the classical FFM to reflect this view of the world by having all cells on the lattice

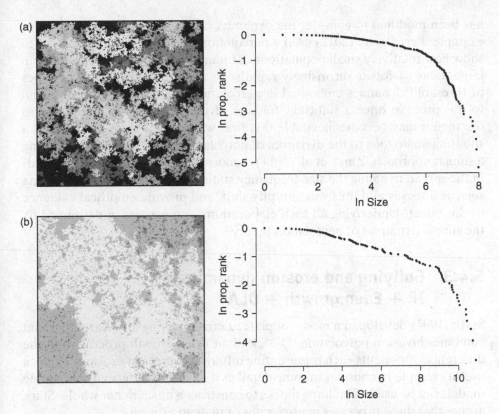

Figure 5.19 Age mosaics and size-frequency statistics produced by two different fire-age models after 300 fire events: (a) flammability is constant with time, $p = 0.23$, and (b) flammability increases with time, $p = 0.025 + 3 \times (\text{age}/1000)^2$ (following Ratz, 1995). In both cases the landscapes are initialised such that all sites have age one. Fires occur at intervals drawn from a Poisson distribution with $\lambda = 10$. Note how spatial structure emerges in both cases (mosaics are colour-coded from young [light] to old [dark]), but there are qualitative differences depending on the fire-age model. Plotting of the fire size-frequency data follows the convention outlined in Figure 5.17.

occupied, keeping track of the time they were last burned and making their flammability dependent on their age. Figure 5.19 shows that such a model not only produces heavy-tailed (but not necessarily power-law) size-frequency statistics, but also results in spatial mosaics of patches of different ages. This spatial organisation is strengthened if we include 'memory' in the model by making flammability a direct function of time since the last fire (Ratz, 1995, Peterson, 2002, Kitzberger et al., 2012).

Many other elaborations can be added to such a basic framework, including directional bias in fire spread as a function of wind or terrain effects, long-distance 'spotting' in fire (non-local fire spread) as occurs via long-distance wind transport of embers and so forth. Indeed the basic process laid out above has been modified to consider the dynamics of fire in 'real' ecosystems. For example, Perry et al. (2012) used a percolation-derived fire spread model to show how relatively small populations of humans could, using fire, transform large areas of forest surprisingly rapidly. Perry et al. argue that the types of threshold dynamics embodied in percolation-derived models are crucial to this process: once a sufficient fraction of the landscape is vulnerable to fire then it may become inevitable that fires will become larger and larger, a situation analogous to the dynamics of percolation clusters around p_c. Using a similar approach, Zinck et al. (2011) demonstrate that at critical thresholds in the spread dynamic the size-frequency statistics of contagious phenomena such as diseases and fires will abruptly shift, and provide empirical evidence of this effect. Underlying all such elaborations, is a process underpinned by the simple dynamics of percolation theory.

5.4.3   Gullying and erosion dynamics: IP + Eden growth + DLA

Stark (1994) developed a model of plateau erosion along an escarpment that 5.21 combines invasion percolation, DLA and an Eden growth process, with the different components each representing different sources of erosion. This is a useful example to consider in some detail as it illustrates how multiple simple models can be used as building blocks to construct a much richer whole. Stark argues that three processes control rates of plateau erosion:

(i) The fundamental strength of the substrate being acted upon.
(ii) Erosion caused by the movement of groundwater through and out of the substrate, or *seepage weathering*.
(iii) Erosion caused by background processes independent of the other two processes, or *background weathering*.

Stark, in turn, uses three different growth processes to represent these three controls on the gross rate of plateau erosion: IP represents the fundamental substrate strength, DLA represents seepage weathering and an Eden process is used to represent background weathering.

The model runs on a lattice with periodic boundary conditions such that the left and the right edge are wrapped, with erosion starting at the bottom edge and moving upward. At the start of each model realisation each site is allocated a substrate strength, analogous to r in the IP process (see Section 5.2.4). At each time step the following occurs (see Stark 1994, page 13959):

(i) A random walker is released from an uneroded site on the plateau and continues walking until it hits an uneroded site neighbouring the eroded edge. The value of r in this site is then reduced by factor γ. This is a modified DLA process and represents seepage weathering.

(ii) An uneroded site neighbouring the eroded edge is selected at random and r is reduced by η. This is a form of Eden growth and represents background weathering.

(iii) The cell at the eroded margin with the lowest value of r becomes eroded. This is the IP process. Note that the seepage and background weathering (DLA and Eden) processes increase the likelihood of sites subsequently eroding due to substrate weakness.

The nature of plateau erosion as simulated is controlled by the values of γ and η (see Figure 5.20). Obviously as they tend towards zero then the importance of seepage and background weathering respectively, become less important. If both are equal to zero then erosion proceeds as an IP process (in other words controlled by the initial random field). If $\gamma \gg \eta \gg 0$ then seepage weathering dominates the process and *vice versa*. A final control on the process is to include noise reduction (as described in Section 5.3.1). This is done by not allowing a cell to erode until it has been hit by a certain number of DLA walkers and/or has been selected a certain number of times in the Eden growth process. Stark (1994) notes that the inclusion of noise reduction dramatically alters the patterns produced by the model and changes its scaling properties.

Although Stark clearly places this multi-facetted growth model in the context of erosion and gullying, it might be used in a range of other settings. For example, in the context of urban growth the Eden growth component could represent the slow expansion of a compact city centre, the DLA component the tendency for cities to grow along corridors formed by arterial

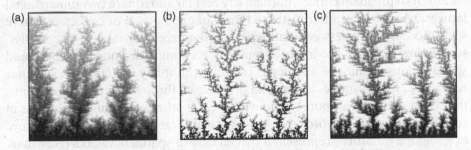

Figure 5.20 Final patterns of plateau erosion produced by the model described by Stark (1994), with (a) $\eta = 0.20$ and $\gamma = 0.05$, Eden growth, representing background weathering, dominant, (b) $\eta = 0.05$ and $\gamma = 0.20$, DLA, representing seepage weathering, dominant and (c) $\eta = 0.25$ and $\gamma = 0.45$, both seepage weathering and background weathering dominant over substrate strength. The lattice ($L = 256$) has periodic left and right boundaries. The underlying random field on which the IP process operates follows U[0,1], but is smoothed to reduce the patchiness of the process.

roads ('ribbon development') and the IP process variability in the desirability of different parts of the landscape for settlement (for example, as a function of local topography). Again, we can see that one of the strengths of starting with rather abstract models of spatial processes is that they can be moulded to fit the specifics of the question at hand and the analytical understanding gained in other contexts remains useful. This is not true when we start with detailed, highly system-specific models.

5.5   Summary

Spatial models derived from either percolation theory or cluster growth approaches (or even both) have been used to simulate and describe phenomena as diverse as opinion forming networks (Shao et al., 2009), the dynamics of ecotones such as treelines (Milne et al., 1996, Zeng and Malanson, 2006), the growth of lichen colonies on rocks (Jettestuen et al., 2010), disease spread (O'Neill et al., 1992b) and groundwater hydrology (Berkowitz and Balberg, 1993). It has not been our intention to review every application of these approaches here. Suffice to say that the very general nature of these models has proved useful across many disciplines and contexts.

A key difference between percolation and cluster growth (or aggregation) models is the role of history and path-dependence in the processes. OP models have no history whereas what happens early in the 'life' of an object growing via clustering determines its final pathway. Asking whether history and associated dynamics such as path-dependence are important may help to determine which of the models discussed in this chapter could be useful in a particular context.

It is worth concluding with a comment on geometry. In this chapter we have considered spread on lattices, particularly regular and square two-dimensional lattices, which are a rather restricted form of geometry. For processes where spread occurs only between entities located adjacent or near to each other, such as low intensity fire in the absence of 'spotting', this model of space and spatial relations may be adequate. However, in many systems this is rather restrictive. For example, the spread of diseases through human populations may have as much (or more) to do with the organisation of human systems in social terms, in other words the social network, rather than spatial structure or geography. Lattices represent a special case of spatial networks (Newman, 2010, Barthélemy, 2011) in which the connections are highly localised and occur only over short distances. Many of the processes discussed in this chapter can also be simulated on more realistic geometries (see, for example, Keeling and Eames, 2005, in the context of disease) and we consider some of these issues in the next chapter.

6
Representing Time and Space

This book is particularly focused on simulation models that produce changes across space through time. Time and space are themselves complicated phenomena, subjects of endless fascination to philosophers, scientists and artists. It is perhaps not surprising then that capturing time and space inside a computer model is not a simple matter. In this chapter, we consider some of the options available, along with their potential implications.

Of course, in the preceding chapters we have already seen many examples of models that represent spatial change through time, but we have not concerned ourselves too much about how these aspects of the models work. This is quite typical of how simpler models are developed, and is appropriate for the very general spatial models that we have focused on. However, what may appear obvious, even trivial choices, such as using a grid of square cells (rather than, for example, hexagons) and having them all update their state at the same time (rather than one at a time), can have profound effects on model outcomes. Furthermore, other approaches to representing time and space can enable richer and more detailed models to be built. This may be particularly important when it comes to building realistic models of particular settings in time and space, which is often necessary when models are intended to inform policy and decision-making.

In some accounts of simulation models (see, for example, Benenson and Torrens, 2004) an emphasis is placed on often-encountered overall 'architectures' such as cellular automata or agent-based models. We are more interested here in the particular effects of particular representational choices and so do not discuss these general categories. Useful overviews of the properties of cellular automata can be found in the references cited in Chapter 3;

Spatial Simulation: Exploring Pattern and Process, First Edition.
David O'Sullivan and George L.W. Perry.
© 2013 John Wiley & Sons, Ltd. Published 2013 by John Wiley & Sons, Ltd.

Grimm and Railsback (2005) provide a thorough account of individual-based models and the recent edited collection by Heppenstall et al. (2012) provides a good overview of agent-based models.

Thus, in this chapter we consider some of the options available for representing time and space in simulation models and consider general properties of these representational choices. A recurring theme, either directly or by implication, is the importance of the *scale* of our chosen model representation. Scale and related key concepts are discussed in Section 2.1.3 and it may be useful to review that material before reading this chapter.

6.1 Representing time

How we represent time in models is less obviously problematic than how we represent space. Time, after all, is just one thing after another, it only 'moves' in one direction and it does not—if we restrict ourselves to Earth's surface!—exhibit the complex geometry that space sometimes does. Ironically, this may make it even more important to think carefully about how time works in a simulation because it can have unexpected implications for model outcomes. There are a number of crucial aspects to consider here.

6.1.1 Synchronous and asynchronous update

A fundamental issue is whether or not all of the state changes in a model occur simultaneously, when they are termed *synchronous*, or if they happen one at a time to each affected component of the model, when they are *asynchronous*. This seemingly innocuous difference can have far-reaching effects on the behaviour of a model. Examples are shown in Figures 6.1 and 6.2, which correspond to the following update timings (or sequences):

(a) Synchronous update with all cells determining their next state at the same time and updating at the same time.
(b) Asynchronous update with cells updating their state one after another, in a random order, but with every cell guaranteed a 'turn' in each round or 'generation' of updating.
(c) Asynchronous update but some cells may not be updated in a generation while others are updated more than once. This is equivalent to repeatedly selecting a single cell at random and updating it.
(d) Asynchronous update, but with the order of cell updating strictly from lower left to upper right, starting at the lower-left corner.

In cases (c) and (d), changes at one location can propagate their effects across the system before other cells have responded to the initial system state, and

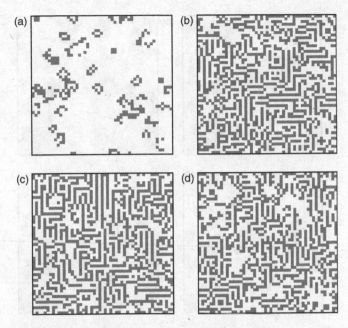

Figure 6.1 Different outcomes after 100 time steps starting from the same random configuration of a toroidally wrapped 51 × 51 cell grid, observing the game of life update rule—see Section 1.3.2. (a) Conventional synchronous update, (b) asynchronous update in random order with every cell guaranteed to update, (c) asynchronous update in random order with no guarantee of update and (d) asynchronous update in order from left to right and bottom to top, starting in the lower-left corner.

this is one reason why we may observe differences in outcomes between the various update methods.

For the game of life, as we can see in Figure 6.1, departing from synchronous update (a) completely changes the nature of the patterns produced by the model, from isolated small clusters to extensive 'maze-like' structures. When we change from random order (b)–(c) to ordered asynchronous update (d) there also appear to be minor differences in terms of the resulting overall pattern density. For a majority rule model, Figure 6.2 demonstrates that it is not a change from synchronous to asynchronous update that makes a notable difference, but a change in the order of asynchronous update: case (d) appears to produce larger contiguous regions of uniform states than do (a)–(c).

6.1

6.2

The important point to note here is that the timing of updates can certainly make a difference *and* that the difference it makes is dependent on the process (Ruxton, 1996). This is a continuing research area in cellular automata (see, for example, Bersini and Detours, 1994, Fatès and Morvan, 2004, 2005, Ruxton and Saravia, 1998, Schönfisch and de Roos, 1999) although interest has been rather sporadic. In any case, the clear dependency of the dynamics on the update mechanism means that it is important to consider which form of

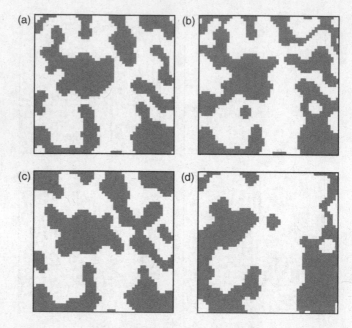

Figure 6.2 Different final states starting from the same random configuration of a toroidally wrapped 51 × 51 cell grid and observing a majority update rule (see Section 3.3.1) where cells adopt the state that is in the majority in their Moore neighbourhood. (a), (b), (c) and (d) are updated as in Figure 6.1, but here the markedly different case is (d).

updating is most appropriate for a particular model and also, where possible, to test what difference (if any) it makes.

As a general rule, either synchronous update or asynchronous random-order updating is preferable, where they are appropriate. However, there are no hard and fast rules, and there are many situations where ordered update makes sense—although the order is more likely to be determined by some attribute other than spatial location, as we used in Figures 6.1 and 6.2. For example, households in an economic model might be given the opportunity to choose a residential location in descending order of their income, as a way to simulate the wider range of choices open to high-income households.

6.1.2   Different process rates

A major advantage of asynchronous update is the ability to include processes that occur at different *rates*, something which is harder to accomplish satisfactorily under synchronous updating. In many cases, particularly interacting particle systems (see Section 3.4), various processes of birth, death, succession and so on occur at different rates λ_i. This is usually taken to mean that state updates are random-order asynchronous, with each of the various possible events that might occur at selected sites occurring with probabilities proportional to their rate parameters. The most straightforward way to implement this effect is to normalise all the rate parameters relative to the fastest rate λ_{max}, so that a slower process with rate λ_i is assigned a relative probability $p_i = \lambda_i/\lambda_{max}$. Then, if a site subject to the fastest process is randomly selected, the relevant event occurs with certainty. If a site subject to a slower process is selected, a uniform random number $x = U(0,1]$ is drawn and the event only occurs if $x \le p_i$, that is, with probability p_i. Over time this will mean that processes with lower rate parameters occur more slowly. Some of the models in Section 3.4 provide examples of this approach.

3.6
and 3.8

6.1.3   Discrete time steps or event-driven time

A different way to think of state updating processes is as a contrast between *discrete time steps*, when state changes are recorded, and *events* which occur in their own time. The approaches are schematically illustrated in Figure 6.3.

The discrete time steps approach effectively assumes that processes are ongoing in the world, and that at some regular time interval their overall effect is recorded. An example might be the seasonal growth and die-back of a resource distributed across a landscape. In each equal time-interval some amount of growth or decline in the resource at each location is possible, governed by various rate parameters. Growth or decline is then applied to every location at each time step—whether synchronously or asynchronously, as discussed above.

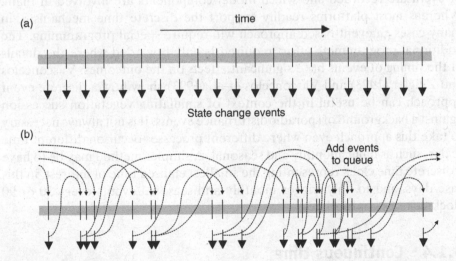

Figure 6.3 Discrete and event-based models of time. (a) The equal interval discrete 'ticks' which trigger state changes in the model. (b) Events triggering the placement of later events on a queue.

An alternative approach is to have a *queue* of events that will happen in the model. An event queue is an ordered list of state changes (or opportunities for state changes) that will occur, where the time interval between events is arbitrary and each event specifies the model component where it will occur. This method makes sense where processes in the system being represented are intermittent or inconsistent in their rate of occurrence, or where the likely time until the next occurrence is related to some aspects of the most recent occurrence. An example might be a detailed model of animal foraging where individuals spend different periods of time browsing the available resources before considering their options for moving on to another location. When an individual arrives at a resource location to start browsing, a 'browse time' can be drawn from a suitable distribution, and the next 'decision time' determined by adding the browse time to the current time and placing an event tagged with this time in the queue. An unpublished model of frugivorous foraging by the native New Zealand kererū described by Pegman (2012) takes exactly this approach. If many individuals' behaviour is modelled in this way, the result is a discontinuous series of times at which events happen. The advantage in model execution terms is that long periods of relative inactivity in the system happen instantaneously in the model as the model clock jumps from event to event. Regularly occurring events in such a model are accommodated by placing the next occurrence of the event onto the queue each time the code for the event is executed.

An event-based representation is more complicated to implement than the discrete, equal time interval approach. It requires maintenance of a time-stamp sorted list of events and some consideration of how different types of event are recorded and which model components are involved in them. Whereas most platforms readily support the discrete time mechanism, in many cases an event-based approach will require special programming. The additional programming effort required is only rewarded where the details of the timing of events have significant effects on the outcomes. Vasconcelos and Zeigler (1993) and Vasconcelos et al. (1993) show how a discrete event approach can be useful in the context of simulating vegetation succession against a background of sporadic disturbance events. It is not always necessary to take this approach even where different processes occur on different time scales, such as daily, monthly and seasonal. A simpler method may be to have a discrete time clock representing the smallest elapsed time of interest, in this case days, and to only trigger monthly and seasonal events every 30 or 90 clock ticks, as appropriate.

6.1.4   Continuous time

The event-based model of time described in the previous section brings us close to a *continuous time* model. In continuous time, changes in model state occur continuously at all times. In truth, it is impossible to implement a continuous time model in a digital computer, which necessarily must update any stored representation of model state from one discrete set of values to a new set. A much older set of methods exists for handling continuous time in the shape of the calculus. This brings us to mathematical models, where mathematical functions represent relationships between model parameters and running a model involves solving a potentially large set of simultaneous differential equations. This is the realm of *systems dynamics modelling*, an important branch of engineering science that is widely used for modelling complex dynamic systems and has its roots in *general systems theory* (von Bertalanffy, 1950). Phillips's MONIAC model (see Section 1.1.2) is a simple example of such a model, albeit an atypical, analogue one. This field is well developed (see, for example, Sterman, 2000, Ford, 2010, Pidd, 2010), with many available tools and software packages. When applied to spatial problems, the systems dynamics approach generally aggregates spatially distributed quantities to whole system quantities. However, this need not be the case, and considerable scope exists for integrating local systems dynamics models with spatially explicit models of the type discussed throughout this book, although such work remains relatively unusual as an overall architecture (Muetzelfeldt and Massheder, 2003). In a similar vein, Vincenot et al. (2011) consider the conditions under which the systems dynamics framework might fruitfully be married with more disaggregated representations such as agent-based models.

6.2   Basics of spatial representation

The field of *geographical information science*, an outgrowth from the technology of *geographical information systems* (GIS), has developed a number of more or less standardised approaches for representing geographical entities in the real world as objects in a computer system. We briefly survey here the options available and their relevance for spatial simulation models. More detailed information can be found in any of the many textbooks in this field. A particularly useful reference, and one which takes a more computationally oriented approach than most, is that of Worboys and Duckham (2004). By partial analogy with the case for representing time, we can think of these as discrete space (raster) and continuous space (vector) approaches, although the correspondence is nowhere near an exact one (see Couclelis, 1992).

6.2.1   Grid or lattice representations

The most basic way to represent two-dimensional spaces in a computer is as a grid of *cells* or *pixels*, a contraction of the phrase 'picture cells'.

As anyone who has more than dabbled in digital photography is aware, many computer images are stored as grids in which each pixel has a particular assigned colour. This grid-based approach is easily extended so that each grid cell has associated with it one or more attributes, each attribute recording some aspect of the phenomenon of interest. Thus a grid might record in each cell, elevation above sea level, the mean annual rainfall, the dominant vegetation cover and land use. Cell attributes may be either qualitative or quantitative. Qualitative attributes may record what type of location the cell represents, in terms relevant to processes of interest. For example, in a model of urban development, land use might be a qualitative attribute and could refer to the zoning of the land each cell represents, whether residential, commercial, industrial or parkland. In a model of animal foraging, a qualitative attribute might tell us whether the cell is grassland or forest. Quantitative cell attributes are probably more commonplace and assign some numerical value to the attribute in question. Zoning might have associated with it some maximum density of development, for example. Other attributes like groundwater permeability, soil pH, mean annual temperature, population density and so on are readily represented as numbers.

We may consider each different attribute as a separate grid in itself, a perspective that lies at the heart of *raster GIS*, which views the world as a set of *layers*, each recording some aspect of the study area (Tomlin, 1990). In simulation models, when we represent multiple attributes in each grid cell, it is probable that there will be interactions between the different attributes. As a result we more often take what might be termed a 'cell-centred' perspective, where all the attributes (or several at a time) of each cell and its near neighbours are relevant to the process represented in the model. The cell-centred view is facilitated by modern programming architectures such as object-oriented design. The contrast between the raster and cell-centred views is schematically illustrated in Figure 6.4. An important feature of the simulation model cell-centred view is its focus on change through time. It has proved difficult to incorporate change over time into GIS, so much so that spatial simulation models have been proposed as a possible way forward in this regard (see, for example, Wesselung et al. 1996; see also O'Sullivan, 2005, for an overview of time in GIS).

From either perspective, it is important that all cell attributes be recorded at the same resolution, that is, for grid cells in each attribute layer to have the same spatial extent. Where some attributes are recorded at lower spatial precision than others there are two possibilities. Most simply, the lower resolution (large cell) attribute is recorded multiple times in all the higher resolution (smaller) cells to which the lower resolution attribute value applies. In more complex models, it may be necessary to employ grids at various different resolutions and keep track of the relationships among grid cells so that large low-resolution cells have pointers to all the higher resolution

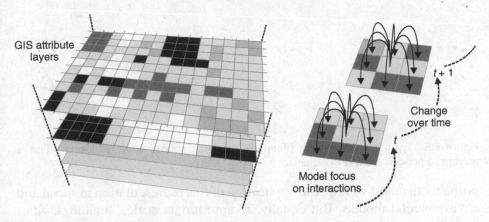

GIS attribute layers

t + 1

Change over time

Model focus on interactions

Figure 6.4   The GIS perspective on grid (raster) data tends to focus on multiple layers of attributes, while the simulation model perspective focuses on cells and their interactions with one another as cell states change through time.

grid cells they contain, and each small 'member' cell also records the larger low-resolution cells to which it belongs. With this scheme, low-resolution 'container' (or 'parent') cells can update their states based on aggregate measures (such as the mean or sum) of the member (or 'children') cells. Meanwhile, the behaviour of member cells may also be affected by the state of the larger cells to which they belong. Some versions of the Schelling model (see Section 3.5) use this approach with low-resolution cells being suburbs or neighbourhoods with fixed boundaries and households attending to both their immediate neighbouring cells, and to the aggregate state of the larger 'suburb' cell in which their current location is found (see, for example, Fossett, 2006).

6.2.2   Vector-based representation: points, lines, polygons and tessellations

A grid-based approach considers geographical space as a continuum and imposes an artificial coordinate system by subdividing it into units. By contrast, vector-based representation treats space as an empty 'box' containing entities at various locations. Depending on the nature of the entities—and on the spatial scale of the model—they may be represented by a variety of geometric object types (Figure 6.5).

A *point* entity $\mathbf{p}_i$ exists at a single location in space defined by a coordinate pair, $\mathbf{p}_i = (x_{i1}, x_{i2})$ in two-dimensional space, or sometimes by a triple, (x_{i1}, x_{i2}, x_{i3}) in three dimensions. A point is an idealised entity with zero extent and so, strictly, cannot represent any actual entity, since all real entities have some spatial extent. However, *at an appropriate scale*, many entities are well approximated by points. The current location of an individual organism is

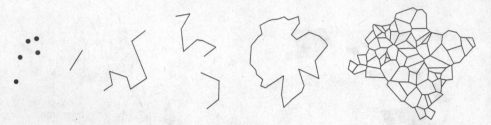

Figure 6.5 Types of vector object (from left to right): points, a line segment, a line, a polyline, a polygon and a tessellation.

probably the most widespread application of this representation in social and environmental models. But equally, at appropriate scales, buildings, cities and other settlements, bus stops, archaeological artifacts, trees, rocks and a wide variety of other entities that might appear in a model can be represented this way (see Section 2.2.2).

Linear phenomena such as roads, paths or rivers require us to record a series of point locations connected together, in a particular order, $\{\mathbf{p}_1, \mathbf{p}_2 \cdots \mathbf{p}_n\}$. Most simply, we have a *line segment* with just two points, one at each end. More likely are more or less complicated *lines* consisting of several line segments. A distinction can be made between lines and *polylines* consisting of several disjoint lines, each of several line segments, although in most cases it is unlikely that this level of complexity will be required. More importantly, the great majority of linear objects are actually links in spatial networks, such as roads in a road network or reaches in a stream network. We briefly consider networks in Section 6.3.3.

Lines may also appear as the boundaries of land parcels, lakes, habitat patches and administrative areas or other regions of space, and these are conveniently represented as *polygon* objects. A polygon is most commonly stored as a sequence of point locations, with the additional proviso that the polygon is closed, which may involve storing the beginning and end location twice or assuming that the last point is connected back to the first. In either case a convention is observed whereby the boundary points are consistently recorded in either a clockwise or anticlockwise direction as this allows easier determination of which parts of the space are inside or outside the polygon, and also calculation of polygon areas.

Isolated individual polygons are unusual. More commonly, a collection of polygons subdivides the modelled space into a *tessellation*, a set of polygons that together cover the whole space without gaps and without overlaps. In a GIS context, this may be called a *polygon layer*, or *coverage*, while in ecology the term *mosaic* refers to a coverage of different habitat patches. Similarly, in urban geography the phrase 'the urban mosaic' refers to how different population mixes in the city may be considered as together forming a coverage of diverse urban neighbourhoods.

When working with a tessellation it may be important to avoid unnecessary (and error-prone) duplicate storage of polygon boundaries. Each line segment is part of the boundary of two polygons, which are adjacent to one another across that line segment. Various data structures have been devised, such as the *doubly-connected edge list* (see Worboys and Duckham, 2004, Chapter 6 for details and references) so that line segments are only stored once, referenced by their start and end points (also stored once only). Polygons are stored as sequences of line segments. This data structure allows for efficient retrieval of the neighbours of each polygon in a tessellation, and may be required even in quite simple spatial models, if frequent access to neighbouring or nearby polygons is required.

The various vector object types capture only the spatial characteristics of the objects they represent. In the same way that a grid representation may record multiple attributes at each cell, vector objects may have attributes associated with them recording characteristics of the entities they represent. Taken together, this approach is the *entity–attribute* data model and is the most widely used general approach to the representation of spatial data in GIS. Although spatial simulation models are not geographical information systems as such, they are often used alongside GIS and it is likely that some appreciation of the entity–attribute data model will be required for successful integration of model results with GIS map outputs. Data imported from GIS are a common starting point for spatial simulation models, and so a good working knowledge of these representational approaches is likely to be useful.

6.3 Spatial relationships: distance, neighbourhoods and networks

Representing the components of a spatial model as a collection of objects is only the first step. Recognising the role of *spatial relationships* among the objects is more important, and the keys to understanding the representation of these are the concepts of distance and neighbourhood, and the closely related networks.

6.3.1 Distance in grids and tessellations

While they are computationally convenient, a potential disadvantage of grid-based models is that they introduce subtle, easily overlooked, distortions of spatial relationships (Figure 6.6). If we measure distance in terms of the numbers of 'steps' required to move from one grid cell to another on a square grid, then there are unavoidable distortions. The key is to recognise that the Euclidean (straight line) distance between a grid cell and its orthogonal neighbours is one distance unit, while the distance to its diagonal neighbours

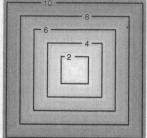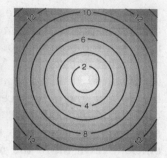

Figure 6.6 How distances measured on a square grid may be distorted. From left to right distances measured by counting moves according to Von Neumann neighbourhoods (orthogonal only), Moore neighbourhoods (orthogonal and diagonal neighbours) and true Euclidean distances. Note how either of the grid-based approaches substantially distorts the actual distance contours.

is $\sqrt{2}$ distance units. If we treat orthogonal and diagonal steps as the same, then movement along the diagonals of the grid is 'accelerated' by a factor of $\sqrt{2}$. On the other hand, if we allow only orthogonal moves, it takes two steps to move from a cell to any of its diagonal neighbours, giving 'Manhattan distances' and so movement in these directions is retarded by a factor of $\sqrt{2}$. These effects may significantly distort the conclusions we draw from a model where movement processes are an important feature. Even where movement processes are not a central concern, the discretisation of space that the grid introduces may be a problem. If some model process uses distances between entities as an input, then *even if we use cell-to-cell Euclidean distances* the fact that on grids only a subset of all possible inter-cell distances actually exist can introduce unexpected effects. To clarify, if grid cells are a single distance unit, then no two objects in the model can be less than one unit apart. Furthermore, objects can only be at separations of 1, $\sqrt{2}$, 2, $\sqrt{5}$, $2\sqrt{2}$ and so on. This effect is more serious at short ranges, but often it is short distances that are of most interest to us!

6.4   Of course, square grids or lattices are only one possibility. Most obviously, hexagonal lattices, where space is subdivided into equal sized hexagons, offer an alternative (for an example, see Phipps, 1989). An advantage of hexagons is that there is no ambiguity about adjacency, since adjacent cells share a common edge and it is impossible for cells to be diagonally adjacent. This removes one disadvantage of square grids since all adjacent cells are now the same distance apart, but it does not resolve the problem completely. Some directions of movement remain favoured over others, and again only certain inter-cell distances exist. A triangular lattice is another possibility, although little used, and it reintroduces the problem of diagonal neighbours. Furthermore, neither option is as computationally straightforward as a square grid

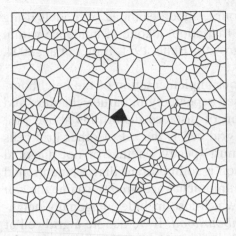 

Figure 6.7 Distance in a Voronoi tessellated space. The tessellation is shown on the left for clarity with the central polygon highlighted, and distance in step-by-step moves from the central polygon outwards is shown on the right, both by grey-scale colouring and by a set of contours at single step intervals. Although distances in each direction are not uniform, on average no direction is favoured, and over large distances this may be preferable to the systematic distortions of regular grid distances (see Figure 6.6).

because the familiar Cartesian grid referencing system falls down, although this difficulty can be overcome with careful programming.

Polygon tessellations offer an alternative to the regular cell structure of grid-based models. Some of the spatial distortions introduced by regular grids **6.5** can be partially solved by using an irregular tessellation, where each polygon is a cell and cell-to-cell moves (although none are precisely identical) are treated as equivalent when handling movement and measuring distance. One approach is to use the Voronoi tessellation (see Section 3.6.2) of a randomly located set of points. Because *on average* there are no favoured directions of travel, for some applications this may be a useful alternative to a regular grid (Schönfisch, 1997, Holland et al., 2007, Etherington, 2012). Figure 6.7 shows how this might work in practice.

6.3.2 Neighbourhoods: local spatial relationships

A central concept in spatial analysis is the neighbourhood (see O'Sullivan and Unwin, 2010, pages 45–46). We can consider the neighbourhood of a location as some specified region of space around it. Alternatively, the neighbourhood of an object is some set of nearby objects. The latter interpretation is usually more relevant for spatial models, although the two ideas are related: if some region of space around an object is its neighbourhood, then any object entering that region may be a neighbour. A good example of models using this idea is provided by forest models that adopt a zone of influence within which trees may affect one another's growth (Bella, 1971, Bauer et al., 2004, Berger et al., 2008).

Grid-based models provide an obvious implementation of the neighbourhood concept, which we have seen used repeatedly in earlier chapters and which is also implicit in the discussion of distance distortions above (see Figure 6.6). The neighbourhood of a grid cell is the set of other grid cells adjacent to it, where, of course, we must specify whether or not diagonal neighbours are included. This distinction is so important that, as mentioned in Section 3.1.1, the orthogonal-only definition has acquired the name the *von Neumann neighbourhood*, while that which also includes diagonally neighbouring cells is called the *Moore neighbourhood*. If we denote a grid cell at coordinates (x, y) as $c_{x,y}$ then the von Neumann neighbourhood is:

$$N_V(c_{x,y}) = \left\{ \begin{matrix} & c_{x,y+1} & \\ c_{x-1,y} & & c_{x+1,y} \\ & c_{x,y-1} & \end{matrix} \right\} \tag{6.1}$$

while the Moore neighbourhood is:

$$N_M(c_{x,y}) = \left\{ \begin{matrix} c_{x-1,y+1} & c_{x,y+1} & c_{x+1,y+1} \\ c_{x-1,y} & & c_{x+1,y} \\ c_{x-1,y-1} & c_{x,y-1} & c_{x+1,y-1} \end{matrix} \right\} \tag{6.2}$$

There is no particular reason to prefer one of these neighbourhood definitions over the other, but it is important to be clear about which is being used in a particular model. These expressions can be written in more compact form to define the cells $c_{i,j}$ that are in the two neighbourhood types as follows:

$$N_V(c_{x,y}) = \{c_{i,j} : |i - x| + |j - y| \leq 1\} \tag{6.3}$$

$$N_M(c_{x,y}) = \{c_{i,j} : \max(|i - x|, |j - y|) \leq 1\} \tag{6.4}$$

and it is a straightforward extension to define von Neumann and Moore neighbourhoods at longer distances. Neighbourhoods based on other distance metrics (see Gatrell, 1983, pages 24–34) are readily defined, the most obvious being a Euclidean (straight line) distance, which yields a neighbourhood N_r at radius r:

$$N_r(c_{x,y}) = \left\{ c_{i,j} : \sqrt{(i - x)^2 + (j - y)^2} \leq r \right\} \tag{6.5}$$

This example shows how the neighbourhood concept is immediately transferrable to a vector representation. The neighbourhood of any object can be defined as those other objects within a specified radius, just as in Equation 6.5. Other approaches are possible, for example we might decide that the k nearest other entities are in the neighbourhood, which has the advantage over a strictly distance based method of guaranteeing that all objects will have *some* neighbours no matter how thinly spread objects are.

This brief discussion only scratches the surface of the possibilities. The important point is that in any spatial modelling exercise *how we define neighbourhoods is a key aspect.* All the processes considered in the models in earlier chapters are formulated with respect to each model entity and its neighbours—in fact, this is what makes the models spatial—and we should take as much care to define neighbourhoods as we do to formulate equations or other descriptions of model processes.

6.3.3 Networks of relationships

A specific kind of relational structure that often appears in spatial models is a *network*. We have already noted that in many cases when linear objects appear in a model they are connected together in a network. Communications networks, infrastructure networks and stream networks are commonplace examples. Network science has seen an explosion of interest in recent years (Newman, 2010), sparked in large part by the rapid development of the Internet, home to multiple overlapping social networks, hyperlink networks and built on electronics and communications networks. These developments have seen much older work on social networks (for a comprehensive survey, see Wassermann and Faust, 1994) gain widespread attention, with particular interest in small world (Watts and Strogatz, 1998) and scale-free networks (Albert et al., 1999). More recently, spatial networks have also attracted considerable interest (Barthélemy, 2011, provides a thorough review), again rediscovering earlier antecedents (Kansky, 1963, Haggett and Chorley, 1969). The analysis of networks relies on a branch of mathematics known as *graph theory* (see Wilson, 1996, Gross and Yellen, 2006), which is foundational to many areas of computer science (Jungnickel, 1999).

In much of this work, how network structures grow and evolve over time based on simple processes is a central concern. There is considerable interest, for example, in how different network growth processes lead to different network *degree distributions*, that is, the distribution of node neighbourhood sizes. Such themes relate closely to the central concerns of this book. However, we are more focused on how processes lead to the emergence of different patterns in fixed spatial structures, while models of network growth focus attention on how processes can alter the spatial structure itself. Readers interested in examples of the latter should consult the now extensive and rapidly growing network literature, starting from the references cited above. An area of emerging interest is *adaptive coevolutionary networks* where network growth affects other processes on the network and *vice versa*. A useful early overview of this field is provided by Gross and Blasius (2008).

For our purposes, the key feature of networks is that they are a highly general structure for representing the relations (spatial or otherwise) among a collection of objects. In a network, the objects are the *nodes* or *vertices* while the relations are *links* or *edges*. An important insight is to realise that a network defines a set of neighbourhoods over a collection of objects, such that each node's neighbours are those other objects to which it is connected by a link. This structure is identical to the description we have set out above of spatial neighbourhoods. Given this fact, it should be apparent that any spatial model defined for a set of objects (be they vector objects or grid cells) with respect to the objects' neighbourhoods *could also be applied to a network.* In effect, many spatial models *imply* an underlying network structure. In the context of grid-based models, this connection has been made explicit (see Couclelis, 1997, Keeling and Eames, 2005, O'Sullivan, 2001), and cellular automata models with irregular connectivity structures have been developed building on this idea (see Semboloni, 2000, Shi and Pang, 2000, Norte Pinto and Pais Antunes, 2010).

The generality of the network concept means that the links between entities may be spatial or aspatial. Where we are primarily concerned with understanding the nature and likely behaviour of systems, the structural characteristics of all the networks of links in a model may be just as interesting as the nature of the links themselves. In this context, Wassermann and Faust (1994) provide good coverage of the different aspects of network structure that can be measured. These can be broadly categorised into *centrality*, *clusters* or *communities*, and *equivalence classes*. Centrality measures identify which nodes (or links) in a network are best connected to everywhere else in the network or the most *accessible*. Where movement processes are at work, we might expect more central locations to be more heavily populated than others. The central cells in any finite lattice model are the most accessible in these terms (obviously so), and this should be borne in mind in analysis. Distinct clusters or communities in a network are subsets of the nodes and links that are more tightly connected to one another than to other parts of the network. We might expect clusters in a network to behave somewhat independently of one another, perhaps supporting different population mixes over time. Numerous methods have been developed in recent years for identifying clusters in large networks (see Fortunato, 2010). Equivalence classes in a network are harder to define. They are (roughly speaking) groups of nodes whose positions in the network are structurally similar. A simple example is the distinctive roles played in some transport networks by hub and spoke nodes.

It would be fair to say that while the long-term potential of developments in network science for improving our understanding of spatial models is clear, considerable work remains to be done. There is also an important role for simple process models such as percolation and random walks in improving our understanding of network structures themselves (see, for example, Newman, 2010, Rosvall and Bergstrom, 2008).

6.4 Coordinate space: finite, infinite and wrapped

So far we have assumed that the model has a Cartesian coordinate system indexing the space, in two or three dimensions. The coordinate system may be integer-valued, meaning that only coordinate values from the set of integers $\mathbf{Z} = \{\ldots -2, -1, 0, 1, 2, \ldots\}$ are allowed, when the space may be denoted $\mathbb{Z}^2$ or $\mathbb{Z}^3$. This definition of the space is effectively a grid-based representation. Where coordinate values are drawn from the set of real numbers, the space is denoted $\mathbb{R}^2$ or $\mathbb{R}^3$ and forms a continuous space.

A common challenge is how to handle the fact that the world is extensive and open, while any model is inevitably a closed system. For example, if we are exploring forest succession at the level of individual trees, then we might wish to build a model extending over many kilometres, including tens of millions of individual trees; Chave (1999) describes efforts to model 2×10^7 trees over an area of $20 \, \text{km}^2$. Parker and Epstein's (2011) EpiSIM framework, introduced in Section 2.1.3, which seeks to represent pandemic spread through the global human population, is another example of this type of effort. Such models are impressive and they certainly have their uses, however, for many purposes, we may want models that are more computationally manageable. Various options are available, and these can have profound effects on the model outcomes and must be considered carefully.

6.4.1 Finite model space

In this approach, we accept the finite limitations of the model, decide on the scope or study region to be covered and construct the model accordingly. In a few specific situations this approach is obvious and preferable. It is usually how we proceed when the model represents an actual as opposed to a generic landscape. If the system under study has clear and distinct boundaries (such as an oceanic island), then provided we account for arrivals and departures, it is sensible to limit the model scope to that area alone.

If the model represents only a portion of a larger landscape, then careful thought must be given to how to deal with the model edges. Issues which must be considered include:

(i) Can new entities arrive from outside, and if so, where will they arrive and at what rate?

(ii) Can entities currently in the model leave by exiting along an edge of the model? If they do, then do they simply disappear or does where they exit the model matter?

(iii) If there are neighbourhood effects in the model, then how are the different neighbourhoods at the edge and the centre of the model to be handled?

The last point may demand particular attention. For example, in a finite Moore lattice, cells in the centre have neighbourhood size $|N|$ of eight, while edge cells have five and corner cells only three neighbours. Such differences in neighbourhood size may affect the operation of transition rules. For example, if a rule includes a critical threshold at say 40% of neighbours for $|N| = 8$ the threshold implies three cells (actually 37.5%), for $|N| = 5$ it implies two (40%), while for $|N| = 3$, it corresponds to only a single cell (33.3%). Whether such differences actually make a difference is hard to predict and it may require careful experimentation to find out. Forsé and Parodi (2010) explore this issue in the context of the Schelling model and suggest that this issue weakens some of the grander claims of emergence made for that system (see also Section 3.5).

6.4.2   Infinitely extensible model space

In some cases, where a developing pattern is growing to fill an empty region of space, it may be possible to devise a model structure that, subject to the limitations of system memory, is infinitely extensible. This approach only records 'occupied' cells or only the locations of objects that currently exist in the model. By only storing locations actually in use, the need to allocate computer memory to large regions where nothing is happening, or where the relevance of what is happening to outcomes of interest is limited, may be avoided. For manageable numbers of objects this can work well, and Wiegand et al. (1999) provide an example of its application in the context of understanding the spatial population dynamics of sparsely distributed *Acacia* trees in the Negev desert. However, if spatial relationships between objects are important to further evolution of the pattern, then this approach can run into computational difficulties without careful thought. For example, if we need to know the distances between objects, so that the configuration of each object's neighbourhood can inform further developments, then with (say) 100 objects there are around 5000 distances to consider. Increasing the number of objects ten-fold to 1000 leads to a hundred-fold increase in the number of distances, and it rapidly becomes computationally intensive to identify the neighbours of an object. Such n^2 scaling effects are commonplace in handling spatial data.

To get around this difficulty requires careful programming of *spatial data structures* such as *quadtrees*. If you need to implement spatial data structures

4.1 and 6.6

the classic work is Hanan Samet's *Design and Analysis of Spatial Data Structures* (1990). Depending on the modelling platform, some spatial data structures may be built in, and it would be sensible to investigate these before building your own from scratch. In such cases, as new objects appear in the model they are added to the data structure, and the complexities of efficiently retrieving other objects nearby are handled by the methods provided by the data structure for such queries. In Section 6.5 we discuss a simple 'trick' that may be useful in some cases without requiring the complexity of fully developed spatial data structures.

6.4.3 Toroidal model space

This is the most widely adopted approach for models where the landscape does not represent a real place and where the aim is the exploration of general system behaviour and properties. The idea of a toroidal coordinate space is that the model 'wraps' around, so that locations at the southern edge of the space, in addition to having neighbours just to their north, also have neighbours to their 'south', which are the locations at the far northern extent of the model. This means, for example, that in a 100×100 toroidally wrapped grid, the von Neumann neighbours of a location $(18, 99)$ at the northern edge of the space are $(18, 98)$, $(17, 99)$ and $(19, 99)$, as we might expect, but also $(18, 0)$, which is ostensibly at the southern edge of the space. The space will usually wrap in both directions (forming a doughnut), although in some situations it may be appropriate for wrapping to occur in one direction only. An example might be simulating deposition of material on a beach, where the onshore and along-shore directions behave very differently and the real-world system is much more extensive in one direction than the other. Likewise, the model of plateau erosion introduced in Section 5.4.3 uses horizontal but not vertical wrapping.

Some of the neighbourhoods from Equations 6.1 to 6.5 are illustrated on toroidal grids wrapped in both directions in Figure 6.8. Toroidal model spaces

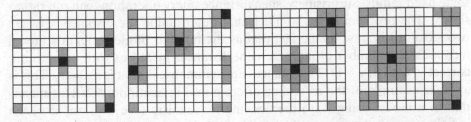

Figure 6.8 Example neighbourhoods in a toroidal space. From left to right the neighbourhoods illustrated correspond to those specified in Equations 6.1, 6.2 , 6.3 (with distance two) and 6.5 ($r = 2.3$), respectively. Note that it would be unusual to toroidally wrap such a small model space as this because it places very large regions of the space in the boundary regions and may make it hard to understand what is happening in the model simply by watching it.

take some getting used to, and the approach is unlikely to work well if the range of local neighbourhoods is large compared to the total model extent. In spite of its widespread use as a proxy for an infinitely extensible space, it is important to keep in mind that toroidally wrapped models are *actually* finite. In some cases the difference between a finite and an infinite space is theoretically important. Many theoretical results refer to the infinite case, which is often easier to analyse mathematically (as noted with respect to both interacting particle systems in Section 3.4 and percolation models in Chapter 5). Generally speaking, however, these disadvantages are heavily outweighed by the convenience of the toroidal space and the removal of edge effects.

6.5   Complicated spatial structure without spatial data structures

It is commonplace in transitioning from simple, abstract models (such as those considered in earlier chapters) to more detailed or empirically grounded simulations for spatial aspects of the representation to become much more complicated. As we have already noted, for serious computation and in large and complicated models, the best approach to dealing with spatial data may be to make use of the many already designed and implemented data structures developed for handling spatial data. A good example is the JTS Topology Suite of Java classes, which implements many standard (vector) spatial data structures (see JTS Topology Suite, 2012).

Even so, there are programming 'tricks' that can be useful without requiring implementation of complex data structures. These revolve around a rule of thumb that if any spatial operation is required frequently during model execution, then it is worth considering whether the results of that operation can be stored for reuse, rather than being recalculated every time.

The most common example is where *distance queries* are required to allow some elements in a model to interact with other model elements that are inside some fixed radius. Surprisingly, such distance query operations can run rather slowly. The problem is not so much that determining the distance between two locations is slow, as the fact that as the number of elements in a model increases, the number of inter-element distances increases with n^2 or, more correctly, with $n(n-1)/2$ if distances are symmetrical, that is $d_{ij} = d_{ji}$. This is another example of the n^2 scaling effect noted in Section 6.4.2. This means that in a model using (say) the flocking mechanism (see Section 4.5) each individual must determine which among all the other individuals are within some range; model efficiency may decrease dramatically as the number of individuals increases. With $n = 100$ individuals, if each must determine which other individuals are within some distance r in order to return a list of nearby individuals, then 99 distance calculations are required for each individual,

4.7

followed by a comparison of the distances to a threshold value. If all 100 individuals perform this calculation every model step, then around 9900 distance calculations are needed. When n increases ten-fold to 1000, the required number of calculations increases one hundred-fold to almost one million distance calculations, dramatically reducing model performance.

A simple approach to speeding things up in this case is to use a filter on candidate nearby individuals so that distances are only calculated for those cases where there is some chance that they are close enough (see Figure 6.9). In most modelling platforms, an underlying grid structure can provide the necessary speed-up without any need to implement a full-blown spatial data structure. Queries that return all those individuals that are on the same grid cell, or perhaps the same and neighbouring grid cells, as the individual making the query are typically quick to run. Since the only individuals that can be in range are probably those on the same or neighbouring grid cells, the distance calculation should only be performed on those individuals and not on all the other individuals in the model.

How well this will work depends in particular cases on factors such as how densely packed individuals are on model grid cells, how large the model grid cells are relative to the rate at which individuals are moving and relative to the model world, how often distances between individuals need to be recalculated and how durable individuals are (if many individuals only live for a single

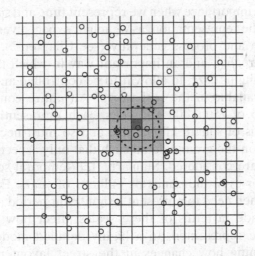

Figure 6.9 Using a local grid as a simple spatial data structure. The 'within range' query indicated by the dashed circle can be speeded up by associating with each cell in the grid a Moore neighbourhood larger than the circular region required and only running the within-range test on objects in that neighbourhood rather than across the whole model space. Here, with only 25 cells in the Moore neighbourhood compared with 400 in the whole model, significant time savings are likely if efficient retrieval of objects in grid cells is supported, as it is on most platforms.

model tick, then many distances may only be calculated once anyway). However, in almost all cases, this type of approach increases model efficiency.

In effect, this method approximates the exact location of point-located entities to grid locations, and makes the reasonable assumption that when we need to determine the near neighbours of an individual they will be those other individuals that share the same approximate grid location. This is a useful approach that can be extended further to what we might term a *zone of influence* method. Assuming that an underlying grid is available as a built-in feature of the modelling platform being used, then at the beginning of a model run zones of influence of each grid cell, that is the other grid cells that might affect each focal grid cell, can be pre-calculated and stored for rapid retrieval either by the grid cells themselves or by individual objects on the grid cell. This approach parallels recently proposed concepts in GIS representation, particularly object fields (Cova and Goodchild, 2002).

6.6   Temporal and spatial representations can make a difference

The goal of this chapter has been to give an overview of some of the key decisions involved in implementing spatial simulation models. It is therefore appropriate to close with some comments (once again!) about scale.

Of fundamental importance when we represent time and space is the choice of some *scale* for the representation, as we might expect given its far-reaching impact on pattern and process generally (see Chapter 2). All models are simplifications of reality, and an important way in which they simplify the world is by changing its 'size' to allow it to be controlled and explored from perspectives unavailable in the real world. An entire continent might be reduced to a map of 250×400 grid cells, each representing an extensive region of hundreds of square kilometres, and so must necessarily abstract away most of the on-the-ground detail such that only aspects relevant to the model's purpose are considered. Similarly, the temporal scope of a model may be decades or even thousands or millions of years. Each 'tick' of the model clock will then need to represent a significant period of real time.

The choices made in terms of these scales affect how every aspect of the system will be represented. In a highly detailed model of pedestrian movement examining how changes in the street layout might affect the numbers of passers-by at a particular corner site, long-term changes to the urban transport system that take decades to accomplish must be handled differently from how they would be in a long-term regional planning model. Equally, if a model is composed of grid cells representing $25\,km^2$ each, and the model clock progresses at one tick per year, it is unlikely that it makes much sense for the model's process aspects to represent individual rabbits feeding. These may seem obvious remarks, but it is surprising how often model builders are vague in response to questions about the scales, both spatial and temporal, of their creations.

Scale then, like all of the representational choices we have discussed in this chapter, makes a difference. Given its fundamental nature and the ensuing effects on other modelling choices, it is important to start out with a clear rationale for the model scale chosen and to work through its implications for other model aspects early in the development process. How these and other modelling decisions play out in practice is the focus of the next two chapters, where we examine different approaches to model analysis (Chapter 7) and present the development and analysis of a model (Chapter 8) composed of building-block models considered in earlier chapters.

7
Model Uncertainty and Evaluation

7.1   Introducing uncertainty

Data, in the form of patterns, play various roles in model design, implementation and evaluation. They can be used to select between alternative model structures and parameterisations, assess the usefulness of a specific model, and inform and guide how a model may be refined in the future. While there is a tendency to focus on the confrontation of a predictive model with empirical data (Mayer and Butler, 1993), this is just one role of data in the iterative process of developing, using and evaluating a simulation model.

A fundamental issue in modelling is *uncertainty*. Briggs et al. (2009, page 191) note that 'ignorance, ambiguity, indeterminacy, variability, unpredictability, error and unreliability are all often used as full or partial synonyms' for uncertainty. While these terms all convey some sense of what uncertainty is, none of them, in isolation, really captures the range of issues that uncertainty encompasses. Perhaps as a result, many authors have attempted to develop typologies of uncertainty (see Regan et al., 2002, Walker et al., 2003, Refsgaard et al., 2007, Parker, 2008, Matott et al., 2009, among many others).

From a modelling perspective typologies of uncertainty tend to emphasise what Regan et al. (2002) call *epistemic uncertainty*, that is, uncertainty in what we know (or believe we know) about a system. Epistemic uncertainty takes many forms. McMahon et al. (2009) include process error, measurement error, random individual and temporal effects, uncertainty about initial conditions and scenario uncertainty, among the many forms of epistemic uncertainty affecting models. In developing their own rather different typology (Walker et al. 2003, page 5) suggest that epistemic uncertainty is anything

Spatial Simulation: Exploring Pattern and Process, First Edition.
David O'Sullivan and George L.W. Perry.
© 2013 John Wiley & Sons, Ltd. Published 2013 by John Wiley & Sons, Ltd.

which, 'provides a deviation from the unachievable ideal of completely deterministic knowledge of the relevant system'. They then distinguish between uncertainties in terms of their location, level and nature. The location of uncertainty relates to where in the model or modelling activity the uncertainty is found—the original question, the parameterisation of the model, the model's structure and so on. The level of uncertainty describes how acute the uncertainty is, from certain epistemic knowledge to complete ignorance. And its nature refers to its origin, for example, natural variability as opposed to measurement error. Finally, Smith (2001) notes that the meaning of uncertainty, and the attention paid to it, varies considerably between disciplines and contexts.

It is important to emphasise that *uncertainty is unavoidable*, even if some forms of it, such as certain kinds of measurement error, can be minimised. Matott et al. (2009) distinguish between *reducible* and *irreducible* uncertainties, arguing that the former are related to errors in knowledge (or ignorance) and data, and that the latter arise from natural variability. Irrespective of the kinds of uncertainty we are faced with, the question for modellers is therefore not how to remove uncertainty, but how to most appropriately represent it in models.

There is considerable current interest in uncertainty in the context of environmental decision-making. This is motivated by a recognition that although there is a pressing need to make forecasts about the environment (Clark et al., 2001), how to effectively represent and communicate uncertainty, in whatever guise, is unclear (Ascough et al., 2008, Polasky et al., 2011). If models are used to support decision-making then there is an obvious need to evaluate them, and for that evaluation to be thorough and reproducible (Bart, 1995, Refsgaard et al., 2007, Matott et al., 2009). The effective communication of the unknown (or even 'unknowable') relates to *linguistic uncertainty* (*sensu* Regan et al., 2002), an issue that has vexed philosophers since the ancient Greeks. While it is beyond the scope of this book to explore this type of uncertainty further, it is worth noting that many debates surrounding models in areas such as climate change science (Stainforth et al., 2005) are arguably as much to do with failures to adequately communicate uncertainty as they are to do with epistemic uncertainty (see Pidgeon and Fischhoff, 2011).

7.2 Coping with uncertainty

One of the important trade-offs that modellers must make is between analytic tractability and realism. One way that this trade-off is practically expressed is in whether a model is deterministic or stochastic. A deterministic model does not represent uncertainty and so for a given set of boundary conditions and input parameters will always produce the same outcomes. Deterministic models embody a 'clockwork universe' view of the world, which holds that given an infinitely well-defined snapshot of the state of a system we should be able to hindcast and forecast the system's (dynamic) behaviour perfectly. This view of the world reached its zenith in the writing and thinking of 18th century mathematicians and scientists such as Pierre-Simon Laplace (1814). Setting aside the existence of deterministic models that lack the quality of predictability (see the discussion of nonlinear and chaotic dynamics in Section 1.3.2), the deterministic approach ignores uncertainty but produces models that are analytically tractable. In other words, it buys tractability, but at the cost of realism.

Stochastic models, on the other hand, include some random component such as, for example, variation in parameter growth rates from year to year in a population model. This stochasticity can arise from many sources, but one way to look at it (see also Bolker, 2008) is in terms of process uncertainty (arising from natural variability in process rates) as opposed to measurement uncertainty (arising from limits on our ability to measure systems). The inclusion of stochasticity typically renders even simple models intractable but arguably increases their realism—after all, few, if any, systems show no variation in time and space, or are perfectly understood and measured.

7.2.1 Representing uncertainty in data and processes

To illustrate some of the issues around uncertainty and model implementation and evaluation we will consider a simple model of the harvesting of a population under logistic growth (see Section 2.2.6). In the simplest case harvesting occurs at a constant rate (so-called fixed-quota harvests), such that H individuals (or units of biomass) are removed every year:

$$N_{t+1} = N_t + rN_t \left(1 - \frac{N_t}{K}\right) - H \tag{7.1}$$

where N_t is the population size at time t, r is the discrete population growth rate, K is the carrying capacity and H is the harvesting rate in each time period.

If we want to manage this hypothetical population so that it is not overharvested then we want to know the *maximum sustainable yield* (MSY). Heino and Enberg (2008) provide an overview of the issues involved in the exploitation of populations, including problems in determining and setting appropriate quotas. The maximum sustainable harvest of a population between times t and $t + 1$ is, at least in theory, the number of individuals that would be added to the population between those times in the absence of harvesting. In other words, if we harvest all the new individuals by setting the harvest H equal to the second term in Equation 7.1, then $N_{t+1} = N_t$ and the abundance will not change. Analysis of the continuous logistic model shows that the population grows most quickly at $N = K/2$, at which point

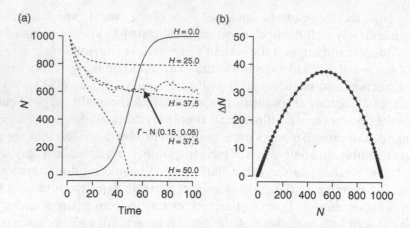

Figure 7.1 Dynamics of the logistic model with fixed-quota harvesting. In all cases $r = 0.15$ and $K = 1000$. (a) Population dynamics with (dashed lines) and without (solid line) harvesting, and with uncertainty in r (see text). (b) The relationship between population growth and population size showing that population growth is highest at $K/2$; with $r = 15$ and $K = 1000$, the deterministic model yields an MSY of 37.5 units per year, which is $rK/4$. Note that with no harvesting ($H = 0$) the population starts at very low N to illustrate the logistic dynamics that occur under these conditions; where $H>0$ it starts at 1000 (the carrying capacity) to illustrate the effects of population reduction.

$rK/4$ individuals are added to the population each time step. This is the MSY under a fixed-quota harvest, and we can arrive at this conclusion by analysis of the deterministic model alone (Figure 7.1). The MSY is very finely balanced: harvesting even fractionally above this value will eventually drive the population to extinction, while values below it are economically suboptimal, although this may be a lesser concern.

The deterministic harvest model is tractable because it is relatively simple and because its parameters r, K and H are assumed to be fixed. In nearly every system of interest, however, uncertainty is pervasive. The population's growth rate (r) is likely to vary between years as a function of environmental stochasticity (for example in climatic conditions). Alternatively, our measurements of the growth rate may be imprecise simply because there are limits to the precision with which we can measure systems. Either way we can represent uncertainty by making the model stochastic, for example we can make r a random variable following a Gaussian distribution with known mean and standard deviation. We can think of the deterministic form of the model as a special case of the stochastic model with the standard deviation of $r = 0.0$.

In the stochastic form of the model the population will *on average* grow at rate $r = 0.15$ (for the parameterisation in Figure 7.1). However, the population will experience some good years and some bad, and because the MSY is estimated on the basis that r is constant once harvesting at the

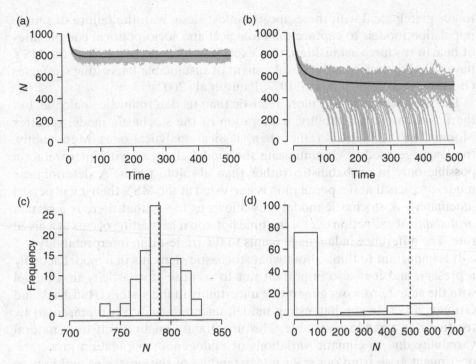

Figure 7.2 Dynamics of the logistic model with fixed harvesting and stochasticity in population growth rate: 100 projections with (a) $r \sim \mathcal{N}(0.15, 0.05)$, $K = 1000$ and $H = 25$, and (b) $r \sim \mathcal{N}(0.15, 0.05)$, $K = 1000$ and $H = 37.5$. Grey lines are individual model runs of the stochastic model, the black lines are the deterministic model projections with the same parameterisation. With r and K at these values, the deterministic model yields an MSY of 37.5 units per year ($rK/4$). (c) and (d) The population size at $t = 500$ for (a) and (b), respectively, with dashed vertical lines representing the deterministic model solutions.

MSY occurs in a few years where growth is below average, an inexorable slide to extinction will begin. Note how in Figure 7.2(a) with variable growth rates a harvest rate of $H = 25$ results in a population trajectory that varies around that seen in the deterministic model. Although good years buffer the population somewhat, extinction will eventually occur because the trajectory is a random walk to a single absorbing state at zero population. As Figure 7.2 shows, in the stochastic model, if harvesting occurs at the theoretically sustainable MSY, extinction occurs over the limited time horizon considered here in the majority of cases (close to 80% in Figure 7.2) and it may even occur if harvesting occurs at levels *below* the deterministic model MSY! In fact, extinction is *inevitable* with stochastic population growth and harvesting occurring at the rate determined by the deterministic model. The problems in assessing fishing activities that uncertainty poses have long been realised (see, for example, the opening sentences of Russell, 1931). Arguably it is failing to adequately deal with these uncertainties, along with the failure of simple population models to capture the biological and socio-political complexities of human resource acquisition, that explains why concepts such as the MSY have failed to underpin the development of sustainable harvesting strategies (Botsford et al., 1997, Larkin, 1977, Pauly et al., 2002).

The stochastic model is more realistic than its deterministic analogue, but there is a cost in tractability: exploration of the stochastic model requires Monte Carlo techniques rather than classical analytical ones. More subtly, the language used to communicate the model changes, with interpretation possible only in probabilistic, rather than absolute terms. A deterministic model predicts that if a population is harvested at the MSY then it will persist indefinitely. A stochastic model only allows us to say that there is a certain *probability* of extinction over some time horizon if harvesting occurs at a given rate. The difference in language points to the trade-off in interpretability.

It is important to think about what stochastic elements in a model actually represent, and it is also important not to confuse stochasticity in a model with the actual *processes* generating uncertainty in the system (Railsback and Grimm, 2012). In the harvesting model, uncertainty in r is represented by treating r as a random variable. This uncertainty might result from natural variability due to climatic variations or endogenous population processes, or it might arise from our poor understanding of the processes and how to represent them in the model, for example the nature of density-dependent effects, or it might be due to measurement error because of practical limits on our ability to estimate population. The uncertainty could be due to any or all of these factors (and more), but we do not explicitly represent any of them. While including all these effects might make a model more realistic, it also makes life much more complicated and, even in simple cases, analysis quickly becomes prohibitively difficult.

7.3   Assessing and quantifying model-related uncertainty

Model evaluation is 'the process of determining model usefulness and estimating the range or likelihood of various interesting outcomes' (Matott et al., 2009, page 1). It is revealing that Matott et al.'s definition emphasises the *usefulness* of a model rather than simply its predictive power. Just as there are many reasons why models are used, there are many ways to evaluate a model's performance, ranging from comparisons of its 'predictions' against some observed pattern, to more qualitative evaluations based on participatory discussions between model builders and other stakeholders (see Table 7.1). At their heart though, *all* of these methods are concerned with identifying uncertainties in the model, whether in the input data (parameterisation) or

Table 7.1 Some commonly used approaches and tools for the evaluation of model outcomes using observed data and patterns (modified from Perry, 2009)

Method	Description and purpose	Section/Reference
Model-based uncertainty		Section 7.3
Error analysis	Analysis of error in model output(s) as a function of the uncertainty associated with each parameter input to the model, including error propagation analysis and error budgeting	Haefner (2005), Jager and King (2004)
Sensitivity analysis	Identification of model components most sensitive to *local* parameter uncertainty	Haefner (2005), Hamby (1994)
Uncertainty analysis	Identification of how uncertainty in multiple (interacting) parameters and their representation will affect a model	Haefner (2005)
Robustness analysis	Analysis of the extent to which different representational decisions influence model dynamics	Levins (1966), Railsback and Grimm (2012)
Confrontational		Section 7.4
Visual 'diagnostics'	Visual comparison of empirical observations and model predictions (e.g. visual inspection for systematic bias via residual plots, etc.)	Mayer and Butler (1993)
Statistical methods	Summary of differences between observations and predictions (non-spatial and spatial difference measures) Quantitative comparison and analysis of predictions and observations (via linear models, correlation, etc.)	Mayer and Butler (1993)
Exploratory/heuristic		Sections 7.6 and 7.7
Pattern-oriented modelling	Use of *multiple* observed patterns to evaluate and refine models and select between alternate representations (drawing on the methods listed above)	Grimm and Railsback (2012), Grimm et al. (2005), Wiegand et al. (2003)
Participatory modelling	Methods of model evaluation that seek to involve all stakeholders in the modelling process from conceptualisation to application This might, for example, involve assessing a model's legitimacy based on whether users believe it adequately represents the system of interest—this may or may not include structural and confirmatory evaluation	Castella et al. (2005), Millington et al. (2011)

in model structure, and with assessing which uncertainties matter the most given the motivation for the model's development and use.

We will only briefly review the various methods outlined in Table 7.1, but careful evaluation of any model is crucial both for learning about the system and in applying that learning more widely. Possibly because of the theoretical context in which they have usually been applied, the types of modelling we have presented in previous chapters have suffered from a lack of this type of evaluation. Haefner (2005) provides a thorough overview of methods for model evaluation and Matott et al. (2009) provide a review of the concepts associated with model evaluation and a catalogue of software tools available to facilitate the process.

7.3.1   Error analysis

Error analysis focuses on the implications of errors in parameter estimation (Haefner, 2005, Jager and King, 2004, Parysow et al., 2000). Thus, we adopt a formal error analysis if we are specifically interested in estimating how much parameter estimation error will influence a given model's dynamics and outcomes. Error analysis is difficult for detailed and/or spatial models, but its underlying aim of understanding how errors combine through a system is important (Heuvelink, 1998, 2006, MacEachren et al., 2005). For example, when data sources with uncertainty attached to them are added together, then the errors in the output could be much greater (*error amplification*) or, if we are lucky, smaller (*error compensation*) than in any of the inputs (Haefner, 2005). In data-rich dynamic models this means that error in the initial conditions can rapidly spread through a model, so that an important component of error analysis is a consideration of *error propagation*. Another method of error analysis sometimes adopted is the development of an *error budget*, which is a formal accounting and partitioning of the error introduced into a model by each of the parameters and their uncertainties (Jager and King, 2004). Again developing an error budget is challenging for complex spatial simulation models and specialised techniques have been developed for this purpose (Parysow et al., 2000).

7.3.2   Sensitivity analysis

Sensitivity analysis (SA), whether formal or informal, is central to most model evaluations and has three broad aims (loosely following Haefner, 2005, page 179ff):

Model 'evaluation' SA serves as a check that the model's behaviour follows our beliefs about how the system works. While our intuition could be faulty, surprising outcomes, such as the model not responding to parameters deemed important, may suggest that the model warrants close checking.

Research design and model confidence Data collection and model development are usually iterative and cyclical. Collecting data is expensive and so needs to be carefully targetted. In the context of improving model parameterisation, priority should be given to bettering the estimation of those parameters identified by SA as sensitive but that are highly uncertain. SA also enables us to establish how precisely each parameter needs to be estimated—robust parameters need less precise parameterisation than sensitive ones. SA also allows the user of a model to determine how much confidence they can have in it, given the level of certainty with which it is parameterised.

Understanding Although SA is often seen as a purely technical exercise, it can play an important role in developing understanding of the system being represented. Models are abstractions and represent model builders' beliefs and understanding of what is important in the system being explored. Via surrogative reasoning, and assuming we have some confidence in the model's representation of the system, SA points us towards the processes and parameters that drive the system dynamics we are interested in. If parameters or processes deemed important seem not to be, or *vice versa*, then this may suggest problems in how the system was conceptualised or in our understanding of it, or both.

SA is often conducted using a local 'one-at-a-time' approach in which each parameter of interest is varied by some amount (say ±10%). Thus, it is a *local* analysis and relevant only close to the parameter values used. A parameter's sensitivity can be expressed by the partial derivative of Y, the state variable, in terms of X_i, the independent (or input) parameter of interest (Fassó and Perri, 2001, Hamby, 1994, 1995):

$$\phi_i = \frac{\partial Y}{\partial X_i}\left(\frac{X_i}{Y}\right) \tag{7.2}$$

where ϕ is the fundamental sensitivity coefficient, which, while it is expressed in different ways in different metrics, provides a measure of the relative change in state variable Y given some change in parameter X_i. The X_i/Y term in Equation 7.2 is a normalising factor. Typically, the response of the state variable of interest is measured as its proportional change, in terms of some baseline value (Y_0), to the proportional change in the input parameter (Hamby, 1994, Haefner, 2005, Soetaert and Herman, 2009). The sensitivity of state variable Y with respect to parameter X_i is given by:

$$S_{Y,X} = \frac{\Delta Y/Y_0}{\Delta X_i/X_0} \tag{7.3}$$

Equation 7.3 (Hamby, 1994) provides an estimate of the derivative in Equation 7.2 , which is often difficult to compute. For this metric, changes in model parameters with $\phi > 1.0$ result in a greater than proportional response in the model and are called *sensitive*. Conversely, those with $\phi < 1.0$ have a less than proportional response and are called *robust*. The use of 1.0 as a 'sensitivity threshold' is arbitrary and different values may be appropriate depending on the context. For stochastic models ϕ will be computed as the average (or median, etc.) response across multiple model realisations.

Drechsler argues that two strong assumptions underpin local sensitivity analysis:

(i) As the parameters are generally varied by a linear proportion, the state variable must be linearly dependent on the parameters.

(ii) As the parameters are varied one at a time, different model parameters do not interact in their influence on the state variable (1998, page 412).

As Drechsler (1998) points out these assumptions are not warranted for most simulation models, meaning that traditional metrics of sensitivity only provide a measure of sensitivity close to the values selected—in other words, they are local, not global. Many different approaches for conducting SA have been developed, including some that attempt to circumvent the problems in local SA highlighted by Drechsler. It is beyond our scope to review them here in detail, and many of them are not well-suited to spatial models, but Helton (1993), Fassó and Perri (2001) and Hamby (1994) all provide brief overviews of a range of approaches to sensitivity analysis, and Saltelli et al. (2000) provide more detail. Examples of local SA are carried out in Section 7.3.6 and in Chapter 8.

7.3.3   Uncertainty analysis

The goal of a local sensitivity analysis is to isolate the parameters to which a model is most sensitive and rank them in terms of the model's sensitivity. However, as Jager and King (2004) point out, sensitivity is a characteristic of a model not its parameters, and the origin of the uncertainty in the parameterisation is not considered – we vary the parameter by ±10% but we do not state what that ±10% represents. Uncertainty analysis (UA) is a more general approach than local SA, and is concerned with estimating how uncertainty in multiple, probably interacting, parameters and their representation will affect a model (Smith, 2001, Jager and King, 2004, Matott et al., 2009, Railsback and Grimm, 2012). There is a large formal literature on uncertainty analysis and, as with SA, many software tools have been developed with which to conduct it (Matott et al., 2009, provides a reasonably current compendium). Semantics aside, UA typically consists of the following steps, which are in the same mould as the Monte Carlo analyses introduced in Section 2.2.3 (following Helton, 1993, Smith, 2001, Railsback and Grimm, 2012):

(i) Select the parameters to be included in the analysis, keeping in mind that computational effort will increase steeply with the number of parameters included.

(ii) Decide on an appropriate probability distribution with which to represent each parameter, and also the parameterisation of that distribution, for example μ and σ for a Gaussian distribution.

(iii) Run the model many times, each time drawing parameters from distributions as appropriate (possibly using an efficient sampling scheme as described below).

(iv) Analyse the output matrix of data using any of a range of visualisation and formal statistical methods.

Many decisions in steps (i)–(iii) affect the outcome of UA, so that for complicated spatial simulation models careful planning is called for (Smith, 2001). In particular, the selection of an appropriate probability distribution to describe each parameter is important. For example, a uniform distribution suggests a higher level of uncertainty than a tightly constrained normal distribution might. As Railsback and Grimm (2012) point out, choosing appropriate probability distributions to describe parameter uncertainty is a key difference between SA and UA. Burmaster and Anderson (1994) provide rules of thumb for conducting detailed Monte Carlo UA. Among their most useful suggestions is not to make it more complicated than necessary! Examples of UA are presented in Section 7.3.6 and in Chapter 8.

Sampling the parameter space The principle advantage of the multivariate UA described above is that it incorporates interaction effects and so allows a more *global* analysis spanning a much wider swathe of the parameter space. A serious problem is that the number of simulations, and hence the computational cost of covering the parameter space, rapidly becomes impractical unless the experiment is carefully designed. Consider a model with three parameters, all in the range zero to one, whose effects we wish to evaluate across a broad range of values in combination with one another. We decide to explore each of the three parameters at values from zero to one in increments of 0.10, giving 11 different settings for each parameter. This results in $11^3 = 1331$ parameter combinations. Assuming the model is stochastic we will need to replicate each of these 1331 settings, say 30 times (more would be better), requiring 1331×30 or 39 930 realisations. If the model takes one minute per run to complete then the total analysis time will be 665 hours, or almost 4 weeks! Keep in mind also that simulation models often have

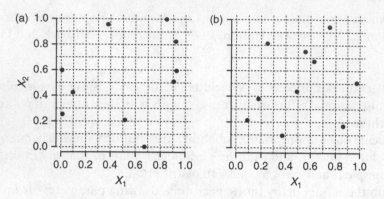

Figure 7.3 Stratified sampling approaches are much more efficient for multi-parameter sensitivity analysis. (a) Ten samples of two variables each following U ~ [0, 1]. (b) Ten samples of the same variables using a simple Latin hypercube scheme (see text). Note the more even distribution of values in (b) and the sampling of the centre and the margins of the parameter space. The Latin hypercube scheme guarantees that there will be an entry in each row and column of the parameter space.

many more than three parameters and that the size of the parameter space increases massively with the number of parameters, so that adding one more parameter in this example would increase the time requirement 11-fold (to around 10 months), and adding two will increase it by the same multiplier again (to over nine years!).

Clearly, sampling a multi-dimensional parameter space calls for careful thought. Imagine we are evaluating two variables and we want to sample them both from $U \sim [0,1]$. Naïve random sampling will be extremely inefficient, with some parts of the parameter space grossly under-sampled relative to others (Figure 7.3). It will take us a very long time (many realisations) to get a sense of how the full range of the distributions behave (Smith, 2001), although the behaviour of the model under 'extreme' parameterisations is often of considerable interest. Various solutions have been proposed for this sampling problem. Among the simplest is the *Latin hypercube*, a form of stratified sampling in which the parameter space is partitioned and then selected from such that each row and column contains at least one sample (McKay et al., 1979, Stein, 1987). Stratified sampling is much more efficient than simple random sampling and guarantees that the edges of the parameter space, which are often of interest during model testing, will be explored.

7.3.4   Analysis of model structural uncertainty

Sensitivity and uncertainty analyses should not be limited to a model's parameterisation and associated uncertainties. Given that there are often multiple ways that a given process could be represented in a model, it is also important to evaluate how decisions about a model's *structure* affect its behaviour. Jansen (1998) calls this *structural uncertainty*. Structural uncertainty is some- times called *model uncertainty*, to discriminate it from parameter uncertainty. As Smith (2001) and Chatfield (2001) note, model uncertainty is often given much less consideration than uncertainty related to model parameterisation, although it is potentially more important (see Chatfield, 1995, for an extended discussion in the context of empirical models). Assessing model uncertainty has been the subject of considerable recent attention in the context of statis- tical model building (Burnham and Anderson, 2002, Link and Barker, 2010) and is beginning to become part of the process of assessing spatial simulation models (Hartig et al., 2011).

Jansen (1998, page 250) notes that structural analyses can be conducted by reducing model complexity by, for example, coarsening the grain at which it operates (implying aggregation of processes), replacing mechanistic with empirical components and so forth. A possibly simpler way to carry out a structural evaluation is by changing small 'submodel' components of a more complicated simulation model (Railsback and Grimm, 2012), a process central to the pattern-oriented modelling framework described in Section 7.6.

As an example, imagine a model of interactions between a predator and its prey that requires you to represent how the consumption rate of prey relates to prey density (the so-called functional response). Ecologists, building on the work of Holling (1959), have developed three types of functional response (Figure 7.4). In a structural analysis of a predator–prey model it might be

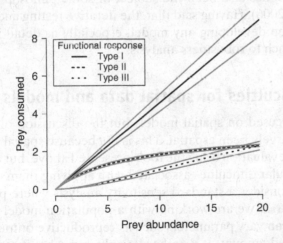

Figure 7.4 The three classical predatory–prey functional responses as proposed by Holling (1959). A structural evaluation of a predator–prey model would likely include a consideration of how the selection of functional response might affect model dynamics. The finer grey lines show the effects of changing one of the parameters in each model by ±10% (a local sensitivity analysis).

important to consider how the model behaves under different functional responses and how sensitive the model is to this decision. If there were data against which the model's dynamics could be judged then the model response may be used (cautiously) to decide which type of functional response is operating in the system of interest, via *reverse inference*. By contrast, a parameter-based sensitivity or uncertainty analysis would fix the type of functional response and focus on how the model responds to changes in the response parameterisation. On the basis of Figure 7.4 it is obvious that in this case the choice of model structure could be at least as influential as its final parameterisation (see also Okuyama, 2009, who shows how the selection of functional responses influences the dynamics of a simple spatial model).

Another similar example is given by the fire spread models presented in Chapter 5 (see Figure 5.19). Changing the relationship between time since fire and flammability represents a structural test of the model and in that case showed that this decision has a marked effect on the model dynamics. Likewise, in the harvest model in Section 7.2.1 the assumption of a fixed-quota could be replaced with a fixed-effort model where a constant proportion, rather than abundance, of the population is harvested.

The endpoint of structural evaluation is to replace the *entire* model with a different representation, a procedure sometimes called *robustness analysis* (Railsback and Grimm, 2012), which Weisberg (2006, page 730) defines as 'the search for predictions common to several independent models'. Robustness analysis is grounded in the strategy of model building advocated by Levins (1966). As Railsback and Grimm (2012) comment, it is formally adopted infrequently and has also been the subject of some philosophical argument (see Weisberg, 2006). Having said that, the iterative testing and checking that takes place when developing any model, especially as detail is added, is an informal approach to robustness analysis.

7.3.5   Difficulties for spatial data and models

This book is focused on spatial models but the discussion of evaluation to this point has largely been aspatial. This is *not* because spatial models do not require careful evaluation using the tools presented above, but rather because there are particular difficulties associated with applying them to spatial data. For example, consider a standard sensitivity analysis where parameters are varied by $\pm 10\%$. If we are working with a population model then it is easy to see how we can vary parameters such as reproductive output or economic growth by a fixed amount. It is rather less obvious what it means to change a map by $\pm 10\%$. Uncertainty in spatial data can take the form of locational error (uncertainty about where an entity is in space), classification error (error in the nature of the entity, such as in misclassification of an aerial image) and errors related to scale (often a result of aggregation). Such errors are usually spatially structured (i.e. spatially autocorrelated) and propagate in complex ways through spatial simulation models (Phillips and Marks, 1996). In the context of categorical maps, neutral models such as those introduced in Section 5.4.1 have a role to play, although there is the added constraint that the final simulated map may have to honour an original data source (Jager et al., 2005). Some headway has been made in developing tools for evaluating and describing the uncertainties associated with spatial data and models (Congalton and Green, 1993, Pontius, 2002), but rigorous evaluation of spatial simulation models remains a real challenge (Jager and King, 2004).

7.3.6   Sensitivity and uncertainty analysis for a simple spatial model

Ruckelshaus et al. (1997) describe a simple spatial model that illustrates how an uncertainty analysis might proceed (see also Mooij and DeAngelis, 1999, Ruckelshaus et al., 1999, South, 1999). Their model considers organisms dispersing through a fragmented landscape and operates as follows:

(i)   A landscape is generated comprising a certain fraction of 'habitat' with habitat patches (sets of contiguous cells) arranged in squares or lines.
(ii) Organisms are released from a habitat patch and search for a new habitat patch following a random walk with step lengths constant over time and turn angles restricted to the eight cardinal directions.
(iii) At each movement there is a probability the organism will die; it will also die if it leaves the simulation arena or fails to find a new patch within some pre-specified number of steps.

This model shares some similarities with Simon's model, discussed in Section 4.3, but is spatially explicit. The model records the success rate across many organisms of those finding habitat as a proportion of the number released, and also the frequency distribution of the number of walk steps successful organisms take before reaching a habitat patch. These two variables index the model's behaviour. Note that although the model is spatially explicit these variables are *not*.

Ruckelshaus et al. (1997) consider three uncertainties: (i) uncertainty in per-step survivorship, (ii) uncertainty in the movement rate (step length) of the organism and (iii) uncertainty in habitat classification. The first two uncertainties can be represented by varying their values around nominal baseline ('best') estimates. Uncertainty in habitat classification is introduced by specifying a certain fraction of habitat cells as 'low-quality habitat' and reducing the probability of an organism stopping in a given patch as a function of the proportion of low-quality habitat in the cells that comprise it (precise

7.1

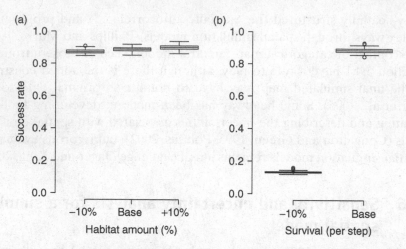

Figure 7.5 Sensitivity of the model described by Ruckelshaus et al. (1997) to changes of ±10% in (a) habitat amount and (b) per-step survival rate. In (a) the base amount of habitat is set to be 20% and in (b) the base survival rate is 0.997. We evaluated sensitivity in habitat amount at ±10% and at −10% for survival rate. Sensitivity was estimated using the average of 100 model runs under each parameter setting.

details are given in Ruckelshaus et al., 1997). All the parameters considered here are notoriously difficult to measure in the field, and the rationale for the original study was to evaluate how these measurement uncertainties might affect predictions from spatially explicit models when used to manage populations in spatially complex settings.

In our analyses we focus on identifying the parameters and processes that most influence success rate of organisms in finding habitat. We start by considering the model's local sensitivity under baseline conditions with no error—this provides a point of departure for subsequent analyses. For these experiments we set the baseline amount of habitat to be 20%, the per-step survival rate to 0.997, the step length to one and the amount of low-quality habitat to zero. A simple sensitivity analysis suggests that, locally at least, the effect of the availability of habitat is minimal compared to changes in the per-step survival rate (Figure 7.5). In terms of the sensitivity metric presented in Equation 7.3, varying habitat amount by ±10% yields values of S of −0.146 and 0.158, respectively. These values suggest the model is *locally* robust (i.e. $S \leq 1.0$) to changes in habitat availability. On the other hand a −10% change in per-step survival rate causes the success to drop to only 15% of its baseline value, giving a sensitivity of $S = 8.5$, which suggests that the model is locally extremely sensitive ($S \gg 1.0$) to changes in this parameter.

If we look at the effect of habitat availability over a wider range, its influence becomes apparent. As Figure 7.6 shows, the success rate of the dispersing organisms is extremely sensitive to changes in the amount of

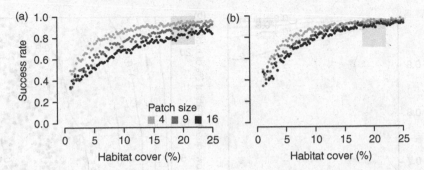

Figure 7.6 Dynamics of the model with the proportion of successful dispersers for (a) square and (b) linear habitat patches as a function of the amount of habitat in the landscape. In these runs per-step mortality and step length are constant and there is no variation in habitat quality. Grey boxes show the area of the parameter space considered by the local SA (Figure 7.5).

habitat (effectively this is a *global* sensitivity analysis). Unsurprisingly, as habitat declines so too does the success rate, but with habitat cover greater than around 15% success rates begin to plateau. The size of the individual patches has some effect, with landscapes compromising smaller patches tending to have more successful dispersers for a given amount of habitat (simply because there are more patches and so more chance of a random walker encountering one). Finally, whether the patches are arranged as squares or lines seems to have little effect (at least qualitatively) and so we restrict further analyses to square patch shapes.

Our uncertainty analysis comprises exploring the effects of uncertainty in survival rates, movement rates and habitat quality as a function of habitat availability. Thus we conduct three two-dimensional Monte Carlo-style analyses with sample values drawn from a random Latin hypercube scheme. In each case we conducted 750 model experiments with habitat availability following $U \sim [1, 25]$, survival rates $U \sim [0.65, 0.998]$, step lengths $U \sim [0.6, 1.4]$ and fraction of *low*-quality habitat (% of total) $U \sim [1, 25]$. We use the dispersers' success rate here as our measure of model behaviour.

The uncertainty analysis (Figure 7.7(a)) shows the overwhelming importance of the per-step survival rate, with values even slightly below the baseline (0.997) resulting in very low success rates; the effect of survival overrides the effects of habitat availability. The effect of the amount of low-quality habitat is less marked, with the gradient along the total amount axis more pronounced (Figure 7.7(b)). In other words, there is minimal evidence of interactions between either habitat amount and survival rate or habitat amount and habitat quality. On the other hand, the effect of step length appears to interact with habitat availability (Figure 7.7(c)), with increased step length boosting success rate even in landscapes with little habitat and *vice versa*. These types of analyses provide important first steps in understanding the dynamics of a model.

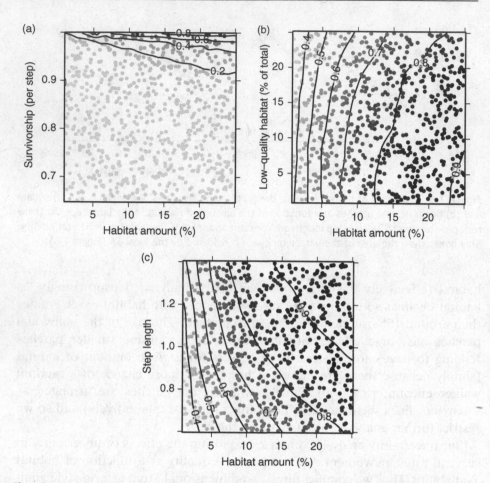

Figure 7.7   Uncertainty analysis of the disperser model for (a) survival × habitat availability, (b) habitat quality × habitat availability and (c) step length × habitat availability. The contours represent success rate. The circles show the points in the parameter space (*n* = 750) sampled using a Latin hypercube with the circles shaded (light to dark) to depict the success rate at that point.

Hearteningly, the direction of the effects that we see in Figure 7.7 are very much in line with our intuitions. Although our intuition might be faulty, checking that the model behaves as we expect, in a qualitative sense, is also an important component of model analysis (a sort of informal 'quality assurance').

Finally, we might consider how important the representation of the dispersal behaviour is. After all, a simple uncorrelated random walk is extremely naïve and not likely to be followed by real organisms. We modified the model structure such that the random walks were correlated, with the organism's turn angle restricted to 45° increments in either direction. This represents

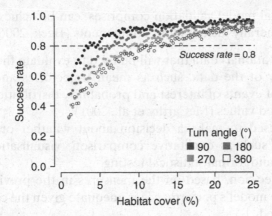

Figure 7.8 Outcomes of modifying the way that the dispersers' movement is represented so that they follow correlated random walks with turn angles constrained to 45° increments within arcs centred on the organism's current heading of 90, 180 and 270 degrees.

a simple model (structural) uncertainty analysis (South, 1999 evaluated the implications of more detailed demographic representations). Constraining the dispersers' turning angle by strengthening the correlation of their walk has a significant effect on success rate, especially as habitat becomes more limiting (Figure 7.8). At low levels of habitat availability more strongly correlated random walks have higher success rates. For example, at a habitat cover of 5% random walkers have a success rate nearly 0.3 lower than those with strongly correlated walks. Likewise to achieve a success rate of around 0.8 random walkers require around three-times as much habitat as do those with tightly constrained correlated walks. This effect is, however, interactive. Above a habitat threshold of around 20% the differences between the walks disappears as the success rate for all converges on 1.0.

This brief example highlights the complexities of even a modest uncertainty analysis of a simple spatial simulation model. Nevertheless, the UA presented here is informative and suggests where we might need to improve the model's representation and parameterisation if we were to further refine it. We build on these sorts of approaches in Chapter 8, where we consider a more complicated simulation model of hunter-gatherer foraging.

7.4   Confronting model predictions with observed data

One of the most routinely conducted evaluations of a model is to assess its predictions against independent observational data. This 'confrontation', which is what traditional model validation comprises, can be achieved using many methods, but generally includes four components (Beck, 2006, page 1276):

(i) The (observational) data that will be used to evaluate the model.

(ii) A summary of the data, such as mean trajectories and patterns, the frequency of events of interest and probability distributions of observed and predicted values (McCarthy et al., 2001).

(iii) Measures used to make a decision about whether or not to accept the model, such as qualitative comparison, visualisations, difference summaries and formal statistical testing.

(iv) Finally, a decision, based on the measures in the previous step, as to whether the model's performance is adequate given the context in which it will be used.

In this skeletal form the process appears logical and straightforward. It has, however, been the subject of considerable debate due, in part, to issues associated with under-determination and equifinality introduced in Section 2.2.6, and also because the approach only makes sense for a limited range of model applications. It suits predictive models but heuristically motivated uses less so, and in many settings observational data may not be available at all. The literature on model evaluation spans a wide diversity of philosophical stances from a position where statistical validation is seen as an absolute necessity to one where it is seen as a logical impossibility (see Rykiel, 1996, Kleindorfer et al., 1998, for reviews). This debate rests on what the purpose of a model evaluation is rather than whether comparing a model with data is useful or not, and in many cases comparison of a model's 'predictions' against some form of observed data is a useful component of model analysis. Mayer and Butler (1993) outline a set of broad approaches to what they call 'statistical validation', including visual diagnostics, summary difference measures and formal statistical tests, and we will consider these below.

7.4.1   Visualisation and difference measures

Visualising a model's outcomes is vital to understanding its dynamics and is also a useful way to detect errors (*visual debugging*, see Grimm, 2002) that may not be obvious otherwise, whether or not a formal comparison between observational and model-derived data is being performed. Exploratory or initial data analysis (Chatfield, 1985, Morgenthaler, 2009) has an important role in any quantitative analysis despite often being neglected. Its general purpose is to 'clarify the general structure of the data, obtain simple descriptive summaries, and perhaps get ideas for a more sophisticated analysis' (Chatfield, 1985, page 214). As Anscombe (1973) comments, visualisation is useful not only to gain a sense of the general trends in the data but also to look beyond these for finer-grained patterns.

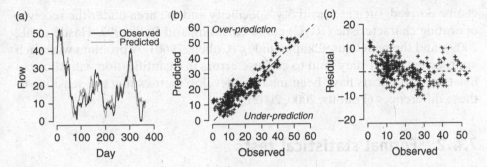

Figure 7.9 (a) Synthetic observed (O) versus predicted (P) data: simply plotting the two variables alongside each other is revealing. In this case there is no evident systematic bias, although the model both over- and under-predicts, and it appears that the predictions are somewhat damped. (b) A plot of observed versus predicted with a line of perfect fit ($P = O$): visually the model over-predicts more often than it under-predicts. (c) A plot of the residuals ($P - O$) versus the observed values, again showing that the model makes both over- and under-predictions, but with no obvious systematic relationship between the residual and the observed values.

As more and more quantitative data become available, effective exploratory analysis becomes ever more important and considerable effort is being put into developing ways of visualising data *and* their underlying uncertainties (Spiegelhalter et al., 2011). While such tools have not been specifically developed for simulation models, it is clear that they have a valuable role to play in evaluating and communicating model-based science. Typical ways of visualising models and data (Figure 7.9) are to plot time series (or maps) alongside each other, to plot predicted and observed values (perhaps accompanied by a perfect-fit line showing observed = predicted) and to plot model residuals (the difference between the predicted and the observed values). All of these methods provide valuable tests of model behaviour and even if 'observed' data are unavailable, visualising model outcomes is informative. At the very least it may help with understanding the model via deductive suggestion.

Visualisation can be supported by simple summary difference measures (Mayer and Butler, 1993), such as the mean absolute error (MAE), the root-mean squared error (RMSE) and the relative RMSE. All such metrics summarise the 'distance' between a set of observations and the corresponding model predictions. In a spatial context similar measures have been developed to measure the concordance between two categorical maps (in our case, one predicted and one observed). A simple approach is to cross-tabulate predicted and observed values to form an *error matrix* (sometimes called a confusion matrix) such that the trace of the matrix is the correct predictions and off-diagonal elements mispredictions (Congalton and Green, 1993, Fielding and Bell, 1997). From such a matrix a number of indices of model performance can be derived, such as sensitivity, specificity and the area under the receiver operating characteristic (ROC) curve (Fielding and Bell, 1997, Hastie et al., 2009), and the widely used kappa index (Cohen, 1960). A problem with such approaches is that they tend to confuse errors in quantification with those in location, and efforts have been made to develop metrics that are sensitive to these differences (Pontius, 2000, 2002).

7.4.2   Formal statistical tests

While visual and summary diagnostics are important tools for understanding and checking a model's behaviour, they usually appear alongside more formal statistical tests such as correlation and various linear models (regression, ANOVA and so on). On the face of it, correlation-based approaches seem ideal for quantifying the association between observations and a model's predictions. However, there are some potential traps with using such metrics as measures of model performance. It has become almost a mantra that 'correlation does not imply causation' or, in our context, that we cannot use the strength of correlation between observed and predicted data to infer a model's 'truth', but there are other technical difficulties associated with the use of correlation measures to assess model performance.

First, as Figure 7.10 shows, the correlation between model prediction and observation is sensitive to the nature of any model error. If the model systematically under- or over-predicts by an approximately constant amount, no matter how large, then the correlation will be unaffected. However, if the timing of the prediction is shifted (the model is 'lagged') the correlation will be low, even if the magnitude of the predictions is approximately the same. A second difficulty is that very different data can give exactly the same

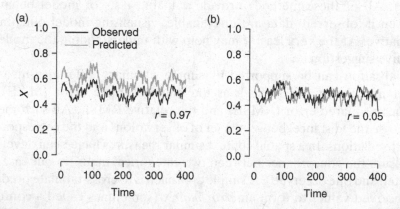

Figure 7.10 Synthetic predicted and observed data where (a) the model constantly over-predicts by about 20% and (b) the model 'lags' in its predictions by about 20%. For (a) $r = 0.97$ but for (b) $r = 0.05$!

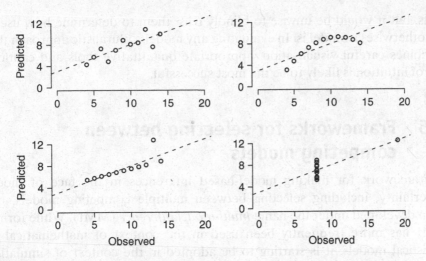

Figure 7.11   The four data sets comprising Anscombe's quartet all have the same correlation coefficient ($r = 0.82$) and RMSE (1.24), despite being very different from each other.

correlation coefficient, as demonstrated by Anscombe (1973) and depicted in Figure 7.11. The question to ask here is whether you, as a modeller, would be equally happy with each of these outcomes? Anscombe's quartet certainly emphasise the importance of visualising model outcomes and not relying on summary statistics alone.

There is an enormous literature on how to evaluate models statistically against observational data (Beck, 1987, 2006, Hamby, 1994, Haefner, 2005, Matott et al., 2009), although as touched on above there are specific problems for spatial models that make direct comparisons difficult. As McCarthy et al. (2001, page 1036) comment, the use of statistical tests of model performance can tend to lead to a view of a model being 'true' once it has passed sufficient tests and 'false' if it fails to do so. The very nature of models suggests, however, that such a simplistic view is to be avoided.

A more subtle difficulty with the use of classical (frequentist) statistics to evaluate a model is that if a model is replicated sufficiently many times then a (null) hypothesis such as 'state variable Y is unaffected by parameter X' can always be rejected. Sufficient replication can always yield a significant difference or effect because a p-value is simply a statement about whether you have enough data to detect an effect rather than one about the size of the effect itself (to paraphrase Bolker, 2008, page 11). In other words, depending on how many model runs you do you can probably get any answer you want! The important questions are about *effect sizes*: if the model and the system are different, or if a parameter makes a difference, is the size of the difference important given the various uncertainties at play? In short, the quantitative tools outlined above are valuable in evaluating models of all sorts, but it would be unwise to blindly trust them to determine how useful (or otherwise) a model is. In evaluating any model a pluralistic approach that combines careful visualisation, appropriate quantitative tools and cautious use of intuition is likely to be the most successful.

7.5 Frameworks for selecting between competing models

A framework for making model-based inferences in the face of model uncertainty, including selecting between multiple competing models, has been developed under the name *multi-model inference* (MMI). While formal MMI has more frequently been used in the context of mathematical or statistical models, it is starting to be adopted in the context of simulation models (see Hartig et al., 2011). Before describing MMI in more detail we must first detour via two related subjects: Occam's razor and likelihood methods of parameter estimation.

7.5.1 Occam's razor

MMI is underpinned by *Occam's* (or *Ockham's*) *razor*, which states that if there are multiple competing models all capable of explaining some phenomena then the simplest is to be preferred. This guiding principle does *not* mean that the simplest is the 'true' or the 'best' model; it is an appeal to heurism (or pragmatism) rather than a strict basis for model selection. Lazar (2010) provides an interesting historical and conceptual overview of the use of Occam's razor, which has been the subject of considerable philosophical debate, and begs the question of what is meant by 'simple' and also why or whether simpler explanations should *a priori* be preferred (Baker, 2011). Nevertheless, the razor is important in model development (see Jansen, 1998, for discussion). Generally, as we have repeatedly argued, it makes more sense to start with a simple—even simplistic—model and slowly add complexity until it is deemed fit for purpose, rather than to start with a detailed model and attempt to remove processes or parameters from it.

In the specific context of MMI, Occam's razor is an important counterweight to goodness-of-fit in evaluating models. On the one hand we should reward models for predictive fit, but on the other we ought to penalise them for buying better fit with too much detail. A fair question is why should we penalise detail? The answer is that a sufficiently complex model can fit any data set it is parameterised on, but it is likely to have low predictive power when a novel prediction is required. Less detailed models may do less well at mimicking the data they were trained on, but often have better capacity

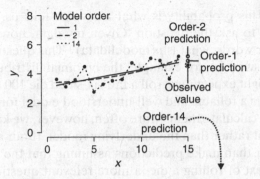

Figure 7.12   Three empirical/statistical models of differing complexity (a linear model, a second-order polynomial and a 14th-order polynomial) fit to data and then used to predict to a new observation (produced from the same underlying mechanism). The most complex model, the 14th-order polynomial, fits the data exactly, with $r^2 = 1.0$. The second-order polynomial has $r^2 = 0.43$, while the simplest model has $r^2 = 0.41$. However, when used to predict the new observation the simplest model performs best (residual $=-0.304$), closely followed by the second-order polynomial (residual $= -0.613$), with the complex model failing abjectly (residual $=$ 2230.98)!

to make new predictions. How important the balance between fitting already observed data compared to predicting new observations is will depend on the context.

The simple example in Figure 7.12 illustrates the nature of this trade-off. Three models of differing complexity are fitted to data and used to predict a previously unobserved data point generated via the same underlying mechanism. While the complex model fits the observed data very well (in fact perfectly) it abjectly fails to predict the new observation, while the two simpler models provide reasonable predictions. The simpler models *underfit* the data: they fail to account for some features in detail. But the complicated model *overfits* the data so that the model structure includes noise (Anderson, 2008). Another way to frame this (following Callender and Cohen, 2006, page 7) is to say that the simple and complicated models 'commit sins of omission and commission by lacking and having features the world does and does not have, respectively'. Finding the appropriate level of 'sin' lies at the heart of model selection and is a concrete example of the trade-off between simplicity and complexity.

7.5.2   Likelihood

Much of conventional statistics is built on assessing how probable is a sample equally or more extreme than the observed data, given some null hypothesis and an underlying model. In other words we are interested in the probability of seeing the observed (or more extreme) data, given some model, that is

p(data|model). This probability is what a p-value tells us. It is often more useful, however, to ask the question 'Given my data, how likely is a given model?', in other words, what is p(model|data)? This question is what *likelihood* helps us answer. So, for example, the binomial distribution tells us how many ones we might expect if we roll a fair six-sided die 100 times. This is fine so long as we have a reliable and well-understood model for the probabilities involved and can calculate them. More often, however, we know the outcome of the experiment rather than the underlying model. If our aim is to improve the model, rather than make predictions assuming that the model is correct, then in the context of rolling a die, a more relevant question is likely to be 'Given 25 die rolls of which seven are ones, what is a good model for this die?'

Unfortunately p(data|model) $\neq p$(model|data). We can, however, estimate the likelihood of a sample coming from a given model. Fisher (1922) showed that this likelihood ($\mathcal{L}$) is proportional to the probability of the sample, X, coming from the model of interest (θ):

$$\mathcal{L}(\theta|X) \propto p(X|\theta) \tag{7.4}$$

For the die example, this result allows us to evaluate $\mathcal{L}$(fair|seven ones) relative to other possible models for the die $\mathcal{L}$(biased|seven ones). It is important to note that this is only useful for comparing specified models: there are many other possible models (e.g. all the binomial distributions where $p_{one} \neq 1/6$) for a biased die, *and* we also need to specify how they are loaded to perform the necessary probability calculations to then determine which of several models for the die is most likely given the observed outcome of seven ones in 25 rolls.

Assuming that two events are independent the probability that they *both* happen is given by the product of their probabilities, that is $p(a \wedge b) = p(a)p(b)$. This generalises to any number (n) of independent events:

$$P_{all\ n} = \prod_{i=1}^{n} p_i \tag{7.5}$$

This result gives us a basis for computing theoretical likelihoods. If we have n independent observations x_i forming a vector X, then the likelihood of model θ is given by:

$$\mathcal{L}(\theta|X) \propto p(X|\theta) = \prod_{i=1}^{n} g(x_i|\theta) \tag{7.6}$$

where $g(x_i|\theta)$ is an appropriate probability distribution or density function. Likelihood is often expressed as a logarithm, the log-likelihood (LL), to avoid computational difficulties in the multiplication of very small values. The LL will be negative because the likelihood is between zero and one, and the most likely model (with highest LL) will have LL closest to zero. The model

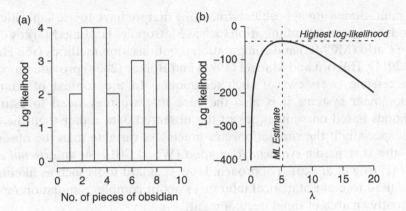

Figure 7.13   (a) Sample data showing the number of pieces of obsidian found in each of 20 plots. These data were simulated from a Poisson distribution with $\lambda = 5.00$, the sample mean is 5.2. (b) Profile plot showing the LL of different values of λ—the value of λ with the highest LL (i.e. that closest to zero) is the maximum likelihood estimate, in this case 5.2, although the flattening of the curve around 5.2 suggests that values close to the MLE are all relatively likely.

parameterisation selected by this procedure is called the *maximum likelihood estimate* (MLE).

So, how does this work in practice? Imagine we have observed the number of worked obsidian shards x_i found across $n = 20$ surveyed plots (Figure 7.13). We assume that the distribution of counts is approximately Poisson and so have predetermined the structure of the model, although in some cases we may want also to assess different probability distributions. The question is, given these data what is the maximum likelihood estimate for the Poisson rate parameter λ? From the Poisson distribution's density function we can determine the likelihood for any chosen value of λ:

$$\mathcal{L} = \prod_{i=1}^{n} \frac{\lambda^{x_i} \exp -\lambda}{x_i!} \tag{7.7}$$

or logarithmically transformed:

$$\log \mathcal{L} = \sum_{i=1}^{n} [x_i \log \lambda - \lambda - \log(x_i!)] \tag{7.8}$$

In a simple case like this it would be possible* to solve the likelihoods analytically and so derive the maximum likelihood estimate of λ, although it is much easier even in simple cases to let software do the work! We can see from Figure 7.13 that the MLE of λ is the same as the mean derived from the data, but this will not always be the case. Furthermore, in most cases we must

---

*In time-honoured fashion, we leave this as an exercise for the reader . . .

solve multi-dimensional problems, meaning that we have to use sophisticated methods of numerical optimisation such as Metropolis-Hastings Markov chain Monte Carlo (MCMC) and computational optimisation methods (see Hartig et al., 2011). Hilborn and Mangel (1997) and Bolker (2008) provide thorough and accessible overviews of such techniques. In the context of complex and nonlinear systems it is also the case that we may need to estimate likelihoods based on simulations of the observed data rather than the data itself, especially if the simulations are much less variable than the observed data—this is termed a *synthetic likelihood* (Wood, 2010) or an *external error model* (Hartig et al., 2011) approach. How standard tools such as likelihood can be used to make statistical inferences about complex simulation models is currently an area of rapid development.

7.5.3   Multi-model inference

To return to the issue of selecting between models, the logic of Occam's razor and the machinery of likelihood calculation underpin a suite of methods that allow us to select between multiple alternative models. In Popper's critical rational view, science proceeds incrementally by testing one hypothesis at a time and provisionally accepting it so long as we find no evidence with which to reject it. This view suggests that science proceeds by *falsification* rather than verification. While this perspective is just about sustainable in settings where clear-cut experiments are possible, as in some branches of physics and chemistry, it is much less useful where tight control and replication are difficult or impossible, and where system dynamics are multi-causal, with the relative importance of different processes varying in space and time (Anderson, 2008). In these circumstances, methods that employ *multiple working hypotheses* (based on Chamberlin, 1897), so that multiple models are allowed to 'compete' with each other, may be more appropriate.

Likelihood provides a basis for selecting among multiple models using measures such as classical likelihood ratios and information-theoretic (IT) criteria such as the AIC *Akaike's* (or simply *An*) *Information Criterion* (Akaike, 1973, 1974):

$$\text{AIC} = -2\log \mathcal{L} + 2k \qquad (7.9)$$

where k is the number of parameters in the model.

Heuristically, AIC rewards for fit, via the likelihood component, and penalises for model complexity via the number of parameters (there are good reasons for the specific selection of $2k$ as the penalty, see Anderson, 2008, Burnham and Anderson, 2002) and so operationalises the principle of parsimony in Occam's razor (Burnham and Anderson, 2004). The $-2 \log \mathcal{L}$ term in Equation 7.9 is the *statistical deviance*. If we compare multiple models using AIC the one with the smallest value is that best *supported* by the data, and so is preferred, and where two models are equally likely, AIC will suggest that we favour the simpler one. It is important to realise that the AIC will *always* identify a best model (one model will always have the lowest AIC) even if *none* of the models under consideration is particularly good, so it remains important to use visualisation methods, measures of fit and intuition to sensibly develop and select models. Intuition and under-standing of the system also play an important role in selecting appropriate models to compare. Furthermore, the absolute value of the AIC is of less interest than the differences in AIC (ΔAIC) across the multiple models being evaluated. We can use these differences to make statements about the the relative support the data provide for different models (such as, for example, model 1 is 12 times better supported by the data than model 2—whether 12 is strong evidence or otherwise is a matter of subjective decision) and to make predictions based on model averages (Burnham et al., 2011). This type of approach emphasises the provisional nature of model selection and model parameterisation. As more data accumulate, the 'best' parameteri-sation will change, and as we understand the system better so too will the 'best' model.

To make this a little more concrete we can return to the example in Figure 7.12. Imagine, using the same data, that we wanted to compare the performance of the simple linear model with a second- and fifth-order polynomial model using a multi-model approach. Table 7.2 shows the typical output of this type of approach. Here we use the AICc, which is a small sample size version of the AIC, to estimate model weights, ω, the relative support of the data for each model (Burnham and Anderson, 2002). ΔAICc is the difference between a given model's AICc score and that of the best supported model. While the fifth-order polynomial has the highest LL it has minimal support from the data relative to the other two models ($\omega = 0.0$). Overall, we can see that despite having higher LL values the data provide less support for models 2 and 5 than they do for model 1 (the linear model).

Table 7.2 Comparison of three models using IT model selection approaches based on data shown in Figure 7.12

Model	df	LL	AICc	ΔAICc	ω
Model 1	3	−12.15	32.702	0.0	0.85
Model 2	4	−11.85	36.214	3.5	0.15
Model 5	7	−10.14	52.956	20.3	0.00

In fact, the ω values suggest the data support model 1 5.67 times (0.85/0.15) more strongly than they do model 2.

It is beyond our scope to provide a thorough overview of model selection. Within an IT framework Millington and Perry (2011) provide a gentle introduction, Burnham et al. (2011) a short synthesis with useful practical advice, and Burnham and Anderson (2002) and Anderson (2008) more rigorous (but still readable) accounts. Link and Barker (2010) provide an overview from a Bayesian perspective. While IT and related Bayesian methods have begun to be widely adopted in the context of empirical (or statistical) models, there is also increasing use of them to select between dynamic simulation models (see, for example, Martínez et al., 2011, discussed in more detail in Section 7.6.1), although this is technically demanding (Hartig et al., 2011). One effort to develop this type of approach is the generalised likelihood uncertainty estimation (GLUE) framework described by Beven and Freer (2001) and Beven (2006), which provides a means of weighting models in model ensembles (assembled on the basis of structure and/or parameterisation) via their likelihoods. There have also been efforts to develop IT-type metrics for evaluating simulation models, where analytical likelihoods are not usually available. For example, Piou et al. (2009) outline an approach derived from the deviance information criteria (DIC) for evaluating competing simulation models in a multi-model framework.

7.6   Pattern-oriented modelling

While widely used, the formal confrontational approaches outlined in Section 7.3 are sometimes either not possible because there is no observational data or are inappropriate because the modelling exercise does not emphasise prediction. Various modelling frameworks have been developed to evaluate models that rely on a more 'experimental' approach. In fact, the literature on model evaluation has begun to emphasise that models are tools for 'experimenting on theories' (Dowling, 1999) and so models should be treated as if they were experiments conducted in a virtual (*in silico*) laboratory, even if important philosophical differences remain between simulations and other forms of experiment (see Guala, 2002, Morrison, 2009, Winsberg, 2009b).

One framework to help design and evaluate models of the type we have emphasised is *pattern-oriented modelling* (POM). POM (Wiegand et al., 2003, Grimm et al., 2005, Grimm and Railsback, 2012) is intended to make use of *multiple* patterns in the system being represented to inform the design, testing and evaluation of detailed simulation models. Grimm and Railsback (2012, page 302) define POM as 'the multi-criteria design, selection and calibration of models of complex systems'. At the heart of the POM approach is the idea that multiple patterns can guide three key phases of the modelling process: defining model structure, selecting between different representations and model calibration. It is important to emphasise that POM relies on more than one pattern. Multiple weak patterns are more informative for model development than a single strong one because they provide more *rejection filters* (*sensu* Hartig et al., 2011). While more than one model structure or calibration might equally well reproduce a single pattern, even a highly distinctive one, using multiple patterns provides a series of filters through which a given model must pass before it is deemed adequate. It may still be that more than one model passes through all the filters, but those remaining should (hopefully) be a greatly reduced subset of the initial ensemble.

Railsback and Grimm (2012) outline a protocol for implementing these steps in the POM approach in the context of agent-based models. In defining model structure POM is intended to ensure that the model has sufficient structural realism to capture the patterns seen in the target system, but not more so—in essence, an application of Occam's razor (see Section 7.5.1). This process may be conceptual, rather than formal, but at the end of it the model structure will have been defined, in the sense that decisions have been made about which processes will be represented in the model (see Grimm et al., 2005, for an example). Having defined the model structure it is then necessary to select between the alternative ways in which each process could be represented. Formal likelihood-based and Bayesian frameworks can be used for this step (Hartig et al., 2011), but it could equally well be a more informal process in which different representations are deemed more or less likely based on how well they reproduce some suite of target patterns. An important decision in this process is also deciding on the summary statistics that will be used to describe the pattern and so, in turn, used to evaluate the model (Wood, 2010, Hartig et al., 2011). As described in Section 2.2.3, we seek a summary that adequately captures the pattern (a sufficient statistic), but we need to remember that in summarising a pattern information will be lost—we should also be aware that the outcomes of this sort of process may be sensitive to the summary statistic used. As Wood (2010) notes it is sensible to use more than one such description, and he provides a brief overview of the types of measures that may be useful in this sort of model evaluation. Finally, the target patterns can be used to calibrate the model. This step may be especially important where the model is to be used predictively but direct observational data are limited. As an example, Rossmanith et al. (2007) adopt a POM approach to estimate demographic parameters in a population viability model of the lesser spotted woodpecker, *Picoides minor*. In this stage, the parameters of interest are varied until an acceptable match with *all* of the target patterns is achieved. Calibration may proceed by formal confrontation, as discussed in Section 7.3, or may be more qualitative. As ever, the most appropriate approach is a function of the purpose of the model.

As Grimm and Railsback (2012) admit, at one level POM is nothing new, it simply formalises a process that modellers and scientists have used for a long time. It does, however, emphasise, and make more systematic, the use of multiple patterns for developing and testing complicated simulation models and so helps to make what is often a difficult and time-consuming process of trial and error less *ad hoc*, particularly when combined with protocols designed to communicate models more effectively (Grimm et al., 2010).

7.6.1   POM case-study: understanding the drivers of treeline physiognomy

Martínez et al. (2011) provide an interesting example of the use of POM approaches coupled with modern computational statistical methods to evaluate and select from alternative simulation models of treeline dynamics. The models used by Martínez et al. (2011) are spatially explicit and individual-based and represent the dynamics of recruitment, establishment, growth and mortality of individual trees across altitudinal gradients on grids of 60×180 m. The key questions of interest were the conditions under which different treeline forms (for example abrupt *versus* diffuse edges) form and the role of krummholz (small and stunted, slow-growing) individuals in this process.

Figure 7.14 outlines the stages in what is a complicated and technical analysis comprising four steps (Martínez et al., 2011, page E141):

(i) Isolating and describing the (multiple) patterns observed in the system which will be used as filters in the POM approach.

(ii) Assessing how well the model reproduces (or fits) those patterns.

(iii) Parameterising the model, in this case using Bayesian methods via MCMC approaches.

(iv) Selecting the model structure(s) and parameterisation(s) that best represents the system.

In a sense the approach outlined by Martínez et al. (2011) is a rather specific type of uncertainty analysis motivated by identifying which regions of a broad parameter *and* model structure space result in model dynamics that map onto a series of specific empirical patterns. While the technical details are of less importance here (Martínez et al., 2011, describe them thoroughly) it is informative to see how complicated the analysis of what is not an overwhelmingly complicated model can become. It is also important to remember the fundamental purpose of the POM approach—to improve the model-based inferences we can make about real systems by making model

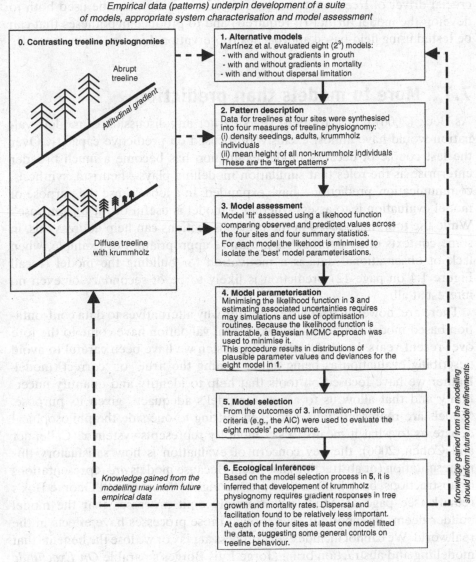

Empirical data (patterns) underpin development of a suite of models, appropriate system characterisation and model assessment

0. Contrasting treeline physiognomies

Abrupt treeline

Altitudinal gradient

Diffuse treeline with krummholz

1. Alternative models
Martínez et al. evaluated eight (2^3) models:
- with and without gradients in growth
- with and without gradients in mortality
- with and without dispersal limitation

2. Pattern description
Data for treelines at four sites were synthesised into four measures of treeline physiognomy:
(i) density seedings, adults, krummholz individuals
(ii) mean height of all non-krummholz individuals
These are the 'target patterns'

3. Model assessment
Model 'fit' assessed using a likelihood function comparing observed and predicted values across the four sites and four summary statistics.
For each model the likehood is minimised to isolate the 'best' model parameterisations.

4. Model parameterisation
Minimising the likelihood function in **3** and estimating associated uncertainties requires may simulations and use of optimisation routines. Because the likelihood function is intractable, a Bayesian MCMC approach was used to minimise it.
This procedure results in distributions of plausible parameter values and deviances for the eight model in **1**.

5. Model selection
From the outcomes of **3**. information-theoretic criteria (e.g., the AIC) were used to evaluate the eight models' performance.

6. Ecological inferences
Based on the model selection process in **5**, it is inferred that developement of krummholz physiognomy requires gradient responses in tree growth and mortality rates. Dispersal and facilitation found to be relatively less important. At each of the four sites at least one model fitted the data, suggesting some general controls on treeline behaviour.

Knowledge gained from the modelling may inform future empirical data

Knowledge gained from the modelling should inform future model refinements.

Figure 7.14 A schematic view of the various stages involved in the detailed POM process that Martínez et al. (2011) conducted to better understand treeline dynamics and physiognomy in the Spanish Pyrenees. Heavy solid lines denote the various steps in the analysis and the dashed lines show how the empirical patterns and the models and model evaluation are interlinked.

design and testing less *ad hoc* than it sometimes becomes. The key outcome in any modelling exercise is improved understanding of the system being considered—models are, after all, tools for surrogative reasoning. By finding the models that best agree with empirical observations and then applying surrogative reasoning the key outcome of the POM exercise conducted by Martínez et al. is that gradients in individual tree mortality and growth are a crucial driver of treeline dynamics. This outcome can then be used both to develop the model further (if required) and to produce hypotheses that can be tested using field-based approaches (observational and/or experimental).

7.7   More to models than prediction

As Beck (2006) notes, 10 to 20 years earlier any discussion of model evaluation would have almost exclusively focused on predictive capacity. Over the last couple of decades model evaluation has become a much broader enterprise as the roles that simulation modelling plays—heurism, synthesis, communication, prediction—have expanded. In a loose sense the purpose of model evaluation is to assess whether a model is useful or 'fit for purpose'. While the tools introduced in the previous sections can help address this, in some contexts other approaches are more appropriate. For example, when lack of observational data is a motivation for building the model (recall Figure 1.4 on page 12) prediction is likely to be of secondary or even no interest at all.

There are, however, deeper reasons for why alternatives to data confrontation based model evaluation and statistical validation have come to the fore over recent years. In the foregoing discussion we have been careful to avoid describing evaluation as being about finding the 'true' or 'correct' model. Rather we have focused on tools that help to identify and quantify uncertainty and that allow us to assess a model's adequacy given its purpose. Models are representations, and even putting to one side the philosophical debate as to what it means to say model *y* represents system *x* (Callender and Cohen, 2006), the key concern of evaluation is how satisfactory this representation (or abstraction) has been. Because models *are* representations or abstractions, by definition they are 'wrong' or 'false' (recall George Box's remark, see page 4). Models deliberately omit processes that the model builder deems unimportant even though those processes have effects in the real world. We cannot include every process at play or we lose the benefits that modelling and abstraction bring (Jorge Luis Borges's parable *On Exactitude In Science* elegantly reinforces this point).

In an influential paper, Oreskes et al. (1994) suggest that while we may be able to *validate* models by checking their internal consistency, we cannot prove them true, that is, *verify* them. Oreskes et al. (1994) argue that successful 'history matching' (*sensu* Beck, 2006) can *confirm* a model, in that it provides some level of empirical support for it, but it does not sanction its truth. They point to three reasons why models cannot be verified:

(i) Models are closed, while real systems are not. Abstraction forces us to leave things out of models, even though these *may* have an effect (so that models are false by definition).

(ii) The data we use in models are 'laden with inferences and assumptions' (Oreskes et al., 1994, page 641), so that in a sense the data we use in models are themselves models and suffer from many of the same problems.

(iii) Non-uniqueness or under-determination means that many different models can produce the same qualitative or quantitative result, so that matching models to data simply means we have isolated one such model. However, we are unable to make claims about its truth because there are many other possible models that could produce the same match with data. Ironically, Oreskes et al. (1994), argue that a *failure* to match observational data may be a better position to be in than having a model that does reproduce some dataset because at least we then know something is wrong!

For models representing environmental and social systems under-determination is most acute via the problem of equifinality (Beven, 2002), introduced in Section 2.2.4. Different models can produce the same outcomes (or predictions), as can different parameterisations of the same model. The latter makes model tuning problematic as it collapses to the problem of which tuning to choose given that many may work equally well. On the basis of the observational data available for model evaluation, equifinality means that the data do not carry sufficient information to allow us to make statements about the model's truth.

Given this rather depressing state of affairs you may wonder why bother at all? Ultimately this brings us back to the other half of George Box's statement from Chapter 1, '[m]odels, of course, are never true, but fortunately it is only necessary that they be useful' (Box, 1979, page 2). The first point to make is that concerns about the truth or otherwise of models become most acute when models are narrowly understood as tools for prediction, to be evaluated solely on these grounds. If we are using models for different purposes then the difficulties are less pressing, although we still need to be aware of the risk of 'affirming the consequent' when making claims about a model on the basis of its ability to produce a desired behaviour: we cannot declare prediction to be unimportant and then go on to make sweeping claims about the real world while side-stepping any formal evaluation of our model!

What Oreskes et al.'s (1994) arguments most emphasise is that the main role of models is as heuristic tools for learning about the world and generating interesting (and testable) hypotheses. In many cases, models may be the only tools available to us for learning (see Chapter 1). In Oreskes et al.'s (1994, page 644) words, '[f]undamentally, the reason for modeling is a lack of full access, either in time or space, to the phenomena of interest.' To be clear, we are emphatically *not* rejecting the place of prediction, rather we, as have many others, are arguing that prediction is just one use, among many, of simulation models. In fact, the dichotomy between models as tools for prediction on one hand and tools for learning on the other is much less troublesome than the academic literature often implies because these goals are complementary. After all, learning about the system should enable better prediction, and efforts to predict, fruitful or otherwise, should mean we learn about the system. A more helpful perspective on quantitative models is as tools for making evidential statements about hypotheses rather than for making predictions (Parker, 2008). In this light, models can be evaluated on the usefulness of those evidential statements and, presumably, a model with little predictive capacity will not consistently provide useful evidence.

As the issues raised by Oreskes et al. (1994), Kleindorfer et al. (1998), Beven (2002) and others have come to the fore, the role of models, and appropriate ways of assessing their adequacy, has become broader and extended far beyond the traditional model *versus* data confrontation (Beck, 2006). On a technical level, simulation modelling is beginning to adopt and adapt new frameworks such as approximate Bayesian computation (ABC), multi-model and Bayesian inference from statistical modelling (Hartig et al., 2011), but this is not the only way evaluation is changing. As models are used more to develop policy and as part of processes such as adaptive management, evaluation methods that encourage the participation of stakeholders (both modellers and non-modellers) have emerged—so-called participatory modelling. In such settings traditional measures of model performance, such as the RMSE or r^2, are generally of much less interest than questions like 'Do we all agree that this model adequately represents the system?' (Castella et al., 2005, Millington et al., 2011).

To summarise, the broader church that is now 'modelling' means that the uses to which models are put are much more diverse than they were two or three decades ago. This diversity of uses, and the recognition of fundamental problems in seeing models in terms of prediction and truth, means that how evaluation is conducted has also diversified. We have only been able to skim the surface of this diversity to provide some orientation; for the details you will need to consult the technical literature. In any case, regardless of the specific methodological details, you will find that the key to successful model evaluation is to keep the *purpose* of the model firmly in mind. Ultimately, the evaluation of a model, which is after all a designed entity, rests on how well it meets the purpose for which it was designed (Beck, 2006, makes a similar argument). The methods we have discussed allow us to formalise this 'usefulness' in some ways, but as Grimm (1999) points out, just as much, or more, is learned in the process of model building than in analysing its outcomes. Equally, while addressing the question of the usefulness of a model may require detailed quantitative analysis, it may be as deceptively simple as asking 'Did I learn anything useful from building this model? And if so, what?'

8
Weaving It All Together

In this chapter we pull together strands from all of the previous chapters to demonstrate our key themes. A central message of this book is that seemingly complicated spatial simulation models can be developed from relatively simple component parts, each of whose general properties and mechanisms are fairly well understood. We have already seen one example of this approach in Stark's (1994) model of plateau erosion combining elements of various different percolation and growth processes (see Section 5.4.3). In this chapter, by working through an example in the context of island resource exploitation by human hunter-gatherers, we show that elements from all three of the building-block process types discussed in Chapters 3, 4 and 5 can be combined in a single model.

Along the way, we also cover practical topics related to model abstraction and conceptualisation, design, development, description and analysis. The best way to develop these skills, as with many practical matters, is *not* by reading a book, but by actually doing them. Even so, there are many lessons to be learned from understanding in some detail the thinking behind a particular model. We are particularly concerned here with the most challenging aspects of developing simulation models, which are related to how we abstract and conceptualise to arrive at a model representation. Having arrived at a plan for our model, how do we then go about building it?

This chapter does not tackle this question from the point of view of a particular programming language or modelling platform. There are good resources available for most programming languages and platforms you might adopt, and also good books to guide you in the use of a number of them (see, for example, Gilbert and Troitzsch, 2005, Soetaert and Herman, 2009, Railsback and Grimm, 2012). Our perspective is a more general one of how to approach the task of turning a model concept into a working prototype and then iteratively refining it until it meets the requirements. We also show how to apply a range of the tools and techniques from Chapter 7 to

Spatial Simulation: Exploring Pattern and Process, First Edition.
David O'Sullivan and George L.W. Perry.
© 2013 John Wiley & Sons, Ltd. Published 2013 by John Wiley & Sons, Ltd.

a specific example. Again, the challenges here are as much about deciding which questions to ask as they are about using particular visualisation and statistical tools.

8.1   Motivating example: island resource exploitation by hunter-gatherers

As people settled the archipelagos of the Pacific during the late Holocene, widespread ecosystem changes ensued. For example, it is believed that New Zealand was settled by the Māori (Polynesians) around 750 years ago (*c.* 1280 AD) and that in just one or two centuries following settlement large areas of forest were burned and a number of bird species became extinct (most famously, the giant flightless ratite moa). The changes that occurred in New Zealand at the time of human settlement are dramatic, but the same dynamic is seen in many other island ecosystems (Anderson, 2002, Rolett and Diamond, 2004). A combination of the introduction of fire to ecosystems where it was previously rare, the hunting of (often) predator-naïve species and the dispersal of invasive species such as the Pacific rat (*Rattus exulans*) proved a devastating combination.

The size of the human populations settling these islands was likely small—Murray-McIntosh et al. (1998) estimate that the initial Māori settlement of New Zealand comprised fewer than 200 people. Despite these small populations, ecosystem changes were rapid. Using a single species population model, Holdaway and Jacomb (2000) suggest that extinction of moa may have occurred in as little as 100 years across all of New Zealand, even under conservative estimates of human population growth rates and resource needs. This analysis supports an 'overkill' model of extinction. Holdaway and Jacomb's model is simple: it ignores other food resources, it is non-spatial and it assumes that the last moa was as easy to catch as the first, implying both that the prey remain naïve in the face of sustained hunting and that there were no spatial refugia. All of these simplifying assumptions are contestable (see Brook and Bowman, 2002, for a more general critique of models of late Quaternary megafaunal overkill). Even so, Holdaway and Jacomb's model strongly suggests that moa populations were fragile to human perturbation, with more recent modelling by Rawlence et al. (2012) demonstrating that moa populations were not in decline prior to human settlement.

It is tempting to wonder why early settlers were so 'destructive', but this question misses the point. Small populations settling in uncertain and unfamiliar conditions must maximise their population growth rate if they are to survive, and this requires maximising resource acquisition. As Anderson (2002) notes, what is from one perspective an 'ecodisaster' also represented an 'ecotriumph' for the humans settling these new environments. A more interesting question then is whether, given the fragility of the ecosystems, the ecological changes and collapses that occurred across the islands of the Pacific were *inevitable* once humans encountered them. It is this question that motivates the model design and analysis that forms the basis of this chapter.

It is also worth thinking about the broader place of models in exploring questions such as these. First, we are interested more in learning about the implications of different behaviours and resource acquisition 'strategies' than we are in prediction *per se*. In the context of moa extinction, for example, we are more interested in how the rate of extinction changes under different resource acquisition or human demographic scenarios than we are in a precise estimate of when extinction occurred. Second, a key advantage of the modelling approach we adopt is that it is both spatial and dynamic. There is a rich palaeoecological and archaeological record describing late Holocene environmental change in New Zealand and elsewhere in the Pacific (Kirch, 2005). This record, however, is inevitably patchy in time and space. It provides a series of largely non-spatial snapshots of environmental conditions, each with their own temporal uncertainty (Brewer et al., 2012). Dynamic spatial simulation models can complement invaluable palaeoenvironmental data sources and help us develop coherent narratives of the *dynamics* in time and space of previous ecosystem changes (Anderson et al., 2006).

8.2 Model description

Among the many challenges of working with simulation models is describing them. Part of the difficulty is that many aspects of a model seem obvious to its builders after the fact and hence hardly worth mentioning, yet may be critical features, not at all obvious to a reader. This is particularly important for the style of models in this book, which are open-ended, incorporate a variety of elements and are not readily described in mathematical equations, in the manner of longer-established, more formalised approaches such as systems dynamics. It has (albeit rather belatedly!) been recognised by the scientific community that standard protocols are required for model description in research papers and reports, so that model-based findings can be replicated by others (see Schmolke et al., 2010).

8.1

An example is the overview, design concepts and details (ODD) protocol set out by Grimm et al. (2006) and subsequently updated by the same group (Grimm et al., 2010). Although ODD was initially proposed for describing agent-based models (ABMs), 'it can help with documenting any large, complex model' (Grimm et al., 2010, page 2760). The model we present here is not large or (particularly) complex, and we consider that adopting the full ODD protocol for its description would lead to an unnecessarily lengthy description (a problem acknowledged by Grimm et al., 2010, page 2766). This is especially true given that the model is freely available for readers to explore at their leisure. In some cases, additional details on the model operation are provided in the model itself, either in comments or in the associated documentation. Even so, we have adopted a stripped-down version of the ODD protocol to ensure a reasonably complete account. Our description is broadly top-down, but somewhat informal compared to what might be expected in a scientific paper.

The difficulties of replication are not solely due to inadequate model descriptions. Even ostensibly identical models differing only in their implementation on different modelling platforms may fail to produce the same results. This is sometimes referred to as the *model docking* problem (Axtell et al., 1996) or model-to-model analysis (Rouchier et al., 2008). Docking refers to aligning different computational models of the *same* conceptual model as a way both to test the conceptual model for ambiguities and to detect bugs in the different implementations. Although the potential advantages of docking experiments are recognised (in particular those related to reproducibility), it is not an approach that has been widely adopted, perhaps understandably, given that it requires a model to be implemented not just once, but several times!

8.2.1 Overview

We have already discussed the overall purpose of this model in Section 8.1. Here we focus on particulars such as the entities represented in the model, their spatial and temporal scale, and the processes in which they engage.

Model environment and model scales   Model time steps represent one month of real time. The main implication of this is that resource and human population changes only occur once every 12 model time steps, that is, on an annual basis. The model space is a 129×129 grid of square cells, with each cell representing approximately one square kilometre of land or sea. The space is *not* toroidally wrapped either north–south or east–west because the model action takes place on an island. A number of different overall island shapes are implemented: a square island extending to one cell from the model edge (that is a 127×127 grid), a circular island of radius 63 cells and an irregularly shaped island whose coastline is formed by a percolation-based erosion process (see Chapter 5). In all cases the island is surrounded by sea so the grid is, effectively, bounded on all four edges. Details of each of these are provided in the model documentation as they are not overly important to the operation of the model. Each grid cell in the model is characterised by state variables as set out below.

On island flag is a boolean true/false flag indicating whether the cell is part of the island or whether it is in the sea. Only locations on the island can be entered by human groups, except when returning to a known hunting ground, when short hops across inlets are permitted.

'Unknownness' is how familiar to the human groups the patch is on a scale from one (well known) to 100 (unexplored). This cell variable is updated as human groups move around in the landscape and affects movement decision-making as detailed in Section 8.2.3 (pages 239ff).

Resource availabilities z_L and z_H record the availability of low- and high-value resources, respectively. Resources are denoted in unspecified units related to their energy yield for the human groups.

Resource capacities k_L and k_H are the maximum resource capacity for each of the two resources. The maximum capacity of the high-value resources in many parts of the model space is zero, when $k_H = z_H = 0$ at all times.

Initialisation of the resource levels and resource capacities is described in Section 8.2.3 on pages 239ff. Initialisation of the resource landscape and its regrowth are controlled by global model parameters:

Landscape initialisation parameters p, p_H, $k_{L,\max}$ and $k_{H,\max}$ control the modified random clusters method (Saura and Martínez-Milán, 2000, see also Section 5.4.1) used to determine which landscape patches support high-value resources. A percolation threshold p is used along with the desired proportion of high-value resource, p_H. $k_{L,\max}$ and $k_{H,\max}$ are the maximum values of the resource capacities k_L and k_H.

Resource demographic parameters $r_L, r_H, \sigma_L, \sigma_H$ and $z_{H,\min}$ control regrowth of the resources as detailed in Section 8.2.3 on pages 239ff and follows logistic growth.

Low value resource diffusion w dictates a secondary landscape process, by which after regrowth low-value resources undergo local averaging, w being the proportion of the resource which each cell shares with its von Neumann neighbours.

Model entities   The model space is populated by groups of hunter-gatherers, with initially only one group present. Groups are mobile agents, who exploit resources in the landscape, moving around as they do so. A group's population may increase or decrease, and groups may split to form new groups when their population increases beyond a threshold level, or a small group may merge with a larger one if it falls below a threshold population size. The current state of a group is recorded in a number of variables:

Population n Detailed demographics are not recorded so this value represents overall group 'capability' and a measure of its resource requirements, based on its population, rather than a literal 'head count'.

Home camp $c = (c_x, c_y)$ is the location of the current home base of the group, around which low-value resource collection occurs and from where hunting trips are mounted. The local area is defined as the cells whose centres are within radius r of the cell centre of the home camp.

Resource collection variables R_T, R_F, R_X and R_K accumulate over the course of a year. These are, respectively, total resources collected R_T, resources collected from local foraging R_F, resources collected by hunting R_X and total resources killed while hunting R_K. R_K may exceed R_X, which allows the possibility of overkill, since a hunting party may kill more high-value resource than it is able to bring back to their home camp. The total resource collected influences decisions about the effort to devote to hunting and also affects reproductive success.

Number of hunts X is the number of hunts undertaken in the current month, which affects the remaining effort available for local low-value resource collection.

Hunt memory H is a set recording previous successful hunting locations and their yields in terms of total resources collected. Each element in H is a pair of values, $(\mathbf{x}_i, Y_i)$. $\mathbf{x}_i$ is the cell encountered during hunt i with the highest remaining z_H when the hunting trip ended and Y_i is the total resource collected during hunt i. We term this pair of values a *hunting spot* or simply *spot*. Each patch of high-value resources may be represented in the memory of one or more groups. H is limited to a length n_H, with the lowest yielding spots 'forgotten' by removal from H if required. n_H is the same for all groups, but each group's memory is exclusive to it: spots are not shared between groups and are lost if a group exits the model.

Search tortuosity s_p controls the random walk process by which groups search for hunting grounds. The tortuosity determines the probability with which hunting parties change direction during searches. When tortuosity is high the search behaviour will be close to a simple random walk on the grid with moves to Moore neighbourhood cells at each step (see Section 4.2), whereas when it is low the walk will be a more direct, correlated random walk (see Section 4.2.3) with fewer changes of direction. Details of the random walk behaviour are set out in Section 8.2.3 on pages 242ff.

Many group behaviours are controlled by global model parameters, as described below.

Maximum and minimum group size $n_{\max}$ and $n_{\min}$ are the maximum population n that a group may attain before it splits, and the minimum population a group can be at before it will merge with a larger group. $n_{\max}$ also affects the 'founder' group size.

Human demographic variables r_G, r_σ, m_{max} **and** Z control population change of groups, as described on pages 243ff. The rate of population change each year is given by a random deviate from $\mathcal{N}(r, r_\sigma)$, where r is the product of the baseline group growth rate r_G and a multiplier based on resource collection relative to the resource requirement per head of Z. The multiplier is limited to a maximum of m_{max} regardless of how well the group has done. Z is the resource required per unit of population per year for the baseline rate of population change r_G to be maintained.

Search controls $s_{p,0}$, $s_{p,min}$, $s_{p,max}$ **and** Δ_s govern the initialisation and subsequent changes in the search tortuosity of groups.

Hunting and gathering parameters f_X, n_X, t_X, Δ_K, Δ_X **and** Δ_L affect the number of hunts undertaken, their effectiveness and the effectiveness of local foraging. Hunt frequency f_X is the number of hunting trips a person can conduct each month. The size of hunting parties, n_X, affects the impact of a hunt in terms of the resource that can be taken and the effort that will remain for local foraging activity. When a group's total population falls below $2n_X$ the hunting party size will actually be $n/2$. Each hunt is modelled as a series of t_X steps across the landscape either in 'search' or 'hunt' mode. This variable can be understood as a *hunting range*, since it is the farthest distance from the home camp that the search process can reach. Finally, the kill rate and take-home rate while hunting are Δ_K and Δ_X. Δ_K is the amount of high-value resource killed during one step in hunt mode per unit of population in the hunting party, while the resource actually returned to the group after a hunt is limited by hunting party size and Δ_X. Details of these variables and their interactions are discussed in Section 8.2.3 on pages 240ff. Δ_L controls the rate at which groups are able to exploit local resources by foraging.

This list of model elements, their state variables and the overall model parameters may seem rather formidable. Models aiming at even a modest level of realism often end up being quite complex regardless of the best attempts at simplification by model builders. During model development and refinement many variables are introduced, but their main role relative to the questions at hand is to establish scenarios (or experiments) where exploration of the effects of specific factors of interest can be investigated. As will be become clear in Section 8.4, many of the parameters listed above are held constant, so are treated as fixed parameters, across model runs. This allows us to investigate (say) high, medium and low population growth scenarios by setting the human demography parameters appropriately. Under any of these broad scenarios a more detailed investigation of the particular effects of (say) the spatial structure of the landscape can then be undertaken.

8.2.2 Design concepts

The model implements and explores the interactions among a number of the simple building-block models that have been discussed throughout this book.

Landscape structure The landscape (or simulation arena) consists of low- and high-value resources, with the latter distributed patchily. Patchiness is controlled by the modified random clusters (MRC) method of Saura and Martínez-Milán (2000) discussed in Section 5.4.1. A question of interest is how different spatial distributions of the high-value resource, even if they are at similar overall levels, affect the long-term ability of the human groups to prosper (compare this with the model of Ruckelshaus et al., 1997, considered in Section 7.3.6). Large patches of high-value resource may sustain a group reliably over a reasonable period of time, but if a group becomes too attached to a particular part of the island it may fail to explore other areas and miss other resource-rich areas. On the other hand, widely distributed small patches of high-value resource may be easily found and rapidly exploited. We therefore expect the spatial structure of the landscape to affect overall outcomes.

Human movement at two scales Human groups are engaged in two distinct movement processes at different temporal but similar spatial scales, both processes being types of random walk (see Chapter 4). Every month groups may relocate their home camp if locally collected resources are in short supply. Relocations may be either random, with a bias towards relatively unknown sites, or focus on known hunting spots, also with a bias to relatively unknown locations. How these relocation strategies perform in different landscapes is potentially of interest. Relocation biased towards known hunting spots appears rational since it enables easier exploitation of known good locations, but it may lead to a failure to exploit all available resources.

The second movement process is hunting search behaviour. This is governed by a biased and correlated random walk. Depending on the success of recent hunting trips, search movement changes. Recent successful hunting leads to movement becoming more tortuous so that local areas are more thoroughly explored. If recent hunting has been unsuccessful, search becomes more direct with longer straight line movement sequences so that a wider area of the landscape is explored. Parameters governing this movement are the hunt 'range' and the initial tortuosity setting after home-camp relocations. We might expect the overall success of groups to be affected by these parameters and their interaction with the landscape structure.

Foraging with memory Group hunting memory introduces an aspect to the hunting and search behaviour that has been of recent interest in the animal and human movement literature (see Boyer et al., 2012, and Sections 4.3 and 4.6).

The persistence of groups in hunting-rich resource locations depends on the length of their memory of hunting spots. Longer memories should mean that a group may stay on the island through leaner times because they still remember 'good' spots to keep them interested. In the long run this may pay off as previously undiscovered resources then have a better chance of being discovered and exploited. A shorter memory might be a good thing too, however, since it allows a group to cut its losses and leave the island in bad times. Bear in mind that this island might be a small part of a larger archipelago, so that leaving if life has become difficult might be a sensible decision, assuming that there are easier pickings elsewhere.

Logistic and exponential growth Resources on the island grow via a logistic growth process with fixed local capacity constraints. For the low-value resource, diffusion means that there will be recovery from localised extinctions. For the high-value resource diffusion does not occur and so extinction is dependent on the efficiency of the human groups' search and relocation strategies with respect to the landscape structure, and on their propensity to kill more resource than can be readily used (overkill, see below).

Human population growth is exponential, with a stochastic component. It is expected that a critical variable controlling overall outcomes will be the initial (random) population of the first human group on the island. A small group may have difficulty getting established, leading to abandonment of the island. Once a human population establishes, rapid exponential population growth is likely and will be accompanied by groups splitting, with an accompanying extension in the spatial range of hunting activity and probably more rapid extinction of high-value resources. Although there is no capacity constraint on human populations, because groups leave the island if hunting has become poor, there is an implicit assumption that in the time frame of the model more sustainable longer-term resource exploitation strategies (such as a balance between hunting and local foraging) are unlikely to be arrived at. This does not preclude the possibility of some scenarios seeing longer survival of high-value resources due to low exploitation rates, and it will be of interest to examine particularly long-lasting scenarios to understand how they differ from rapid extinction cases.

'Overkill' Human groups have differing abilities to kill the high-value resource and to make use of it, with the former higher than the latter. The model assumes that hunting groups do not pay any attention to the long-term implications of their actions, simply killing as much high-value resource as they can. The palaeoecological and archaeological records suggest that this is (sadly) a reasonable assumption. If more is killed than can be used, the immediate resource needs of the group will be met, but any overkill is lost from the system. It is of interest to explore how overkill rates in particular model runs affect the time to extinction of the high-value resource or before the last group leaves the island. It may be that if no overkill or very low levels of overkill occur that extinction is unlikely, which would suggest that such extinctions need not have been inevitable.

8.2.3   Details

8.1

Here we depart from the ODD protocol by providing only a sketch of the detailed workings of the model. The full description and the model itself are available online at http://patternandprocess.org. This is in keeping with common practice in the contemporary research literature on simulation models, of providing model code and supplementary descriptive materials to accompany published articles.

In spite of the large number of parameters listed, the overall behaviour of the model is straightforward. The easiest way to get to grips with it is to consider a flowchart of the sequence of events in each model time step (Figure 8.1). Developing such 'high-level' schematic views is a good way to conceptualise a model both pre- and post-implementation. The basic sequence is that groups assess their situation and relocate if necessary, then

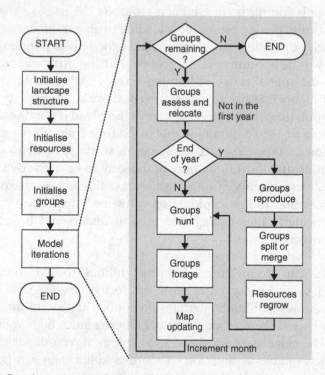

Figure 8.1   A flowchart overview of the model. The left-hand diagram shows model initialisation and scheduling. The right-hand panel shows details for each model iteration.

hunt and gather, and then the collective landscape map is updated. During the relocation step a group may leave the island if hunting is not going well or relocate its home câmp if local gathering is not going well. To ensure that groups spend at least one year on the island, this relocation dynamic only occurs after the first year has ended. Every 12th iteration, in other words once a year, additional operations occur: human population growth, group splitting or merging, and landscape resource regrowth. Each of these steps is discussed in more depth below.

Landscape initialisation   The key landscape feature of interest is the spatial distribution of high-value resource locations where $k_H > 0$. This distribution is initialised using the percolation-based MRC method (see Section 5.4.1) with percolation threshold p and assigning the percolation clusters created so that the required proportion of grid cells are high-value sites. Low-value resource capacities are set by local averaging of an initial random allocation. After capacities have been set, initial low- and high-value resource allocations are randomly determined. Ten rounds of the landscape regrowth process (see below) complete landscape initialisation.

Landscape regrowth   Resource regrowth is modelled using independent logistic growth models with capacity constraints (see pages 50ff), one for each resource type. Two minor adjustments are that the low-value resource diffuses by a local averaging process each step, which means that local extinctions eventually recover and that the high-value resource has a minimum sustainable level $z_{H,\min}$. If z_H is below this level after regrowth, extinction occurs and is set to zero.

Collective 'map' of the island   All groups in the model maintain a *shared* map of the island. This is represented by maintaining an 'unknownness' index for each grid cell. This index ranges from one (very familiar) to 100 (completely unknown). All cells visited by a group during a month have the index set to one, and at the end of a time step all cells except the current home camp have the index incremented by one. Additionally, at the end of a month all cells perform a weighted local average operation over the von Neumann neighbours, which makes the 'memory' of locations that have not been visited recently fade and spreads the effect of known locations so that, for example, cells adjacent to a recently travelled route become better known. The map of the island is held in common by all groups and plays a role in search and relocation behaviour by encouraging groups to explore less well-known territory.

Initialisation of groups   A single group is placed at a random location on the island edge, but one that is adjacent to a high-value resource. This represents

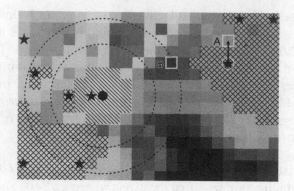

Figure 8.2 Group relocation of the home camp. Grid cells are coloured from light (well-known areas) to dark (unknown areas), reflecting the collective map of the island. Hunting grounds are cross-hatched and hunting spots are black stars. The current home camp is the black hexagon. Relocation with respect to hunting spots chooses the hunting spot in the least explored location and then moves from there to the nearest site with no high-value resource present (site A). Random relocation selects the least well-known location that is between two and three times the local area radius distant from the current home camp, as indicated by the dotted circles (site B)—ties are broken by random selection. The single-hatched region is the local area of the current home camp.

first arrival on the island on the shoreline, near useful resources. The group's initial population size n is drawn from a Poisson distribution with mean $\lambda = n_{max}/2$. Hunting memory H is initially empty and the search tortuosity of the group is set to $s_{p,0}$.

Group assessment of situation and relocation   Each month after the first year of model time, groups first decide if they wish to leave the island or relocate. If no hunting spots with a non-zero yield are in the memory, then the group leaves the island. If the resources available in the local area around the home camp are poor then the group relocates either at random or by selection of a location near a spot in the hunt memory. These mechanisms are shown in Figure 8.2. In either case the selection of a site favours less well-known locations in the collective map of the island.

Hunting and gathering   Hunting and associated decision-making occur each month. The sequence of operations for each group is shown in Figure 8.3. Each groups carries out all of these operations before any other group gets its 'turn' in random order, so that no group is consistently favoured. Figure 8.4 shows an example of one (busy!) month of hunting activity.

Deciding how many hunting trips   The number of hunting trips a group will undertake in a given month is determined via an estimation of likely success by consulting the hunting memory and assessing the proportion of

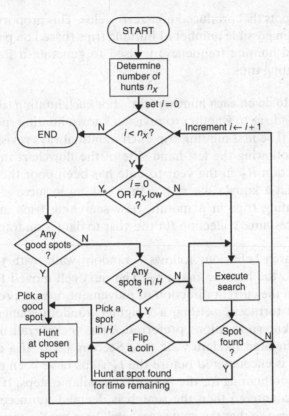

Figure 8.3 A flowchart of the hunting process: n_x = number of hunts, R_x = amount of resource collected in the year to date and H = collection of hunting spots in the memory.

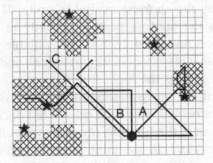

Figure 8.4 Hunting and search activity during a single month. The first of five hunting trips in this month revisited a known spot to the north-east, travelling there by a direct route (A). Subsequent hunting resulted in storage of a new spot in memory. Note that this is not where the hunting ground was entered, but the best site encountered during the hunt. Five subsequent trips were searches; since this group is on target for resource collection it can conduct search trips, even though several known good spots are already in memory. One search yields a new hunting ground to the north-west (B). Note that one unsuccessful search to the north-west comes very close to a hunting ground but still misses it (C).

remembered spots that produced non-zero yields. This proportion multiplied by the maximum possible number of hunting trips (based on party size, group population and hunting frequency) is used to generate a Poisson random number of hunting trips.

Deciding what to do on each hunting trip   For each hunting trip undertaken, a decision is made as to whether to revisit a known hunting spot or to search for a new one. The first hunting trip each month always revisits an old spot if one exists, following the left-hand side of the flowchart in Figure 8.3. If resource collection R_X in the year to date has been poor then the decision will be to revisit a known site rather than risk an unsuccessful search. For successive hunting trips in a month, new searching trips are undertaken provided that resource collection for the year to date is on track.

Searching   Search behaviour follows a random walk, with the number of steps limited to the hunting range t_X. The next cell moved to each step is based either on the current direction of movement or with probability given by the current tortuosity setting, a weighted random change of direction with less well-known locations preferred. Thus the overall movement is a biased and correlated random walk (see Section 4.2.3). If a cell with high-value resource is encountered before the t_X steps have been taken then the group switches to hunting for the remaining available steps. If no high-value resource is encountered then the search is deemed 'unsuccessful' and the tortuosity reduced so that the next search will be more far-ranging.

Hunting   If hunting is based on a known spot in the memory then the group moves there by the shortest available route. If hunting is initiated during a search then the hunting party is already in the hunting ground and hunting commences immediately. For the available number of steps, the group exploits high-value resources in its current cell and moves to whichever of the Moore neighbours has the highest remaining z_H. When the available steps have been used the group instantaneously returns to home camp.

The amount of resource killed in each cell visited during the hunt is based on the hunting party size, the hunting range (less is taken each step if the hunting range is longer) and the availability of resources. The total resource taken is limited by the ability to take kill back to home camp, Δ_X and may be less than the total kill. When the hunt ends, the grid cell visited during the hunt with the highest *remaining* z_H value is stored in the hunt memory, along with the total kill for the hunt, and the search tortuosity is increased so that the next search undertaken will be a more thorough local exploration.

Local foraging   Local foraging occurs in the local area around the home camp. The effort available for foraging is based on the group population and

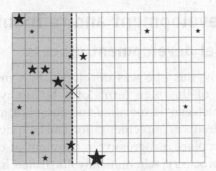

Figure 8.5 Splitting hunting spots when a group splits. The hunting spots (stars) are sized to show which have the higher yields. The yield-weighted centre location marked by the large $\times$ divides the hunting spots across the shorter axis of the overall distribution. Here, because the orientation of the distribution is east–west, the split is from north to south. One group will get more spots, while the other gets fewer, but these include the richest spot. Each group will then relocate with respect to their new hunt memory (see Figure 8.2).

on the amount of hunting undertaken this month—more hunting leaves less effort for foraging. Foraging involves randomly selecting cells in the local area with high levels of the low-value resource z_L.

Human demography Once a year, the human population is adjusted using an exponential growth model with noise. A mean birth rate is determined, derived from the baseline birth rate r_G and a multiplier reflecting the success of the previous year's resource collection. The mean birth rate is used to define a Gaussian distribution from which the actual birth rate is drawn, and the resulting population change is rounded to the nearest whole number. Negative birth rates are possible, so that group populations may fall as well as rise.

If group population falls to $n \leq n_{min}$ then the group merges with the nearest other group (if any). If there is only one group and it falls below a minimum threshold size then the island is abandoned (this prevents small groups persisting indefinitely). Merging groups' hunting memories are combined, and if the merged memory exceeds the maximum allowed then the required number of spots with the lowest yields are forgotten. If after reproduction and group merges the population of any group exceeds the maximum allowed then the group splits into two groups. Each member of the population is assigned with equal probability to one of the two groups. Figure 8.5 shows how the hunt memory of the original group is split between the new groups by finding a yield-weighted centre location of the spots in memory and dividing the spots east–west or north–south at this location, depending on the longer axis of the area of the island covered with hunting spots.

8.3   Model development and refinement

8.3.1   The model development process

As with many models, this one was developed in a modular fashion. The easiest way to think of this is in terms of a hierarchy of levels of model operation, each of which can be specified and designed, and for each of which an initial (perhaps primitive) implementation can be quickly developed. With all the components in place, progressive refinement of each aspect can then be undertaken. The process is schematically illustrated in Figure 8.6. There is no set order in which to do this, still less a 'correct' way to do it. In some cases the best way to make progress is to tackle those components for which the chosen representations are best understood first. This can build momentum and provides a sense that progress is being made, which is important to the success of any project: it is much easier to give up on a project when no progress has been made than on one where significant elements are in place and just a few more components are needed.

Of course, tackling the more difficult aspects cannot be put off indefinitely, since these are often the parts of the model most central to its operation. It is also important to find out at an early stage just how hard those difficult elements *really* are, so that realistic plans can be made for how much time to dedicate to them, or even to reconsideration of the design. Note also that if the

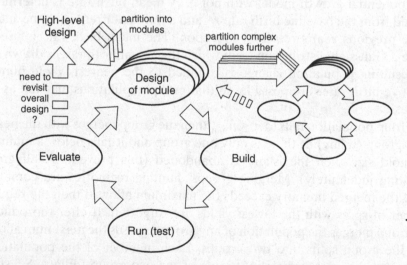

Figure 8.6   The model development process. High-level design decomposes the model into component modules. Each is designed, built, tested, evaluated and redesigned separately. Complex modules may require further decomposition. The design–build–test–evaluate cycle is iterative and may be carried out several times. Reconsideration of the high-level design may also be required from time to time.

more difficult model components are inessential parts of the whole then you should go back to the drawing board: there is little point in over-elaborating secondary elements of the model. The use of tools such as sensitivity and uncertainty analysis, even on prototype models, can help to identify these unimportant components. For example, in the model we are discussing we might incorporate complicated seasonal variation in the regrowth of resources and in the behaviour of the hunter-gatherer groups. If the aim of the exercise was to explore the relative impacts of climatic variation and human numbers on newly colonised places, this would be a reasonable plan. On the other hand, if the focus is on how different resource exploitation strategies affect survival times of high-value resources then climatic variation is of secondary importance and investing a lot of effort in this aspect makes little sense—a richer representation of climate can always be included later, if desired.

There are limits to how much can be simplified and abstracted away, however, and most simulation models inevitably include some feature that is difficult for, or at any rate is unfamiliar to, the model builder at the outset of the process. Even making a start on these model features can be difficult (we speak from bitter experience . . .). The best advice we can offer is that you should aim to start from the simplest possible implementation and add features rather than the other way around. This pragmatic application of Occam's razor may appear obvious advice—after all we usually make complicated artifacts from simple ones—but it is tempting, and with current development tools surprisingly easy, to try to assemble all the elements at once. A much more reliable approach is, as for the model as a whole, to break the model's components into pieces, decide a logical order in which to implement them and then work on them one at a time, testing each new feature as it is developed. For example, our model includes movement behaviour in three different modes (relocation, search and resource exploitation). In this case a possible approach is to build a model of each movement mode on its own, without the distraction of mode-switching. Once the distinct movement behaviours are working as intended, they can be combined. The goal is *never* to be faced with a model-building or programming task that is so complicated that it becomes intractable.

A good rule of thumb is that if you find yourself writing more than a dozen lines of program code or adding more than a couple of boxes and arrows to your model, and you have not tried running the model since the last time that you knew it worked, then you are trying to do too much at once! It is easy to get into a habit of writing large slabs of model code without testing, only to spend the following week laboriously rewriting every single line just to get it to run at all. Remember that professional programmers spend more of their time designing and testing code rather than actually writing it. It is likely to be more productive (and more enjoyable) to develop the model by the progressive addition of no more than a few lines at a time, testing as you go. You may even discover by this approach that some planned elaborations of your model are unnecessary.

8.3.2   Model refinement

Model development eventually gives way to model refinement. In practice it is difficult to tell when the transition occurs. Arguably, as soon as even the most primitive implementations of all the model components have been developed, a whole model implementation exists and the refinement process begins. In any case, model refinement is really a continuation of the development process, but with a stronger emphasis on testing, evaluation and (re)design.

It is particularly important at this stage not to be afraid to break what you have already built and start over again. This applies to the whole model as much as to any of its parts, but it is much more likely that you will need to completely change components rather than the whole model. This is another good reason to pursue a modular approach, since redesign of one component is then less likely to have far-reaching knock-on effects on all the other components. A modular approach also facilitates assessment of model uncertainty (Section 7.3.4), which is an important part of the refinement process. A pattern we have seen repeated many times (in our own work and in student projects) is that the model goes through stages of added complication followed by consolidation, when the model implementation becomes simpler. After the fact, a lot of work that has been done may appear redundant, as it is removed from later versions of the model by the refinement process. It is more useful to see this as an inevitable aspect of how complicated artifacts are created, rather than as wasted effort. Often, it is only by adding complications that we are able to see simpler alternatives.

This pattern of progress carries with it two important consequences. The first we have referred to several times already: the necessity of regularly running and testing your model. It is only by ongoing testing that avenues for simplification are likely to be identified and progress made. Interactive exploration and testing should inform refinement of each model component, usually one at a time in isolation from one another. During this process you will learn a lot about your model and start to understand which are the critical parameters, and hopefully also what are the interesting behaviours and patterns it generates. This overall 'feel' for the model is an important stage in model development and analysis, as it will inform many subsequent design and analysis decisions.

The second consequence of 'progress by redesign' is the importance of documenting what you do during the modelling process. Model documentation consists of two aspects: in-code commenting and 'laboratory notes'. Comments in the model itself are there so that at a later date you (or someone else) can easily figure out what the model does and why it is implemented in the way that it is. Things which appear obvious when you wrote them may not be so apparent when you return later to revise or add new features. Much time can be saved if code is well annotated with comments. Most model builders recognise the need for comments and while they probably fail (like most programmers) to adequately comment their creations, they at least recognise this failing. A potentially more serious (and more common) failure is neglecting to keep sufficiently detailed laboratory notes, by which we mean a record of design decisions made as the model proceeded to its final form. As is apparent from the foregoing account, model development involves a process of trial and error, with significant progress often only made after *omitting* features or departing from more obvious approaches. By their nature, what is left out of the final version of a model cannot easily be documented by comments alone, and it is here that detailed notes on decisions through the development process can be invaluable. Unfortunately, the modelling community as a whole has yet to seriously tackle this problem. An automated solution might involve building a model in a *version control system*, which forces the user to make notes associated with significant changes, but as yet such practices are not part of the everyday routines of scientists in this field (see Schmolke et al., 2010).

8.4   Model evaluation

We now present a preliminary analysis of the hunter-gatherer foraging model described above, using some of the approaches introduced in Chapter 7. We start by looking at the model's dynamics under 'baseline' conditions before moving onto a local univariate (one-at-a-time) sensitivity analysis (SA; Section 8.4.2) and then a multivariate uncertainty analysis (UA; Section 8.4.3). As the model description has shown, even a relatively simple question—how might humans optimally forage in a new environment?—quickly results in a complicated model with many components and parameters. Fundamentally model evaluation is concerned with exploring why a model behaves the way it does and what controls that behaviour. We do not present an exhaustive analysis of all the parameters and their combinations here. Our intent is to illustrate the types of approaches that might be used to evaluate a model of moderate complexity. Furthermore, our analyses emphasise the model's temporal dynamics and uncertainty in its parameterisation rather than its structure.

8.4.1   Baseline dynamics

A first step in analysing *any* model is to explore its dynamics under so-called 'baseline' conditions, which usually encompass our current best estimates of the model's parameters and our current understanding of the processes of interest. This 'baseline' scenario is also our 'best' representation of the processes that we have included in the model, and as such provides a point of comparison for all subsequent analyses.

The model developed in this chapter contains 32 parameters in five broad classes:

(i)   Control over the initial spatial structure of the island.
(ii) Human demography.
(iii) Resource demography.
(iv) Resource exploitation.
(v)   Search strategy.

Our estimates of these parameters (see Table 8.1) represent plausible values, informed by previous studies where possible. Estimating some of these parameters is extremely difficult given that we are seeking to recreate systems that no longer exist or have lost important components. In this situation modelling is potentially useful in allowing us to make inferences about the importance of various processes and parameters, and may even enable us to estimate plausible values for key parameters, an approach sometimes called *reverse inference*.

Figure 8.7 shows a 'typical' model run and exhibits interesting temporal dynamics (see also Plates 9 to 13). Periods of high resource availability are coupled with periods of population growth, but there are also times where resources are harder to come by and population growth slows (or even becomes negative), with several groups leaving the island before it is finally abandoned after around 230 years. Having interesting dynamics does not, of course, constitute a model analysis, but it is a good start! It is also reassuring, although again it is not a formal analysis, that the model behaves as we would expect. For example, population dynamics and resource acquisition are clearly linked. Such informal model assessments via visual debugging (Grimm, 2002) are an important part of model testing during development.

How to measure a model's behaviour is critical to any analysis and requires that we decide which state variables will be used to describe the dynamics. There is no single 'right' answer to this question. The appropriate variables will partly be a function of the questions being asked, and it is usually sensible to use multiple, ideally uncorrelated, measures. Possible measures include the value of a specific variable at a given point in space and time, mean values (over space and/or time), extreme values, ratios and so forth.

Here we are interested in the rate at which humans could have reduced high-value resources on encountering them, and in how the demography and behaviour of human populations, together with spatial patterns of resource

Table 8.1 Baseline parameter values for the agent-based foraging model, subdivided into the five main model 'components'. Those parameters marked [C] are categorical. Abbreviated codes for each parameter used in some of the later figures are also shown. Note that min-viable-human-pop was treated as a fixed parameter (with a value of five) to prevent model runs persisting almost indefinitely with a group size of just one

Component	Parameter	Code	Baseline value
Spatial structure (5)			
	percolation-threshold p	*p.thresh*	0.5
	proportion-high-resource p_H	*phr*	0.15
	no-singleton-patches?	[C]	TRUE
	island-type	[C]	'irregular'
	wiggly-edge	*w.edge*	0.28
Human demography (5)			
	min-viable-human-pop n_{min}	*min.via*	5
	max-group-size n_{max}	*mx.grp*	30
	r-humans r_G	*r.hum*	0.015
	r-humans-sd r_σ	*r.hum.sd*	0.1
	max-birth-rate-multiple m_{max}	*mx.birth*	3
Resource demography (8)			
	r-high r_H	*r.high*	0.05
	r-hi-sd σ_H	*rh.sd*	0.1
	max-high-k $k_{H,max}$	*mx.high.k*	2.5
	min-sustainable-h $z_{H,min}$	*min.sus.h*	0.1
	r-low r_L	*r.low*	0.2
	r-low-sd σ_L	*rl.sd*	0.2
	max-low-k $k_{L,max}$	*ms.lo.k*	0.5
	diffusion-rate w	*diff.rt*	0.1
Resource exploitation (9)			
	resource-per-head Z	*res.ph*	1
	hunt-kill-per-head Δ_K	*kill.ph*	5
	hunt-take-per-head Δ_X	*take.ph*	0.1
	hunt-party-size n_X	*h.size*	6
	hunt-range t_X	*h.rng*	16
	max-hunts-per-month f_X	*mx.hpm*	4
	hunt-memory-length n_H	*h.mem*	15
	gather-per-head Δ_L	*gat.ph*	0.05
	nearby-range r	*nr.rng*	2.3
Search behaviour (5)			
	initial-search-tortuosity $s_{p,0}$	*init.tort*	0.1
	search-adjust Δ_s	*s.adj*	0.05
	max-tortuosity $s_{p,max}$	*mx.tort*	0.95
	min-tortuosity $s_{p,min}$	*mn.tort*	0.05
	relocate-near-hunting?	*rnh* [C]	TRUE

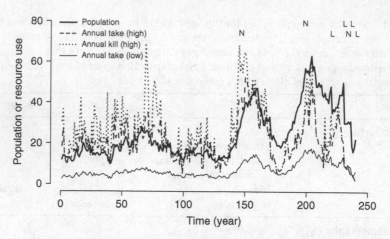

Figure 8.7 Typical model dynamics under baseline conditions. The population (black line) changes as a function of the amount of high-value resource being exploited (taken and killed—dashed lines), which changes as groups of humans find new areas locally rich in resources. The difference between the kill and the take is the overkill, which also varies over time (maximum of slightly under 120% in this example). Letters above the lines denote times when new groups form as existing groups split up (N) and when groups leave the island due to a lack of hunting success (L).

availability, might have influenced this rate. Thus, the $n = 16$ state variables we initially considered as useful descriptors relate to:

population dynamics mean and maximum population size (*mn.pop* and *mx.pop*), year in which the maximum population size was attained (*t.mx.pop*) and the length of time the population persisted for (*t.persist*)

high-value resource take mean and maximum take per year (*mn.take* and *mx.take*), and proportion of the initially available high-value resource taken at the end of the model run (*pr.take*)

high-value resource kill mean and maximum kill per year (*mn.kill* and *mx.kill*), and maximum overkill (*mx.okill*)

low-value (local) resource exploitation mean low take (*mn.low*), and median and maximum low:high take ratio (*med.ratio* and *mx.ratio*)

temporal dynamics of resource exploitation year at which 50% of the initial high-value resource had been exploited (*t50*), and year in which the maximum take and kill occurred (*t.mx.take, t.mx.kill*).

We also record demographic conditions and events such as the initial population size (*init.pop*), group mergers and splits, and groups leaving the island for whatever reason. If our model implementation and evaluation were motivated by other questions then different state variables might be more appropriate.

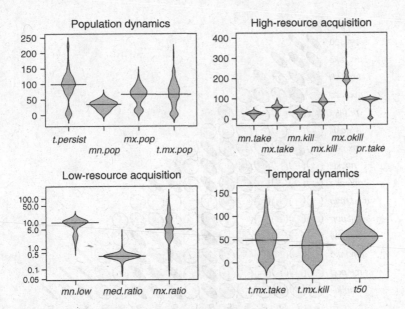

Figure 8.8 Beanplots (smoothed density of observations) showing variation for each output variable across 100 simulations. Thick horizontal lines are group medians. This graph shows which model outputs are most variable under the same parameter conditions. Note that *pr.take* is expressed here as a percentage and may exceed 100 due to resource regrowth over time. We have omitted model runs (22 of 100) where less than 50% of the high-value resource was exploited from the plot of *t50*.

Another important step in understanding a model's behaviour is to look at the variability between model runs under the same conditions. Because our model is stochastic its analysis requires synthesising multiple model runs. The question 'How many runs is sufficient?' has no clear answer, but within reason the more the better, with 30 often considered a minimum*. If you are exploring a large multivariate parameter space then computational expense also becomes a consideration (see Section 7.3.2). To look at variability in model behaviour under fixed conditions we ran the model $n = 100$ times under the baseline parameter set (Table 8.1). Note, however, that this type of analysis does not depict all of the variability that may occur under identical model parameterisations. In this case, for example, it does not consider the effects of different initial settlement locations on islands with identical initial resource conditions.

Looking at the distribution of each of the model outputs and then assessing the extent of any correlation between them is another useful step in under-standing any model's behaviour. As is clear from Figure 8.8, under baseline conditions the model's behaviour varies even under the same initial param-eterisation because the state variables all show some scatter. Furthermore,

---

*This is presumably a misguided appeal to the central limit theorem.

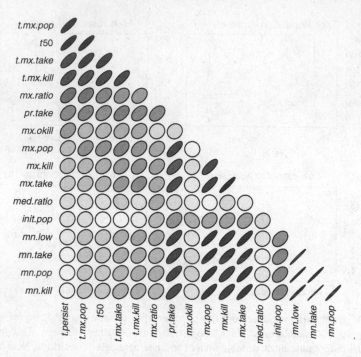

Figure 8.9 Pairwise correlations (non-parametric Spearman's rank) between the 16 state variables initially considered as useful descriptors of model behaviour; the shape of the ellipse depicts the scatter and directionality in the relationship and the grey-scaling the (absolute) correlation (dark = high to light = low).

some of the state variables are not unimodal (e.g. *t.persist*), which suggests that qualitatively different types of model behaviour can occur under the same model parameterisation. For example, the time the population persists for shows peaks at around 25 and around 100 years, reflecting less and more successful (and prolonged) colonisations, respectively.

Correlations between the state variables are shown in Figure 8.9. Evaluating highly interdependent outputs is not likely to be informative: if we have two strongly correlated variables *a* and *b* we (usually) gain little new insight in analysing *b* after we have analysed *a*. We can see that the various temporal measures are strongly positively correlated (upper left of Figure 8.9). Likewise the mean take, kill and population size are also strongly positively correlated (lower right of Figure 8.9). Thus there is probably little to gain by including all of these variables in further analysis. On the other hand, this correlation is again heartening as we would intuitively expect these variables to be closely linked to each other.

Classes of model behaviour Another way to think about the model's behaviour is to consider the types, in a qualitative sense, of dynamics it produces—in other words, can we classify model dynamics into groups such as 'failed settlement' and 'successful settlement'? Under baseline conditions, 22% of simulations resulted in the population exploiting less than 50% of the available resource and in 15% the population persisted for less than 25 years. We might regard these as 'failed' settlements. Depending on the questions being asked it may be appropriate to use derived measures such as these as an index of model behaviour.

There are also more technically sophisticated ways to summarise the model's behaviours. The 100 baseline runs and 17 state variables (the 16 in Figures 8.8 and 8.9, and the initial population size [*init.pop*]) comprise a matrix that can be used to measure the 'distances' between model runs. To measure these inter-run distances we first standardise each of the 17 parameters using its mean and standard deviation because the parameter values span very different ranges. We then build a pairwise distance matrix, with matrix elements being the Euclidean distances between each pair of model runs. Finally, we use an ordination method—here non-metric multi-dimensional scaling (nMDS), but other methods such as principal components analysis (PCA) may also be appropriate—to identify groupings in the data. It is beyond our scope to provide a detailed description of multivariate analysis (see Jongman et al., 1999, Quinn and Keough, 2001, Gotelli and Ellison, 2004, among many others for that), but, in essence, points representing model runs that are near one another in the nMDS space are more similar to each other than those further apart. Finally, we overlay vectors on the nMDS to form a 'biplot' showing how position in the ordination space relates to the state variables. Such a two-dimensional depiction of the multi-dimensional ($n = 17$) model space loses information, but few of us are able to visualise that many dimensions!

Figure 8.10 suggests two broad categories of model dynamic. Axis 1 broadly represents the success, or otherwise, of the human population. The small, diffuse cluster towards the right of the ordination space are 'failed' colonisations (such as (d) in Figure 8.11), while the more compact cluster towards the left comprises the more successful settlements, although there is considerable variation within this group (compare (a) and (b) in Figure 8.11) and some model realisations (Figure 8.11 (c)) are intermediate between groups. The biplot (Figure 8.10) suggests that populations that failed to establish had a higher ratio of low-value to high-value resource use, which makes sense because they presumably did not find areas of high-value resource. The median ratio of low-value to high-value resource use is somewhat orthogonal to all of the others (note the direction of the vector in Figure 8.10), suggesting that it may be a useful descriptor of the model's dynamics.

Axis 2 of the ordination space broadly relates to the time-scaling of the model run, with length (*t.persist*) increasing towards the top of the space (along with *t.mx.take*, *t.mx.kill*, etc.); these variables are negatively correlated with the mean and maximum take and kill. This pattern suggests that there

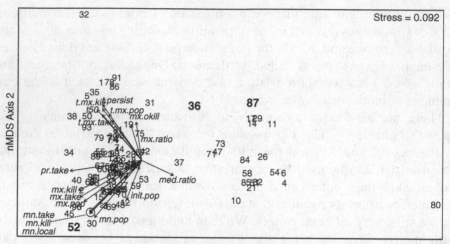

Figure 8.10 nMDS plot showing the multi-dimensional structure of the 100 model runs based on 17 state variables (the 16 in Figure 8.8 and the initial population size [*init.pop*]). Numbers on the plot represent individual model realisations, with realisations close together more similar than those further apart. The distance between model realisations is measured by Euclidean distance for a centred and scaled matrix of the state variables by run. The low stress (<0.10) suggests that the nMDS is a fair depiction of the underlying data and that distances between pairs in the matrix and in the nMDS space are strongly correlated. Arrows show the direction and strength of correlations between the position in parameter space and model outcomes. Individual model runs in Figure 8.11 are those in larger, bold typeface.

are two types of 'successful' occupations: (i) those where the population grows rapidly, acquires high levels of resource quickly and leaves the island soon after colonisation, and (ii) those where population growth is slower, acquisition rates slower and occupation time longer.

Having got a feeling for how the model behaves and the range of dynamics it produces under baseline conditions the logical next step is a sensitivity and/or uncertainty analysis.

8.4.2   Sensitivity analysis

Sensitivity analysis approach   Sensitivity analysis (SA; Section 7.3.2) seeks to identify those components of a model that most influence model outcomes. We evaluate *local* sensitivity by systematically varying the parameters of interest, one by one, by a fixed amount and assessing whether the proportional change in the state variable(s) of interest is more (sensitive) or less (robust) than that of the input parameter. Typically we use a sensitivity metric, such as that in Equation 7.3, that scales the proportional change in the variable to the proportional change in the parameter, so that a value of one represents a

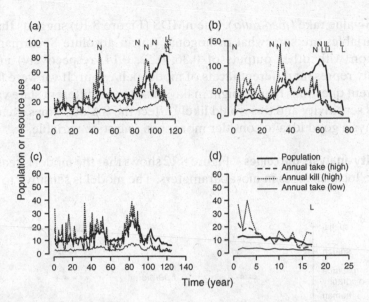

Figure 8.11 The range of dynamics produced by the model under baseline conditions: (a) a 'typical' model realisation (run 74), (b) rapid growth followed by abrupt collapse (52), (c) prolonged occupation but little population growth and no new groups initiated (36), and (d) a failed settlement (87). Run numbers correspond with those on Figure 8.10. Note that the scaling of both axes varies between graphs.

perfectly proportional change, while values greater or less than one represent greater or less than proportional changes, respectively. While there are shortcomings with a local SA (see Section 7.3.2), it is a useful first step in understanding model behaviour and also serves as a good logic check. In our case the model is stochastic and we need to run the model many times for each parameter value and compare the mean response with the mean value of the state variable of interest under baseline conditions. In the analyses here we manipulate each of the input parameters in Table 8.1, except for min-viable-human-pop (n_{min}), by ±20% and conduct 30 replicate model runs for each.

Some of the parameters listed in Table 8.1 cannot be changed by ±20% as they are boolean (true/false) or categorical values. In these cases we have performed 30 model runs using each of the values the parameter can take, for example the relocate-near-hunting *rnh* parameter was set to true (the baseline condition) and false. This can be seen as a form of robustness analysis (Section 7.3.4)—it is a change in the model's *structure*—but, semantics aside, it is important to evaluate non-quantitative parameters in a sensitivity framework.

An important decision in any sensitivity analysis is which state variables to assess—16 is rather unwieldy! Considering the analyses of the baseline model dynamics presented above we will use the time the population persists for (*t.persist*), the mean *per annum* take (*mn.take*) and the median *per annum* low:high-value take (*med.ratio*). The nMDS (Figure 8.10) suggests that these three variables are somewhat orthogonal (mean absolute Spearman's rank correlation with other outputs of 0.36, 0.52, 0.14, respectively) and they intuitively represent different facets of model behaviour. If we were focusing on different questions then it might make sense to use different state variables to assess sensitivity and this would likely affect the SA outcomes. In any case it is always a good idea to consider more than one state variable.

Sensitivity analysis outcomes   Figure 8.12 shows that the model is reasonably robust to local changes in most parameters. The model is sensitive to at least

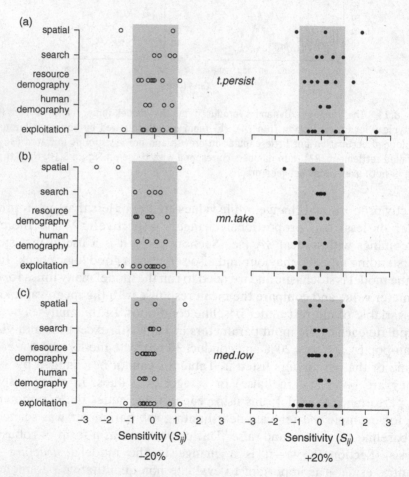

Figure 8.12   Stripcharts showing the sensitivity index values (Equation 7.3) for each of the non-categorical model parameters for (a) persistence time (*t.persist*), (b) the mean take (*mn.take*) and (c) the median low:high-value resource take (*med.low*). Values greater than |1.0| indicate a proportional change in the state variable larger than that in the input parameter. The grey area bounds the area where sensitivity indices are less than |1.0| where the model is robust to changes in those parameters.

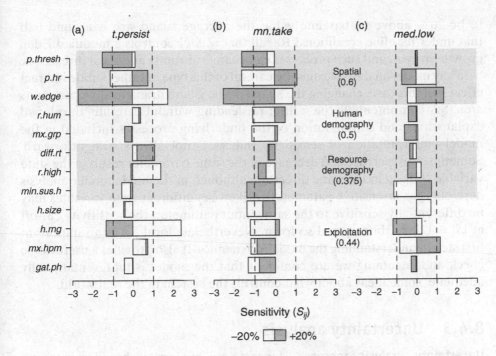

Figure 8.13 Sensitivity index values for those parameters where $|S_{ij}| > 1.0$ in at least one analysis across the three state variables assessed for either a 20% increase *or* decrease. Parameter abbreviations follow codes in Table 8.1 and values in parentheses show the proportion of parameters in each category that were 'sensitive' (see Figure 8.12). The three graphs show S_{ij} for (a) persistence time (*t.persist*), (b) the mean take (*mn.take*) and (c) the median low:high-value resource take (*med.low*). Note that the model never responded more than proportionally to a change in any of the search behaviour parameters.

one parameter (i.e. $|S| > 1.0$) for all groups except those related to search behaviour, with none of those showing $|S| > 1.0$. None of the changes in the categorical input parameters yielded a significant response (not shown).

In short, the initial availability of the high-value resource, and to a lesser extent how it is spatially structured, the level at which the high-value resource can no longer be sustainably hunted and the dynamics of the hunting activity (group size, frequency of hunting activity) are the parameters to which the model is most sensitive (Figure 8.13). If population growth rate falls and groups are smaller this reduces the mean annual take. The model is also extremely sensitive to changes in the *w.edge* parameter, and in fact changes in this parameter yielded the highest mean value of $|S|$. This effect may seem somewhat strange—why should the wiggliness of the island's coastline have such a significant effect? The answer is that as *w.edge* increases—so that the coastline becomes more crenellated—the size of the island markedly decreases with flow-on effects for the human population. When *w.edge* is set to be 20% above its baseline value the average island size is around half that under baseline conditions! Recall that *w.edge* controls a modified Eden growth process, and such processes can change abruptly at critical thresholds.

In any model analysis we must be careful of this type of rather subtle indirect effect—in this case, changing the shape of the island dramatically changes its area. SA can potentially be a little misleading without carefully developed explanations and consideration of the underlying processes included in the model—interpretation of sensitivity indices is not always straightforward. Sometimes an increase and decrease in the same parameter result in the state variable shifting in the same direction, although in the model under analysis here only for insensitive parameters. Likewise, different state variables may be differentially sensitive to the same input parameter, the sensitivity cut-off at $|S| = 1.0$ is arbitrary and so forth. Nevertheless local SA is an important first step in understanding the model's dynamics. It also provides a useful logic check and, assuming we are confident that the model is representationally adequate, some ideas about what controls the system in the real world.

8.4.3   Uncertainty analysis

Uncertainty analysis approach   Uncertainty analysis is broader in scope than the local sensitivity analysis presented above and can easily become a massive undertaking. Here we focus on trying to understand better how the resource exploitation strategy that human groups use affects the population and hunting dynamics of the settler. While the fate of megafauna such as moa *may* have been sealed as soon as humans settled oceanic islands, the rate and nature—slow and steady or abrupt collapse—at which extinction occurred are matters of ongoing debate and can be difficult to assess on the basis of the palaeoecological and archaeological records alone. As we have seen, under baseline conditions this model produces a range of dynamics: failed settlements, prolonged settlements with steady resource acquisition before abandonment and short settlements with rapid growth and resource use before collapse (Figures 8.10 and 8.11). Examples of all of these types of trajectory are known to have occurred.

To explore the effect of the nine parameters (Table 8.1) related to exploitation strategies in more detail we conducted an uncertainty analysis. As described in Section 7.3.3 simple random samples from the parameter space are inefficient in exploring the edges or extremes of that space, especially where it is highly multi-dimensional. Thus, our uncertainty analysis was conducted based on 2000 model runs in which the nine parameters were simultaneously varied using a Latin hypercube sampling strategy as per Stein (1987) and Stocki (2005).

Synthesising the wealth of data produced by this sort of analysis—a 2000×9 input matrix and a 2000×17 output matrix—poses challenges and necessitates the use of some sophisticated analytical methods. We used boosted regression trees (BRTs) to evaluate which exploitation parameters had the largest effect on the time the population persists for (*t.persist*), the mean per annum take (*mn.take*), the median per annum low:high-value take (*med.ratio*) and the proportion of model runs in which 50% of the initially available high-value resource was taken (a measure of 'success'). BRTs are a machine-learning type statistical tool that have good explanatory power for the sorts of multivariate and interactive data we are concerned with (see Elith et al., 2008, for a detailed review). Such methods are increasingly used to analyse complex simulation models (see, for example, Esther et al. 2010). The BRT analysis: produces a quantification of the relative importance of each parameter in explaining the state variables' behaviour, based on the extent to which inclusion of the parameter improves the BRT's performance; provides partial dependence plots, showing the relationship between input parameters and outputs; and identifies important parameter interactions. While the predictions of a BRT model can be evaluated against a battery of model performance metrics (see Fielding and Bell, 1997, for a thorough review), we do not use such metrics here—our approach is explanatory rather than predictive.

Uncertainty analysis outcomes   For three of the four state variables explored, the range over which hunting occurs, *h.rng*, has the highest relative influence (Figure 8.14). As the range over which hunting increases the maximum population size achieved, the time the population persists for and the 'success' of the settlement all increase (Figure 8.15). This is an intuitive dynamic—increasing the hunting range means that more of the landscape is explored and so patches of high-value resource are more likely to be found. The amount of resource gathered per head, *gat.ph*, is also an important parameter, with maximum population size, the median ratio of low-to-high-value resource used and the likelihood of a successful settlement all increasing with this parameter, again a somewhat intuitive outcome.

The maximum number of hunts conducted per month, *mx.hpm*, influences the maximum population size and the median ratio of low-to-high-value resource—in both cases negatively. This might seem counter-intuitive. Surely in terms of resource exploitation, the more hunting the better? On the other hand, there is no guarantee that any given hunting trip will be successful and effort expended in hunting is unavailable for local gathering activity. Thus, over-investing in hunting as a strategy may be a suboptimal investment of effort, and this is what the model suggests. In terms of establishing a population and then ensuring that it grows, the 'best' strategy in terms of resource exploitation may be one that balances the exploitation and use of low- and high-value resources, with a median ratio of 1:1 appearing to be optimal (although we note that the model is not calibrated in terms of effort per kJ and so this interpretation requires caution). This balance is probably

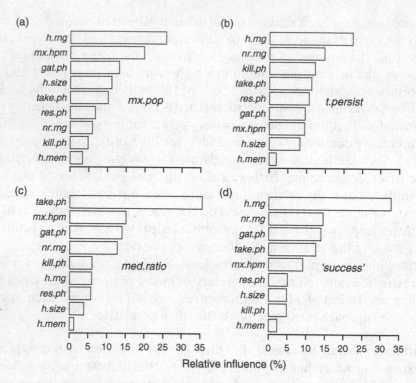

Figure 8.14 Relative influence (%) of the exploitation parameters on (a) *mx.pop*, (b) *t.persist*, (c) *med.ratio* and (d) proportion of successful colonisations where more than 50% of high-value resource was taken.

weighted in favour of the high-value resource more than might have been the case in many prehistoric settings.

Finally, the tree-based structure of the BRT models lets us explore potential interactions between parameter values. Such interactions are often very important in driving a model's behaviour, but the one-at-a-time approach we have adopted to this point, whether local, as in the sensitivity analysis, or global, ignores them. Table 8.2 shows that there are some significant and strong interactions between parameters, especially those that have the strongest independent (main) effect. In each case, however, parameters that in isolation appear to have little influence do, in tandem with some other parameter, feature in a signification interaction. For example, for 'success' the parameters *max.hpm* and *take.ph* have only moderate relative influences, but together feature in one of the strongest interactions.

In summary, the most successful exploitation strategy, in terms of achieving maximum population size and persistence, is one where there is a balance between hunting for the higher value resource and local foraging for the lower value resource. On the other hand, history suggests that this does *not* seem

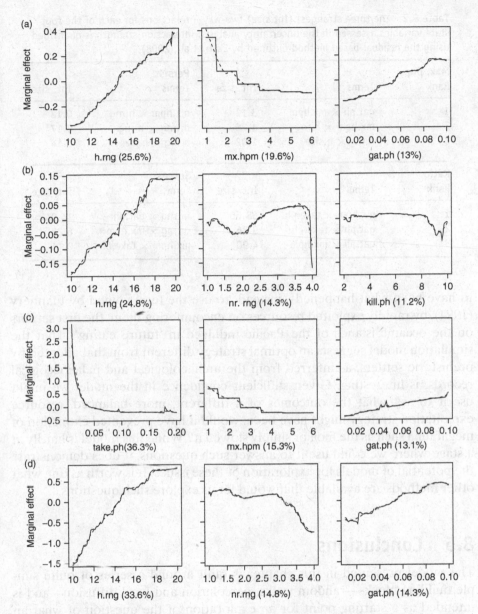

Figure 8.15 Partial dependence plots for the three most influential parameters for (a) *mx.pop*, (b) *t.persist*, (c) *med.low* and (d) proportion of successful colonisations. Partial dependence plots show the marginal effect of changes in the parameter holding all other parameters fixed. The solid line is the fit directly from the BRT model and the dashed line a smoothed form of that fit. Numbers in parentheses on the x-axis labels are the relative influence of that parameter (see Figure 8.14).

Table 8.2   The three strongest (Int.size) two-way interactions for each of the four state variables assessed in the uncertainty analysis; interaction strength is quantified using the residual-based method outlined by Elith et al. (2008)

Max. pop.			Persist	
Rank	Terms	Int. size	Terms	Int. size
1	gat.ph × mx.hpm	3.17	mx.hpm × h.rng	0.18
2	mx.hpm × h.rng	1.26	nr.rng × h.rng	0.17
3	h.rng × h.size	1.02	h.rng × h.size	0.14
Ratio			**Success**	
Rank	Terms	Int. size	Terms	Int. size
1	mx.hpm × take.ph	38.65	nr.rng × gat.ph	14.94
2	gat.ph × take.ph	9.36	nr.rng × mx.hpm	6.22
3	gat.ph × mx.hpm	4.90	mx.hpm × take.ph	0.14

to have been what happened—instead, to use the term coined by Flannery (1994), by rapidly exploited resources on encountering them, the first settlers on the oceanic islands of the Pacific indulged in 'future eating'. That the simulation model suggests an optimal strategy different from that adopted by prehistoric settlers, as inferred from the archaeological and palaeoecogical records, is interesting. Given sufficient confidence in the model we might use it to ask what the outcomes of a different, more balanced, resource exploitation strategy might have been: would it have prevented extinction of megafauna such as the moa or simply slowed it? While this model is hardly at a stage where we could use it to answer such questions, it does demonstrate the potential of models for exploration of these issues. It is worth asking what other methods are available that would let us explore such questions.

8.5   Conclusions

The model presented in this chapter is built around a series of quite simple building blocks—random walks, percolation and local diffusion—and is intended as a starting point for an exploration of the question of what an optimal foraging or resource acquisition strategy in a new environment might be. Despite the simplicity of the question and its building blocks, the model as a whole is structurally complex and produces a rich range of (plausible) dynamics in space and time.

As the analyses above show, evaluating a model of even moderate complexity is a difficult and technically demanding task. In seeking to understand any model a fundamental question is whether an appropriate balance between what is included and what is omitted has been achieved. Answering this question is not easy. Methods such as pattern-oriented modelling (Section 7.6) are potentially helpful in deciding how structurally complex models such as this one really need to be. POM emphasises the use of empirical data ('patterns') as filters to select between alternate model structures and parameterisations. We have only informally adopted a POM approach here by exploring the parameterisations under which different types of model dynamics arise. In doing this, we have largely side-stepped model structural uncertainty, but such uncertainty is likely to be high in the sort of exploratory framework we are using.

This model is representationally rather simplistic, and it is always worth considering how it might be refined. A key area where improvements could be made is in the representation of the resources. As it stands resources are represented by abstract numerical values representing the relative value or yield of these resources to the human groups, and are expressed loosely in terms of undefined units of energy. This avoids having to represent dietary selection in terms of calorific values and so forth, but comes at a cost. For example, in developing the model we had to spend considerable time in 'tweaking' the baseline parameterisation until the resource acquisition components of the model produced plausible dynamics. It might be possible to calibrate the energetic component of the model against the palaeoecological record and so make it more empirically grounded, but this would not be straightforward.

Representing food resources as energy units also makes mapping back to the fundamental units of the population (individuals) difficult, not least due to discretisation issues (you can have 1.5 units of energy but not 1.5 individuals!). On the other hand, representing individual animals would also lead to problems. The simplest way to directly represent populations might be to keep track of the number of individuals in each grid cell and assume no movement between cells. Such an approach could be made more detailed if it was deemed necessary to represent, for example, individuals of different ages separately from each other. Technically it is not difficult to see how this sort of model structure could be achieved, but clearly the model would become more complex, with an added demographic component, and costs would be incurred in uncertainty and interpretability.

A further step towards realism in representing the prey species would be to develop a disaggregated agent-based approach (see Section 2.2.6). Wainwright (2008) provides an example of how such a model might be structured in the context of trying to understand how prehistoric peoples and their activities, including animal husbandry, might have influenced landform change (for example via soil erosion). Despite its appeal, this type of disaggregated representation carries with it some difficult scaling issues. If we are representing individual organisms, presumably we want to represent their movements, otherwise a static model in which aggregated numbers are used would be sufficient. Representing movement brings to the fore difficult questions related to the spatio-temporal grain of the model. How far would an organism move in a month (the current model time step)? Already the human movement is handled in a generalised way, with only localised resource collection in an area around the home camp and longer range hunting. This approach does not imply that only these human movements would have occurred in such settings, rather that at the level of detail in this model these are the movements that matter. Similar decisions about the movement of the prey species and its relevance to the questions of interest would have to be addressed and would raise further difficult questions about chance encounters between humans and prey, along with our usual concerns about spatial and temporal scale, grain and extent. Representing the system in this way might actually be necessary in some contexts, the interactions of small populations being a situation where highly localised, individual-level heterogeneity can make important differences. As always, we must recognise that realism and detail come at a cost: our goal should be to build models structurally realistic enough for our purpose, but not more so.

9
In Conclusion

To conclude, we return to our three major themes: (i) the utility of a 'building-block' approach to model development, (ii) the challenges of inferring process from pattern in simulation models and, more generally, (iii) the need for careful evaluation of spatial simulation models.

9.1 On the usefulness of building-block models

One of the key ideas of this book is that defining and understanding some basic building-block spatial models can assist in the development of more complex models. We believe that when setting out to implement a potentially complex model, rather than starting with the specifics of the system of interest, it makes sense to begin by thinking about the system in a more general way. In identifying the building blocks we use we have been motivated by identifying models that reproduce general classes of pattern rather than by a desire to capture in detail the physical processes that underpin them. This is not to argue that mechanistically motivated spatial modelling is unimportant, but rather that starting with a more phenomenological approach is (at least pragmatically) easier. In any case as our phenomenological building blocks become more complex the distinction between phenomenological and mechanistic models becomes blurred.

In our view the strength of using building blocks is that their behaviour and dynamics are fairly well understood, so that their use facilitates building more complex models by grounding model development in an extensive body of theory. The use of well-understood components should, in turn, mean that more informed, less *ad hoc* decisions about model structure can be made. We are drawing here on a fundamental assumption of the systems

Spatial Simulation: Exploring Pattern and Process, First Edition.
David O'Sullivan and George L.W. Perry.
© 2013 John Wiley & Sons, Ltd. Published 2013 by John Wiley & Sons, Ltd.

dynamics approach: that the behaviour of the components in isolation is better understood than is the system as a whole.

We do not claim to have presented a comprehensive typology or encyclopaedia of spatial processes or models. The material presented in Chapters 3 to 5 should also have made it clear that the building-block models are not independent of each other. There are, for example, deep and important links between the interacting particle systems of Chapter 3, the random walks of Chapter 4 and the ordinary percolation models of Chapter 5. We have also committed some 'sins of omission', two of which warrant a brief comment. First, reaction–diffusion systems are a fundamental way of exploring pattern formation in complex systems, and have a long and rich history in this regard, but one that we touch on only briefly (Section 3.7.1). Their development and analysis, however, requires a level of mathematical comfort that we considered somewhat challenging given the goals of this book—we encourage interested readers to learn more about them, nevertheless. Second, we do not cover networks in any great depth. As noted in Section 6.3.3, networks are a general data structure rather than a spatial process *per se*. The lattice spaces we generally focus on here can be seen as a specific (densely but only locally connected) form of network. How processes related to random walks and percolation behave on networks has been the subject of considerable recent interest and attention. Understanding how these processes behave on a lattice provides a sense of how they will behave in other spatial contexts. Again we encourage the interested reader to pursue this large and growing literature more closely.

9.2   On pattern and process

Understanding the relationship between data and theory has been a persistent theme in the philosophy of science in Western thought during the last four or five centuries (Stanford, 2009). The general question is, 'Given the data we have available to us, what position should we adopt with respect to some theory?' This is a difficult question because the data available to us are often consistent with more than one theory or hypothesis. In other words, the theories or hypotheses that we evaluate may be empirically equivalent or equifinal, making it difficult (if not impossible) to select between them on the basis of data alone. Rosindell et al. (2012, page 204) use the analogy of children mixing paint to reinforce these points. We can be sure that if children mix together paints of different colours they will end up with a (generally unappealing) brownish colour. But that colour can be achieved in many different ways: different combinations of colours mixed in different amounts all result in more or less the same hue. There is no way to work out how a particular brown colour (the pattern) emerged without careful observation of what each child actually did (the process).

Under-determination, equifinality and the relationships between process and pattern pose challenges across the disciplines that make up the social and environmental sciences (see, for example, Gifford-Gonzalez, 1991, Beven, 2002, O'Sullivan, 2004, Vellend, 2010, among many others). We are concerned here with the implications of equifinality for the inferences we make using models (Beven, 2002). It is just about always the case that we want to use the kinds of models considered in this book to make statements about how a system arrived at its current state. This need arises because it is almost always easier to collect data describing snapshot patterns than it is to monitor long-term processes. Equifinality matters because it implies that the capacity of a model to reproduce some pattern does *not* mean that it is 'correct' (let alone 'true')—there will be other candidate models that reproduce the pattern just as well. Even if we settle on a single model structure, different parameterisations can yield the same outcomes. Recall, for example, in Section 4.2.4 the difficulties of discriminating between random walks with drift and correlated random walks, and Lévy flights from multi-scale walk processes. In short, as Millington et al. (2011) argue, mimetic accuracy does not guarantee structural realism, and these concerns are potentially even more acute for the phenomenologically motivated models considered here. While these issues are fundamental, patterns can, nevertheless, be used to inform our understanding of environmental and social systems (McIntire and Fajardo, 2009). Furthermore, spatial simulation models of the type we present here can be used to explore how effectively pattern can 'reveal the qualitative or quantitative nature of underlying generative processes' (Cale et al., 1989, page 501).

One more model (the last one!) reemphasises these points. Cale et al. (1989) describe a simple model of competition between two plant species. The model runs on a two-dimensional lattice in discrete time. Each species is defined by two parameters describing competition (c_1 and c_2) and reproduction (r_1 and r_2), both in the range zero to one. Each generation, individuals of each species s survive (reproduce) with probability r_s and they remain in place, otherwise the cell they occupy becomes empty. Species compete for empty cells in random order, with each species occupying each empty cell with probability c_s.

It is apparent that the same outcomes may occur under many parameterisations of this model. Figure 9.1(a) and (b) show that the ratios of the species abundances can be similar even under very different values of c_a/c_b and r_a/r_b. This is true even for c_a/c_b despite its strong effect on the species abundances. Thus, even in this simple system, inferring processes (the reproduction and competition dynamics) from final conditions (the relative abundance of the species) is almost impossible. As Cale et al. (1989) point out, we cannot even reliably discriminate cases where the species are the same or neutral with respect to each other from those where they are different. As Figure 9.1(c) shows, relative abundances statistically no different from a null model of equal proportions occur across a wide range of the parameter space. Such problems will be even more acute in systems where multiple processes, whose importance varies in time and space, are at play. We can, however, use models

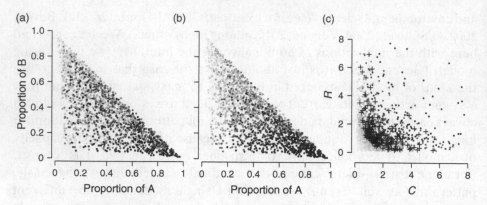

Figure 9.1   Uncertainty analysis of the model described by Cale et al. (1989) with c_a, c_b, r_a and r_b all independently drawn from U ∼ [0.1, 0.9]. Plots of proportional abundance of species A and B with grey-scaling denoting (a) $C = c_a/c_b$, (b) $R = r_a/r_b$, with data divided into ten percentile groups, and (c) a plot of R vs C, with shading showing abundance of species A (light to dark) and the crosses outcomes statistically indistinguishable from an equal proportion null model. The model was run on a lattice of 100×100 cells for 200 generations (sufficient for dynamics to stabilise).

of this sort to assess how strongly linked are pattern and process, which is in itself a useful outcome.

9.3   On the need for careful analysis

The difficulties of inferring process from pattern highlight the need for careful model analysis and evaluation. Despite the issues raised in the previous section, we are *not* rejecting the importance of the use of data and patterns to test and evaluate models via some form of model-data confrontation. Rather, we are arguing that the ability of a model to match some target pattern is a weak test and should be undertaken alongside other evaluation methods (as introduced in Chapter 7). The patterns produced by dynamic spatial models can be beguiling, and it is easy to be convinced that reproducing some facet of the real world 'proves' your model—it is important, however, not to be blinded by mimetic accuracy! Recognising this issue sets in train what Railsback and Grimm (2012, page 61ff) term 'moving from animation to science'.

Nevertheless, patterns remain important guides or filters in model design and evaluation. We have discussed pattern-oriented modelling (POM) at various stages and used it informally to guide model design and evaluation in Chapter 8. For us, the strength of the POM approach is the use of *multiple* patterns to guide decisions about model design and to find levels of model complexity sufficient to answer the questions of interest but not more so. While POM can be used in a predictive context it is also well suited to heuristically motivated modelling, especially when employed in a framework where different model structures are considered as competing multiple working hypotheses (*sensu* Chamberlin, 1897). Others have developed approaches that support the effective use of pattern to unravel underlying processes. For example, McIntire and Fajardo (2009) consider that the inference-making process requires the explicit statement of *a priori* hypotheses, grounded in appropriate theory, followed by careful analysis. Developing testable upfront hypotheses about how a system (or model) might behave is important as it should limit the development of *ad hoc* (or retroductive) explanations of the origins of some pattern of interest. Because we understand what to expect from building-block models—a correlated random walk or a contact process, for example—their use is effectively an explicit hypothesis test, and this is another benefit of our approach.

Finally, the careful application of modern statistical methods (see Hartig et al., 2011) is an important component of model evaluation, as we have seen in Chapters 7 and 8. However, as it has become more obvious that models are valuable tools for learning, so a wider range of tools are being developed to support model evaluation and communication. For example, there has been considerable interest in the use of models to develop and explore alternative futures in the forms of scenarios (Coreau et al., 2009, Thompson et al., 2012), and in this setting a model's ability to reproduce a specific pattern may be less important than its capacity to integrate and synthesise data and facilitate communication. In such settings rather different metrics are needed to evaluate a model's 'usefulness', even if some form of model-data confrontation remains helpful. New methods to communicate the rich dynamics of complex disaggregated simulation models have begun to appear, for example Millington et al. (2012) highlight the use of narratives (supported by careful quantitative analysis) as tools to improve understanding and communication of models.

A move away from 'history matching' (Beck, 2006) towards more qualitative approaches may seem to encourage an 'anything goes' mentality in model evaluation. After all, at a rather trivial level any given model can be described as 'useful' in some way. We believe, however, that seeing models as tools with which to make evidential statements and to evaluate hypotheses (Parker, 2008), and assessing them in that light, is *more* rigorous than simply quantifying their imitative capacity. As Parker (2008, page 381) argues, this view means that 'Instead of always focusing on the question of whether simulation results are true of a real-world target system, we can shift attention to the question of which range of hypotheses about that target system can be rejected or accepted in light of the production of those results by that model.' As we have argued from the outset, models are tools for learning about the real world. A model's capacity to generate testable hypotheses and to aid in their evaluation is, therefore, surely its most fundamental test.

References

Square bracketed number(s) at end of each reference denotes the page(s) on which it is cited.

Adler, L. and Lev, U. (2003) Bootstrap percolation: visualizations and applications. *Brazilian Journal of Physics* **33**, 641–644. [138]

Adler, P.M. and Thovert, J.F. (1999) *Fractures and Fracture Networks* Theory and Applications of Transport in Porous Media. Kluwer Academic. [87]

Akaike, H. (1973) Information theory and an extension of the maximum likelihood principle. In *2nd International Symposium on Information Theory* (ed. Petrov, B.N. and Csaki, F.), pp. 267–281. Akademiai Kiado. [220]

Akaike, H. (1974) A new look at the statistical model identification. *IEEE Transactions on Automatic Control* **19**, 716–723. [220]

Albert, R., Jeong, H. and Barabási, A-L. (1999) Diameter of the World-Wide Web. *Nature* **401**, 130–131. [183]

Alexandrowicz, Z. (1980) Critically branched chains and percolation clusters. *Physics Letters A* **80**, 284–286. [149, 152, 153]

Allen, P.M. (1997) *Cities and Regions as Self-organizing Systems: Models of Complexity*. Gordon Breach Science Publishers. [22]

Anderson, A.J. (2002) Faunal collapse, landscape change and settlement history in Remote Oceania. *World Archaeology* **33**, 375–390. [230]

Anderson, D.R. (2008) *Model-based Inference in the Life Sciences: A Primer on Evidence*. Springer. [217, 220, 222]

Anderson, N.J., Bugmann, H., Dearing, J.A. and Gaillard, M-J. (2006) Linking palaeoenvironmental data and models to understand the past and to predict the future. *Trends in Ecology and Evolution* **21**, 696–704. [231]

Andersen, A.N., Cook, G.D. and Williams, R.J. (eds) (2003) *Burning Issues in Savanna Ecology and Management*. Springer. [17]

Anscombe, F.J. (1973) Graphs in statistical analysis. *The American Statistician* **27**, 17–21. [212, 215]

Arthur, W.B. (1994) *Increasing Returns and Path Dependence in the Economy* Economics, Cognition, and Society. University of Michigan Press. [157]

Ascough, II, J.C., Maier, H.R., Ravalico, J.K. and Strudley, M.W. (2008) Future research challenges for incorporation of uncertainty in environmental and ecological decision-making. *Ecological Modelling* **219**, 383–399. [194]

Spatial Simulation: Exploring Pattern and Process, First Edition.
David O'Sullivan and George L.W. Perry.
© 2013 John Wiley & Sons, Ltd. Published 2013 by John Wiley & Sons, Ltd.

Axelrod, R. (1984) *The Evolution of Cooperation*. Basic Books. [93]

Axtell, R., Axelrod, R., Epstein, J.M. and Cohen, M.D. (1996) Aligning simulation models: a case study and results. *Computational & Mathematical Organization Theory* **1**, 123–141. [232]

Baas, A.C.W. (2002) Chaos, fractals and self-organization in coastal geomorphology: simulating dune landscapes in vegetated environments. *Geomorphology* **48**, 309–328. [91]

Bak, P., Chen, K. and Tang, C. (1990) A forest-fire model and some thoughts on turbulence. *Physics Letters A* **147**, 297–300. [162]

Baker, A. (2011) Simplicity. In *The Stanford Encyclopedia of Philosophy* (ed. Zalta, E.N.) Summer 2011 edn. [216]

Ball, P. (2009) *Nature's Patterns: A Tapestry in Three Parts*. Oxford University Press. [xxi, 27, 30, 91]

Banerjee, S., Carlin, B.P., and Gelfand, A.E. (2003) *Hierarchical Modeling and Analysis for Spatial Data* C&H/CRC Monographs on Statistics & Applied Probability. Chapman and Hall/CRC. [30]

Barabási, A-L. and Stanley, H.E. (1995) *Fractal Concepts in Surface Growth*. Cambridge University Press. [151]

Bard, J.B.L. (1981) A model for generating aspects of zebra and other mammalian coat patterns. *Journal of Theoretical Biology* **93**, 363–385. [90]

Barnes, T.J. (2004) A paper related to everything but more related to local things. *Annals of the Association of American Geographers* **94**, 278–283. [57]

Bart, J. (1995) Acceptance criteria for using individual-based models to make management decisions. *Ecological Applications* **5**, 411–420. [194]

Barthélemy, M. (2011) Spatial networks. *Physics Reports* **499**, 1–101. [168, 183]

Barthélemy, M. and Flammini, A. (2008) Modeling urban street patterns. *Physical Review Letters* **100**, 138702. [87]

Bartumeus, F., da Luz, M.G.E., Viswanathan, G.M. and Catalan, J. (2005) Animal search strategies: a quantitative random walk analysis. *Ecology* **86**, 3078–3087. [105]

Batschelet, E. (1981) *Circular Statistics in Biology*. Academic Press. [105]

Batty, M. (2005) *Cities and Complexity: Understanding Cities with Cellular Automata, Agent-Based Models, and Fractals*. MIT Press. [22, 64, 95, 129, 135]

Batty, M. and Longley, P. (1994) *Fractal Cities: A Geometry of Form and Function*. Academic Press. [34, 157]

Batty, M. and Xie, Y. (1996) Preliminary evidence for a theory of the fractal city. *Environment and Planning A* **28**, 1745–1762. [94]

Batty, M. and Xie, Y. (1997) Possible urban automata. *Environment and Planning B: Planning and Design* **24**, 175–192. [94]

Batty, M. and Xie, Y. (1999) Self-organized criticality and urban development. *Discrete Dynamics in Nature and Society* **3**, 109–124. [94]

Batty, M., Couclelis, H. and Eichen, M. (1997) Urban systems as cellular automata. *Environment and Planning B: Planning and Design* **24**, 159–164. [94]

Batty, M., Longley, P. and Fotheringham, S. (1989) Urban growth and form: scaling, fractal geometry, and diffusion-limited aggregation. *Environment and Planning A* **21**, 1447–1472. [155, 157]

Bauer, S., Wyszomirski, T., Berger, U., Hildenbrandt, H. and Grimm, V. (2004) Asymmetric competition as a natural outcome of neighbour interactions among plants: results from the field-of-neighbourhood modelling approach. *Plant Ecology* **170**, 135–145. [182]

Beck, B. (2006) Model evaluation and performance. In *Encyclopedia of Environmetrics* (ed. El-Shaarawi, A.H. and Piegorsch, W.W.) John Wiley & Sons, Ltd. [212, 215, 226, 228, 269]

Beck, M.B. (1987) Water quality modeling: a review of the analysis of uncertainty. *Water Resources Research* **23**, 1393–1442. [215]

Beckage, B., Gross, L.J. and Kauffman, S. (2011) The limits to prediction in ecological systems. *Ecosphere* **2**, 125 doi:10.1890/ES11-00211.1. [22]

Begon, M., Mortimer, M. and Thompson, D.J. (1996) *Population Ecology: A Unified Study of Animals and Plants*, 3rd edn. Blackwell Science. [70]

Bella, I.E. (1971) A new competition model for individual trees. *Forest Science* **17**, 364–372. [182]

ben-Avraham, D. and Havlin, S. (2000) *Diffusion and Reactions in Fractals and Disordered Systems*. Cambridge University Press. [110, 138, 141, 142, 143, 157]

Benenson, I. and Torrens, P. (2004) *Geosimulation: Automata-Based Modelling of Urban Phenomena*. John Wiley & Sons. [169]

Benguigui, L. (1995) A new aggregation model. Application to town growth. *Physica A: Statistical Mechanics and its Applications* **219**, 13–26. [152]

Benguigui, L. (1998) Aggregation models for town growth. *Philosophical Magazine Part B* **77**, 1269–1275. [152]

Benhamou, S. (1994) Spatial memory and searching efficiency. *Animal Behaviour* **47**, 1423–1433. [116, 120]

Benhamou, S. (2004) On the expected net displacement of animals' random movements. *Ecological Modelling* **171**, 207–208. [107]

Benhamou, S. (2007) How many animals really do the Lévy walk? *Ecology* **88**, 1962–1969. [119]

Berg, H.C. (1993) *Random Walks in Biology*. Princeton University Press. [99]

Berger, U., Piou, C., Schiffers, K. and Grimm, V. (2008) Competition among plants: concepts, individual-based modelling approaches, and a proposal for a future research strategy. *Perspectives in Plant Ecology, Evolution and Systematics* **9**, 121–135. [182]

Berkowitz, B. and Balberg, I. (1993) Percolation theory and its application to groundwater hydrology. *Water Resources Research* **29**, 775–794. [168]

Berkowitz, B. and Ewing, R.P. (1998) Percolation theory and network modeling applications in soil physics. *Surveys in Geophysics* **19**, 23–72. [138]

Berlekamp, E.R., Conway, J.H. and Guy, R.K. (2004) *Winning Ways for Your Mathematical Plays*, 2nd edn. A. K. Peters Ltd. [21, 70]

Bersini, H. and Detours, V. (1994) Asynchrony induces stability in cellular automata based models. In *Proceedings of the 4th International Workshop on the Synthesis and Simulation of Living Systems Artificial Life IV* (ed. Brooks, R.A. and Macs, P.) MIT Press–pp. 382–387. [171]

Beven, K. (2002) Towards a coherent philosophy for modelling the environment. *Proceedings of the Royal Society of London. Series A* **458**, 2465–2484. [52, 227, 228, 267]

Beven, K. (2006) A manifesto for the equifinality hypothesis. *Journal of Hydrology* **320**, 18–36. [222]

Beven, K. and Freer, J. (2001) Equifinality, data assimilation, and uncertainty estimation in mechanistic modelling of complex environmental systems using the GLUE methodology. *Journal of Hydrology* **249**, 11–29. [222]

Bian, L. (2004) A conceptual framework for an individual-based spatially explicit epidemiological model. *Environment and Planning B: Planning and Design* **31**, 381–395. [135]

Binmore, K.G. (1994) *Game Theory and the Social Contract: Playing Fair*. MIT Press. [93]

Blöschl, G. and Sivapalan, M. (1995) Scale issues in hydrological modelling: a review. *Hydrological Processes* **9**, 251–290. [34, 36]

Blythe, R.A. and McKane, A.J. (2007) Stochastic models of evolution in genetics, ecology and linguistics. *Journal of Statistical Mechanics: Theory and Experiment*, P07018. [82]

Bolker, B.M. (2008) *Ecological Models and Data in R*. Princeton University Press. [38, 39, 195, 215, 220]

Bolker, B.M., Pacala, S.W. and Neuhauser, C. (2003) Spatial dynamics in model plant communities: what do we really know? *American Naturalist* **162**, 135–148. [74]

Bonnell, T.R., Sengupta, R.R., Chapman, C.A. and Goldberg, T.L. (2010) An agent-based model of red colobus resources and disease dynamics implicates key resource sites as hot spots of disease transmission. *Ecological Modelling* **221**, 2491–2500. [127]

Boots, B. and Shiode, N. (2003) Recursive Voronoi diagrams. *Environment and Planning B: Planning and Design* **30**, 113–124. [88]

Börger, L., Dalziel, B.D. and Fryxell, J.M. (2008) Are there general mechanisms of animal home range behaviour? A review and prospects for future research. *Ecology Letters* **11**, 637–650. [129]

Boswell, G.P., Britton, N.F. and Franks, N.R. (1998) Habitat fragmentation, percolation theory and the conservation of a keystone species. *Proceedings of the Royal Society of London. Series B* **265**, 1921–1925. [159]

Botsford, L.W., Castilla, J.C. and Peterson, C.H. (1997) The management of fisheries and marine ecosystems. *Science* **277**, 509–515. [198]

Bovet, P. and Benhamou, S. (1988) Spatial analysis of animals' movements using a correlated random walk model. *Journal of Theoretical Biology* **131**, 419–433. [105]

Box, G.E.P. (1979) Some problems of statistics and everyday life. *Journal of the American Statistical Association* **74**, 1–4. [4, 227]

Boyer, D. and Walsh, P.D. (2010) Modelling the mobility of living organisms in heterogeneous landscapes: does memory improve foraging success? *Philosophical Transactions of the Royal Society of London. Series A* **368**, 5645–5659. [120, 126, 127, 129]

Boyer, D., Crofoot, M.C. and Walsh, P.D. (2012) Non-random walks in monkeys and humans. *Journal of the Royal Society Interface* **9**, 842–847. [108, 128, 236]

Boyer, D., Ramos-Fernández, G., Miramontes, O., Mateos, J.L., Cocho, G., Larralde, H., Ramos, H. and Rojas, F. (2006) Scale-free foraging by primates emerges from their interaction with a complex environment. *Proceedings of the Royal Society of London. Series B* **273**, 1743–1750. [116, 117, 127]

Brewer, S., Jackson, S.T. and Williams, J.W. (2012) Paleoecoinformatics: applying geohistorical data to ecological questions. *Trends in Ecology & Evolution* **27**, 104–112. [231]

Briggs, D., Sabel, C.E. and Lee, K. (2009) Uncertainty in epidemiology and health risk and impact assessment. *Environmental Geochemistry and Health* **31**, 189–203. [193]

Broadbent, S.R. and Hammersley, J.M. (1957) Percolation processes I. Crystals and mazes. *Mathematical Proceedings of the Cambridge Philosophical Society* **53**, 629–641. [137]

Brockmann, D., Hufnagel, L. and Geisel, T. (2006) The scaling laws of human travel. *Nature* **439**, 462–465. [110, 129]

Brook, B.W. and Bowman, D.M.J.S. (2002) Explaining the Pleistocene megafaunal extinctions: models, chronologies, and assumptions. *Proceedings of the National Academy of Sciences of the United States of America* **99**, 14624–14627. [230]

Brown, D.G., Page, S., Riolo, R., Zellner, M. and Rand, W. (2005) Path dependence and the validation of agent-based spatial models of land use. *International Journal of Geographical Information Science* **19**, 153–174. [95]

Brown, J.H., Gupta, V.K., Li, B-L., Milne, B.T., Restrepo, C. and West, G.B. (2002) The fractal nature of nature: power laws, ecological complexity and biodiversity. *Philosophical Transactions of the Royal Society of London. Series B* **357**, 619–626. [34]

Bruch, E.E. and Mare, R.D. (2006) Neighborhood choice and neighborhood change. *American Journal of Sociology* **112**, 667–709. [85]

Buchanan, M. (2000) *Ubiquity: The Science of History … or Why The World is Simpler Than We Think*. Weidenfeld and Nicholson. [21]

Bunde, A., Herrmann, H.J., Margolina, A. and Stanley, H.E. (1985) Universality classes for spreading phenomena: a new model with fixed static but continuously tunable kinetic exponents. *Physical Review Letters* **55**, 653–656. [149, 152, 153]

Burks, A.W. (1970) Von Neumann's self-reproducing automata. In *Essays on Cellular Automata* (ed. Burks, A.W.) University of Illinois Press pp. 3–64. [18]

Burmaster, D.E. and Anderson, P.D. (1994) Principles of good practice for the use of Monte Carlo techniques in human health and ecological risk assessments. *Risk Analysis* **14**, 477–481. [203]

Burnham, K.P. and Anderson, D.R. (2002) *Model Selection and Multimodel Inference: A Practical Information-Theoretic Approach*, 2nd edn. Springer. [205, 220, 221, 222]

Burnham, K.P. and Anderson, D.R. (2004) Multimodel inference: understanding AIC and BIC in model selection. *Sociological Methods and Research* **33**, 261–304. [221]

Burnham, K.P., Anderson, D.R. and Huyvaert, K. (2011) AIC model selection and multimodel inference in behavioral ecology: some background, observations, and comparisons. *Behavioral Ecology and Sociobiology* **65**, 23–35. [221, 222]

Buss, L.W. and Jackson, J.B.C. (1979) Competitive networks: nontransitive competitive relationships in cryptic coral reef environments. *American Naturalist* **113**, 223–234. [76]

Buttell, L., Cox, J.T. and Durrett, R. (1993) Estimating the critical values of stochastic growth models. *Journal of Applied Probability* **30**, 455–461. [72]

Byrne, D. (1998) *Complexity Theory and The Social Sciences: An Introduction*. Routledge. [22]

Cai, A., Landman, K. and Hughes, B. (2006) Modelling directional guidance and motility regulation in cell migration. *Bulletin of Mathematical Biology* **68**, 25–52. [109]

Cale, W.G., Henebry, G.M. and Yeakley, J.A. (1989) Inferring process from pattern in natural communities. *BioScience* **39**, 600–605. [267, 268]

Callender, C. and Cohen, J. (2006) There is no special problem about scientific representation. *Theoria* **55**, 7–25. [217, 226]

Cantrell, R.S. and Cosner, C. (2003) *Spatial Ecology via Reaction-Diffusion Equations*. John Wiley & Sons. [136]

Carlson, J.M. and Doyle, J. (1999) Highly optimized tolerance: a mechanism for power laws in designed systems. *Physical Review E* **60**, 1412–1427. [164]

Carlson, J.M. and Doyle, J. (2002) Complexity and robustness. *Proceedings of the National Academy of Sciences of the United States of America* **99**, 2538–2545. [164]

Carpenter, S.R. (1996) Microcosm experiments have limited relevance for community and ecosystem ecology. *Ecology* **77**, 677–680. [17]

Castella, J.C., Trung, T.N. and Boissau, S. (2005) Participatory simulation of land-use changes in the northern mountains of Vietnam: the combined use of an agent-based model, a role-playing game, and a geographic information system. *Ecology and Society* **10**, 27. [199, 228]

Castellano, C., Fortunato, S. and Loreto, V. (2009) Statistical physics of social dynamics. *Reviews of Modern Physics* **81**, 591–646. [80, 138]

Casti, J.L. (1997) *Would-be Worlds: How Simulation is Changing the Frontiers of Science*. John Wiley & Sons. [15]

Caswell, H. (1976) Community structure: a neutral model analysis. *Ecological Monographs* **46**, 327–354. [158]

Chamberlin, T.C. (1897) The method of multiple working hypotheses. *Journal of Geology* **5**, 837–848. [220, 269]

Chatfield, C. (1985) The initial examination of data. *Journal of the Royal Statistical Society. Series A.* **148**, 214–253. [212]

Chatfield, C. (1995) Model uncertainty, data mining and statistical inference. *Journal of the Royal Statistical Society. Series A.* **158**, 419–466. [205]

Chatfield, C. (2001) Model uncertainty. In *Encyclopedia of Environmetrics* (ed. El-Shaarawi, A.H. and Piegorsch, W.W.) John Wiley & Sons, Ltd. [205]

Chave, J. (1999) Study of structural, successional and spatial patterns in tropical rain forests using TROLL, a spatially explicit forest model. *Ecological Modelling* **124**, 233–254. [185]

Chesson, P. (2000) Mechanisms of maintenance of species diversity, *Annual Review of Ecology and Systematics* **31**, 343–366. [xxi]

Chopard, B. and Droz, M. (1998) *Cellular Automata Modeling of Physical Systems*. Cambridge University Press. [18]

Chopard, B., Luthi, P. and Droz, M. (1994) Reaction-diffusion cellular automata model for the formation of Leisegang patterns. *Physical Review Letters* **72**, 1384–1387. [91]

Christaller, W. (1966) *Central Places in Southern Germany*. Prentice-Hall. Translation by C. W. Baskin. [48]

Christensen, K. and Moloney, N.R. (2005) *Complexity and Criticality*. Imperial College Press. [137, 138, 140, 141, 142]

Cilliers, P. (1998) *Complexity and Postmodernism: Understanding Complex Systems*. Routledge. [22]

Clark, J.S. (1988) Why trees migrate so fast: confronting theory with dispersal biology and the paleorecord. *American Naturalist* **152**, 204–224. [110]

Clark, J.S. (2009) Beyond neutral science. *Trends in Ecology & Evolution* **24**, 8–15. [82]

Clark, J.S. (2012) The coherence problem with the Unified Neutral Theory of Biodiversity. *Trends in Ecology & Evolution* **27**, 198–202. [82]

Clark, J.S., Carpenter, S.R., Barber, M., Collins, S., Dobson, A., Foley, J.A., Lodge, D.M., Pascual, M., Pielke, R., Pizer, W., Pringle, C., Reid, W.V., Rose, K.A., Sala, O., Schlesinger, W.H., Wall, D.H. and Wear, D. (2001) Ecological forecasts: an emerging imperative. *Science* **293**, 657–660. [194]

Clark, J.S., Fastie, C., Hurtt, G., Jackson, S.T., Johnson, C., King, G.A., Lewis, M., Lynch, J., Pacala, S., Prentice, C., Schupp, E.W., Webb III, T. and Wyckoff, P. (1998) Reid's paradox of rapid plant migration: dispersal theory and interpretation of paleoecological records. *BioScience* **48**, 13–24. [110, 136, 155]

Clark, P.J. and Evans, F.C. (1954) Distance to nearest neighbor as a measure of spatial relationships in populations. *Ecology* **35**, 445–453. [43]

Clark, W.A.V. (2006) Ethnic preferences and residential segregation: a commentary on outcomes from agent-based modeling. *Journal of Mathematical Sociology* **30**, 319–326. [85]

Clark, W.A.V. and Fossett, M. (2008) Understanding the social context of the Schelling segregation model. *Proceedings of the National Academy of Sciences of the United States of America* **105**, 4109–4114. [85]

Clarke, K.C. and Gaydos, L.J. (1998) Loose-coupling a cellular automaton model and GIS: long-term urban growth prediction for San Francisco and Washington/Baltimore. *International Journal of Geographical Information Systems* **12**, 699–714. [94]

Clarke, K.C. and Olsen, G. (1996) Refining a cellular automata model of wildfire extinction and propagation. In *GIS and Environmental Modeling: Progress and Research Issues*. (ed. Goodchild, M.F., Steyaert, L.T., Parks, B.O., Johnston, C.A., Maidment, D., Crane, M. and Glendinning, S.) GIS World Books pp. 333–338. [155]

Clarke, K.C., Hoppen, S. and Gaydos, L.J. (1997) A self-modifying cellular automaton model of historical urbanization in the San Francisco Bay area. *Environment and Planning B: Planning & Design* **24**, 247–261. [94]

Clauset, A., Shalizi, C.R. and Newman, M.E.J. (2009) Power-law distributions in empirical data. *SIAM Review* **51**, 661–703. [26]

Clifford, P. and Sudbury, A. (1973) A model for spatial conflict. *Biometrika* **60**, 581–588. [78]

Clover, C. (2004) *The End of the Line: How Overfishing Is Changing the World and What We Eat*. Ebury. [14]

Coco, G., Huntley, D.A. and O'Hare, T.J. (2000) Investigation of a self-organization model for beach cusp formation and development. *Journal of Geophysical Research* **105**, 21991–22002. [91]

Codling, E.A., Plank, M.J. and Benhamou, S. (2008) Random walk models in biology. *Journal of the Royal Society Interface* **5**, 813–834. [99]

Cohen, J. (1960) A coefficient of agreement for nominal scales. *Educational and Psychological Measurement* **20**, 37–46. [214]

Condit, R., Ashton, P.S., Baker, P., Bunyavejchewin, S., Gunatilleke, S., Gunatilleke, N., Hubbell, S.P., Foster, R.B., Itoh, A., LaFrankie, J.V., Seng Lee, H., Losos, E., Manokaran, N., Sukumar, R. and Yamakura, T. (2000) Spatial patterns in the distribution of tropical tree species. *Science* **288**, 1414–1418. [39]

Congalton, R.G. and Green, K. (1993) A practical look at the sources of confusion in error matrix generation. *Photogrammetric Engineering and Remote Sensing* **59**, 641–644. [207, 213]

Conlisk, J. (1996) Why bounded rationality? *Journal of Economic Literature* **34**, 669–700. [112]

Conradt, L. (2012) Models in animal collective decision-making: information uncertainty and conflicting preferences. *Interface Focus* **2**, 226–240. [128]

Cook, M. (2004) Universality in elementary cellular automata. *Complex Systems* **15**, 1–40. [21]

Coreau, A., Pinay, G., Thompson, J.D., Cheptou, PO. and Mermet, L. (2009) The rise of research on futures in ecology: rebalancing scenarios and predictions. *Ecology Letters* **12**, 1277–1286. [13, 269]

Costopoulos, A. (2001) Evaluating the impact of increasing memory on agent behaviour: adaptive patterns in an agent based simulation of subsistence. *Journal of Artificial Societies and Social Simulation* **4**, http://jasss.soc.surrey.ac.uk/4/4/7.html. [128]

Costopoulos, A. and Lake, M. (eds) (2010) *Simulating Change: Archaeology into the Twenty-First Century*. University of Utah Press. [51]

Couclelis, H. (1985) Cellular worlds: a framework for modelling micro-macro dynamics. *Environment and Planning A* **17**, 585–596. [93, 94]

Couclelis, H. (1988) Of mice and men: what rodent populations can teach us about complex spatial dynamics. *Environment and Planning A* **20**, 99–109. [93]

Couclelis, H. (1992) People manipulate objects (but cultivate fields): beyond the raster-vector debate in GIS. In *Theories and Methods of Spatio-Temporal Reasoning in Geographic Space* (ed. Frank, A.U., Campari, I. and Formentini, U.) vol. 639 of *Lecture Notes in Computer Science* Springer-Verlag pp. 65–77. [175]

Couclelis, H. (1997) From cellular automata to urban models: new principles for model development and implementation. *Environment and Planning B: Planning and Design* **24**, 165–174. [95, 184]

Courtat, T., Gloaguen, C. and Douady, S. (2011) Mathematics and morphogenesis of cities: a geometrical approach. *Physical Review E* **83**, 036106. [87]

Cova, T.J. and Goodchild, M.F. (2002) Extending geographical representation to include fields of spatial objects. *International Journal of Geographical Information Science* **16**, 509–532. [190]

Coveney, P. and Highfield, R. (1995) *Frontiers of Complexity: The Search for Order in a Complex World*. Ballantine Books. [21]

Cox, J.T. and Griffeath, D. (1986) Diffusive clustering in the two dimensional voter model. *The Annals of Probability* **14**, 347–370. [80]

Crawley, M.J. and May, R.M. (1987) Population dynamics and plant community structure: competition between annuals and perrenials. *Journal of Theoretical Biology* **125**, 475–489. [74]

Czirók, A. and Vicsek, T. (2000) Collective behavior of interacting self-propelled particles. *Physica A: Statistical Mechanics and its Applications* **281**, 17–29. [124]

Davis, B. (1990a) Reinforced random walk. *Probability Theory and Related Fields* **84**, 203–229. [108, 129]

Davis, M. (1990b) *City of Quartz: Excavating the Future in Los Angeles*. Verso. [36]

de Gennes, P.G. (1976) La percolation: un concept unificateur. *La Recherche* **7**, 919–927. [142, 161]

de la Torre, A.C. and Mártin, H.O. (1997) A survey of cellular automata like the "game of life". *Physica A: Statistical Mechanics and its Applications* **240**, 560–570. [70]

Delcourt, H.R., Delcourt, P.A. and Webb, III, T. (1983) Dynamic plant ecology: the spectrum of vegetational change in space and time. *Quaternary Science Reviews* **1**, 153–175. [36]

Dhar, D. (1985) The vibrational spectrum of Eden clusters in many dimensions. *Journal of Physics A: Mathematical and General* **18**, L713. [149]

Dhar, D. and Ramaswamy, R. (1985) Classical diffusion on Eden trees. *Physical Review Letters* **54**, 1346–1349. [149]

Diamond, J.M. (1983) Laboratory, field and natural experiments. *Nature* **304**, 586–597. [17]

Diggle, P.J. (2003) *Statistical Analysis of Spatial Point Patterns*, 2nd edn. Edward Arnold. [41, 43, 44]

Dijkstra, E.W. (1959) A note on two problems in connection with graphs. *Numerische Mathematik* **1**, 269–271. [145]

Doebeli, M. and Hauert, C. (2005) Models of cooperation based on the Prisoner's Dilemma and the Snowdrift game. *Ecology Letters* **8**, 748–766. [93]

Dowling, D. (1999) Experimenting on theories. *Science in Context* **12**, 261–273. [14, 222]

Drechsler, M. (1998) Sensitivity analysis of complex models. *Biological Conservation* **86**, 401–412. [202]

Drossel, B. and Schwabl, F. (1992) Self-organized critical forest fire model. *Physical Review Letters* **69**, 1629–1632. [162]

Du, Q., Faber, V. and Gunzburger, M. (1999) Centroidal Voronoi tessellations: applications and algorithms. *SIAM Review* **41**, 637–676. [88]

Durrett, R. (2009) Coexistence in stochastic spatial models. *The Annals of Applied Probability* **19**, 477–496. [73, 74, 75, 76]

Durrett, R. and Levin, S.A. (1994a) The importance of being spatial (and discrete). *Theoretical Population Biology* **46**, 363–394. [74, 75, 76]

Durrett, R. and Levin, S.A. (1994b) Stochastic spatial models: a user's guide to ecological applications. *Philosophical Transactions of the Royal Society of London. Series B.* **343**, 329–350. [70, 71, 72, 80]

Durrett, R. and Levin, S.A. (1998) Spatial aspects of interspecific competition. *Theoretical Population Biology* **53**, 30–43. [78]

Durrett, R. and Schinazi, R. (1993) Asymptotic critical value for a competition model. *The Annals of Applied Probability* **3**, 1047–1066. [74]

Durrett, R. and Swindle, G. (1991) Are there bushes in a forest? *Stochastic Processes and their Applications* **37**, 19–31. [74]

Ebrahimi, F. (2010) Invasion percolation: a computational algorithm for complex phenomena. *Computing in Science & Engineering* **12**, 84–93. [145, 146]

Economou, A.D., Ohazama, A., Porntaveetus, T., Sharpe, P.T., Kondo, S., Basson, M.A., Gritli-Linde, A., Cobourne, M.T. and Green, J.B.A. (2012) Periodic stripe formation by a Turing mechanism operating at growth zones in the mammalian palate. *Nature Genetics* **44**, 348–351. [90]

Eden, M. (1961) A two-dimensional growth process. In *4th Berkeley Symposium on Mathematical Statistics and Probability* vol. IV (ed. Neyman, J.), University of California Press. pp. 223–239. [148, 149]

Edmonds, B. and Hales, D. (2005) Computational simulation as theoretical experiment. *Journal of Mathematical Sociology* **29**, 209–232. [14]

Edwards, A.M. (2011) Overturning conclusions of Lévy flight movement patterns by fishing boats and foraging animals. *Ecology* **92**, 1247–1257. [119]

Edwards, A.M., Phillips, R.A., Watkins, N.W., Freeman, M.P., Murphy, E.J., Afanasyev, V., Buldyrev, S.V., da Luz, M.G.E., Raposo, E.P., Stanley, H.E. and Viswanathan, G.M. (2007) Revisiting Lévy flight search patterns of wandering albatrosses, bumblebees and deer. *Nature* **449**, 1044–1048. [119]

Edwards, S.F. and Wilkinson, D.R. (1982) The surface statistics of a granular aggregate. *Proceedings of the Royal Society of London. Series A.* **381**, 17–31. [148, 151]

Elith, J., Leathwick, J.R. and Hastie, T. (2008) A working guide to boosted regression trees. *Journal of Animal Ecology* **77**, 802–813. [259, 262]

Englund, G. and Cooper, S.D. (2003) Scale effects and extrapolation in ecological experiments. *Advances in Ecological Research* **33**, 161–213. [32, 34, 36]

Enquist, B.J. and Niklas, K.J. (2002) Global allocation rules for patterns of biomass partitioning in seed plants. *Science* **295**, 1517–1520. [37]

Epstein, J.M. and Axtell, R. (1996) *Growing Artificial Societies: Social Science from the Bottom Up*. Brookings Press & MIT Press. [15]

Érdi, P. (2008) *Complexity Explained*. Springer. [22]

Esther, A., Groeneveld, J., Enright, N.J., Miller, B.P., Lamont, B.B., Perry, G.L.W., Blank, F.B. and Jeltsch, F. (2010) Sensitivity of plant functional types to climate change: classification tree analysis of a simulation model. *Journal of Vegetation Science* **21**, 447–461. [259]

Etherington, T.R. (2012) Least-cost modelling on irregular landscape graphs. *Landscape Ecology* **27**, 957–968. [181]

Etienne, R.S. and Rosindell, J. (2011) The spatial limitations of current neutral models of biodiversity. *PLoS ONE* **6**, e14717. [82]

Evans, K.M. (2003) Larger than Life: threshold-range scaling of Life's coherent structures. *Physica D: Nonlinear Phenomena* **183**, 45–67. [70]

Fassó, A. and Perri, P.F. (2001) Sensitivity analysis. In *Encyclopedia of Environmetrics* (ed. El-Shaarawi, A.H. and Piegorsch, W.W.) John Wiley & Sons, Ltd. [201, 202]

Fatès, N.A. and Morvan, M. (2004) Perturbing the topology of the game of life increases its robustness to asynchrony. In *Cellular Automata* (ed. Sloot, P., Chopard, B. and Hoekstra, A.) vol. 3305 of *Lecture Notes in Computer Science* Springer pp. 111–120. [171]

Fatès, N.A. and Morvan, M. (2005) An experimental study of robustness to asynchronism for elementary cellular automata. *Complex Systems* **16**, 1–27. [171]

Fielding, A.H. and Bell, J.F. (1997) A review of methods for the assessment of prediction errors in conservation presence/absence models. *Environmental Conservation* **24**, 38–49. [213, 214, 259]

Fisher, R.A. (1922) On the mathematical foundations of theoretical statistics. *Philosophical Transactions of the Royal Society of London. Series A* **222**, 309–368. [218]

Fisher, R.A. (1937) The wave of advance of advantageous genes. *Annals of Eugenics* **7**, 355–369. [136]

Fisher, R.A. (1966) *The Design of Experiments* 8th edn. Oliver and Boyd. [57]

Flannery, T.F. (1994) *The Future Eaters: An Ecological History of the Australasian People*. Reed Books. [262]

Flügge, A.J., Olhede, S.C. and Murrell, D. (2012) The memory of spatial patterns: changes in local abundance and aggregation in a tropical forest. *Ecology* **93**, 1540–1549. [39, 46, 49, 53, 54, 56]

Ford, A. (2010) *Modeling the Environment*, 2nd edn. Island Press. [175]

Forsé, M. and Parodi, M. (2010) Low levels of ethnic intolerance do not create large ghettos: a discussion about an interpretation of Schelling's model. *L'Année Sociologique* **60**, 445–473. [186]

Fortin, M-J. and Dale, M.R.T. (2005) *Spatial Analysis: A Guide for Ecologists*. Cambridge University Press. [31]

Fortunato, S. (2010) Community detection in graphs. *Physics Reports* **486**, 75–174. [184]

Fortune Staff (1952) The MONIAC: Economics in thirty fascinating minutes. *Fortune* March, 100–101. [7]

Fossett, M. (2006) Ethnic preferences, social distance dynamics, and residential segregation: theoretical explorations using simulation analysis. *Journal of Mathematical Sociology* **30**, 185–273. [85, 177]

Frean, M.R. and Abraham, E. (2001) Rock-scissors-paper and the survival of the weakest. *Proceedings of the Royal Society of London. Series B* **268**, 1323–1327. [76]

Frenken, K. and Boschma, R.A. (2007) A theoretical framework for evolutionary economic geography: industrial dynamics and urban growth as a branching process. *Journal of Economic Geography* **7**, 635–649. [157]

Fritz, H., Said, S. and Weimerskirch, H. (2003) Scale-dependent hierarchical adjustments of movement patterns in a long-range foraging seabird. *Proceedings of the Royal Society of London. Series B*. **270**, 1143–1148. [111]

Gardner, M. (1970) Mathematical games: the fantastic combinations of John Conway's new solitaire game 'life'. *Scientific American* **223**(4), 120–123. [18, 19, 69]

Gardner, R.H., Milne, B.T., Turner, M.G. and O'Neill, R.V. (1987) Neutral models for the analysis of broad-scale landscape pattern. *Landscape Ecology* **1**, 19–28. [158, 159]

Gatrell, A. (1983) *Distance and Space: A Geographical Perspective*. Oxford University Press. [182]

Gautestad, A.O. and Mysterud, I. (2010) The home range fractal: from random walk to memory-dependent space use. *Ecological Complexity* **7**, 458–470. [127]

Gelfand, A.E., Diggle, P.J., Fuentes, M. and Guttorp, P. (eds) (2010) *Handbook of Spatial Statistics*. CRC Press. [31, 41]

Getz, W.M. and Saltz, D. (2008) A framework for generating and analyzing movement paths on ecological landscapes. *Proceedings of the National Academy of Sciences of the United States of America* **105**, 19066–19071. [127]

Gifford-Gonzalez, D. (1991) Bones are not enough: analogues, knowledge, and interpretive strategies in zooarchaeology. *Journal of Anthropological Archaeology* **10**, 215–254. [267]

Gilbert, N. and Troitzsch, K.G. (2005) *Simulation for the Social Scientist*, 2nd edn. Open University Press. [229]

Gleick, J. (1987) *Chaos: Making a New Science*. Viking Penguin. [22]

Goering, J. (2006) Schelling redux: how sociology fails to make progress in building and empirically testing complex causal models regarding race and residence. *Journal of Mathematical Sociology* **30**, 299–317. [85]

González, M.C., Hidalgo, C.A. and Barabási, A-L. (2008) Understanding individual human mobility patterns. *Nature* **453**, 779–782. [110, 126, 129]

Goodchild, M.F. (2004) The validity and usefulness of laws in geographic information science and geography. *Annals of the Association of American Geographers* **94**, 300–303. [57]

Goodchild, M.F. and Mark, D.M. (1987) The fractal nature of geographic phenomena. *Annals of the Association of American Geographers* **77**, 265–278. [34]

Gotelli, N.J. and Ellison, A.M. (2004) *A Primer of Ecological Statistics*. Sinauer Associates. [253]

Gould, H. and Tobochnik, J. (2010) *Statistical and Thermal Physics with Computer Applications*. Princeton University Press, Princeton. [141]

Gouyet, J-F. (1996) *Physics and Fractal Structures*. Masson; Springer. [34, 137, 141, 142, 149, 151, 152, 157]

Grassberger, P. (2003) Critical percolation in high dimensions. *Physical Review E* **67**, 036101. [141]

Greenberg, J.M. and Hastings, S.P. (1978) Spatial patterns for discrete models of diffusion in excitable media. *SIAM Journal on Applied Mathematics* **34**, 515–523. [91]

Grégoire, G., Chaté, H. and Tu, Y. (2003) Moving and staying together without a leader. *Physica D: Nonlinear Phenomena* **181**, 157–170. [124, 125]

Grim, P., Mar, G. and Denis, P.S. (1998) *The Philosophical Computer: Exploratory Essays in Philosophical Computer Modeling* Bradford Books. MIT Press. [93]

Grimm, V. (1999) Ten years of individual-based modelling in ecology: what have we learned and what could we learn in the future? *Ecological Modelling* **115**, 129–148. [52, 228]

Grimm, V. (2002) Visual debugging: a way of analyzing, understanding and communicating bottom-up simulation models in ecology. *Natural Resource Modelling* **15**, 23–38. [212, 248]

Grimm, V. and Railsback, S.F. (2005) *Individual-based Modeling and Ecology*. Princeton University Press. [51, 170, 222]

Grimm, V. and Railsback, S.F. (2012) Pattern-oriented modelling: a 'multi-scope' for predictive systems ecology. *Philosophical Transactions of the Royal Society of London. Series B.* **367**, 298–310. [15, 52, 199, 222, 224]

Grimm, V., Berger, U., Bastiansen, F., Eliassen, S., Ginot, V., Giske, J., Goss-Custard, J., Grand, T., Heinz, S.K., Huse, G., Huth, A., Jepsen, J.U., Jørgensen, C., Mooij, W.M., Müller, B., Pe'er, G., Piou, C., Railsback, S.F., Robbins, A.M., Robbins, M.M., Rossmanith, E., Rüger, N., Strand, E., Souissi, S., Stillman, R.A., Vabø, R., Visser, U. and DeAngelis, D.L. (2006) A standard protocol for describing individual-based and agent-based models. *Ecological Modelling* **198**, 115–126. [231]

Grimm, V., Berger, U., DeAngelis, D.L., Polhill, J.G., Giske, J. and Railsback, S.F. (2010) The ODD protocol: a review and first update. *Ecological Modelling* **221**, 2760–2768. [224, 231, 231]

Grimm, V., Revilla, E., Berger, U., Jeltsch, F., Mooij, W.M., Railsback, S.F., Thulke, H-H., Weiner, J., Wiegand, T. and DeAngelis, D.L. (2005) Pattern-oriented modeling of agent-based complex systems: lessons from ecology. *Science* **310**, 987–991. [15, 30, 52, 199, 223]

Grimmett, G.R. (1999) *Percolation*, 2nd edn. Springer. [137, 138, 155]

Gross, J.L. and Yellen, J. (2006) *Graph Theory and Its Applications* Discrete Mathematics and its Applications. Chapman & Hall/CRC. [183]

Gross, T. and Blasius, B. (2008) Adaptive coevolutionary networks: a review. *Journal of the Royal Society Interface* **5**, 259–271. [183]

Guala, F. (2002) Models, simulations, and experiments. In *Model-Based Reasoning: Science, Technology, Values* (ed. Magnani, L. and Nersessian, N.) Kluwer pp. 59–74. [222]

Gueron, S. and Levin, S.A. (1993) Self-organization of front patterns in large wildebeest herds. *Journal of Theoretical Biology* **165**, 541–552. [125, 128]

Haefner, J. (2005) *Modeling Biological Systems: Principles and Applications*, 2nd edn. Springer. [199, 200, 201, 215]

Hägerstrand, T. (1968) *Innovation Diffusion as a Spatial Process*. University of Chicago Press. [93]

Haggett, P. and Chorley, R.J. (1969) *Network Analysis in Geography*. Edward Arnold. [183]

Haklay, M. Singleton, A. and Parker, C. (2008) Web mapping 2.0: the neogeography of the geoweb. *Geography Compass* **2**, 2011–2039. [14]

Haklay, M., Schelhorn, T., O'Sullivan, D. and Thurstain-Goodwin, M. (2001) 'So go down town': Simulating pedestrian movement in town centres. *Environment and Planning B: Planning and Design* **28**, 343–359. [130]

Halley, J.M., Hartley, S., Kallimanis, A.S., Kunin, W.E., Lennon, J.J. and Sgardelis, S.P. (2004) Uses and abuses of fractal methodology in ecology. *Ecology Letters* **7**, 254–271. [34]

Hamby, D.M. (1994) A review of techniques for parameter sensitivity analysis of environmental models. *Environmental Monitoring and Assessment* **32**, 135–154. [199, 201, 202, 215]

Hamby, D.M. (1995) A comparison of sensitivity analysis techniques. *Health Physics* **68**, 195–204. [201]

Harris, R., Sleight, P. and Webber, R. (2005) *Geodemographics, GIS and Neighbourhood Targeting*. John Wiley & Sons. [58]

Harris, T.E. (1974) Contact interactions on a lattice. *The Annals of Probability* **2**, 969–988. [71]

Harsch, M.A. and Bader, M.Y. (2011) Treeline form–a potential key to understanding treeline dynamics. *Global Ecology and Biogeography* **20**, 582–596. [136]

Hartig, F., Calabrese, J.M., Reineking, B., Wiegand, T. and Huth, A. (2011) Statistical inference for stochastic simulation models—theory and application. *Ecology Letters* **14**, 816–827. [45, 205, 216, 220, 222, 223, 228, 269]

Hastie, T., Tibshirani, R. and Friedman, J.H. (2009) *The Elements of Statistical Learning Data Mining, Inference, and Prediction* 2nd edn. Springer. [214]

Hastings, A., Cuddington, K., Davies, K.F., Dugaw, C.J., Elmendorf, S., Freestone, A., Harrison, S., Holland, M., Lambrinos, J., Malvadkar, U., Melbourne, B.A., Moore, K., Taylor, C. and Thomson, D. (2005) The spatial spread of invasions: new developments in theory and evidence. *Ecology Letters* **8**, 91–101. [136]

Havlin, S. and ben-Avraham, D. (2002) Diffusion in disordered media. *Advances in Physics* **51**, 187–292. [142, 143, 157]

Heath, D.F. (1967) Normal or Log-normal: appropriate distributions. *Nature* **213**, 1159–1160. [26]

Heino, M. and Enberg, K. (2008) Sustainable use of populations and overexploitation. *Encyclopedia of Life Sciences (ELS)* John Wiley & Sons, Ltd. [195]

Helbing, D., Farkas, I.J. and Vicsek, T. (2000) Simulating dynamical features of escape panic. *Nature* **407**, 487–490. [125, 130]

Helbing, D., Molnár, P., Farkas, I.J. and Bolay, K. (2001) Self-organizing pedestrian movement. *Environment and Planning B: Planning and Design* **28**, 361–383. [51]

Helbing, D., Schweitzer, F., Keltsch, J. and Molnár, P. (1997) Active walker model for the formation of human and animal trail systems. *Physical Review E* **56**, 2527–2539. [129]

Helton, J.C. (1993) Uncertainty and sensitivity analysis techniques for use in performance assessment for radioactive waste disposal. *Reliability Engineering & System Safety* **42**, 327–367. [202, 203]

Hemelrijk, C.K. and Hildenbrandt, H. (2008) Self-organized shape and frontal density of fish schools. *Ethology* **114**, 245–254. [125]

Heppenstall, A.J., Crooks, A.T., Batty, M. and See, L.M. (eds) (2012) *Agent-Based Models of Geographical Systems*. Springer. [170]

Herrmann, H.J. (1986) Geometrical cluster growth models and kinetic gelation. *Physics Reports* **136**, 153–224. [148, 149, 152]

Herrmann, H.J., Hong, D.C. and Stanley, H.E. (1984) Backbone and elastic backbone of percolation clusters obtained by the new method of 'burning'. *Journal of Physics A: Mathematical and General* **17**, L261–L266. [143, 147]

Heuvelink, G.B.M. (1998) *Error Propagation in Environmental Modelling With GIS.* Taylor & Francis. [38, 200]

Heuvelink, G.B.M. (2006) Analysing uncertainty propagation in GIS: why is it not that simple? In *Uncertainty in Remote Sensing and GIS* (ed. Foody, G.M. and Atkinson, P.M.) John Wiley & Sons. [200]

Hilbert, M. and López, P. (2011) The world's technological capacity to store, communicate, and compute information. *Science* **332**, 60–65. [14]

Hilborn, R. and Mangel, M. (1997) *The Ecological Detective: Confronting Models with Data.* Princeton University Press. [220]

Hoffmann, G. (1983) Optimization of Brownian search strategies. *Biological Cybernetics* **49**, 21–31. [103]

Holdaway, R.N. and Jacomb, C. (2000) Rapid extinction of the moas (Aves: Dinornithiformes): model, test, and implications. *Science* **287**, 2250–2254. [230]

Holland, E.P., Aegerter, J.N., Dytham, C. and Smith, G.C. (2007) Landscape as a model: the importance of geometry. *PLoS Computational Biology* **3**, e200. [181]

Holland, J.H. (1998) *Emergence: From Chaos to Order.* Perseus. [22]

Holling, C.S. (1959) The components of predation as revealed by a study of small-mammal predation of the European Pine Sawfly. *Canadian Entomologist* **91**, 293–320. [205]

Holmes, E.E., Lewis, M.A., Banks, J.E. and Veit, R.R. (1994) Partial differential equations in ecology: spatial interactions and population dynamics. *Ecology* **75**, 17–29. [136]

Hubbell, S.P. (2001) *The Unified Neutral Theory of Biodiversity and Biogeography.* Princeton University Press. [82]

Huth, A. and Wissel, C. (1992) The simulation of the movement of fish schools. *Journal of Theoretical Biology* **156**, 365–385. [125]

Illian, J., Penttinen, A., Stoyan, H. and Stoyan, D. (2008) *Statistical Analysis and Modelling of Spatial Point Patterns.* John Wiley & Sons. [41, 42, 43, 48]

Illius, A.W. and O'Connor, T.G. (2000) Resource heterogeneity and ungulate population dynamics. *Oikos* **89**, 283–294. [128]

Jager, H.I. and King, A.W. (2004) Spatial uncertainty and ecological models. *Ecosystems* **7**, 841–847. [199, 200, 202, 207]

Jager, H.I., King, A.W., Schumaker, N.H., Ashwood, T.L. and Jackson, B.L. (2005) Spatial uncertainty analysis of population models. *Ecological Modelling* **185**, 13–27. [207]

James, A. and Plank, M.J. (2007) On fitting power laws to ecological data. *ArXiv e-prints.* (http://arxiv.org/abs/0712.0613) [163]

James, A., Plank, M.J. and Edwards, A.M. (2011) Assessing Lévy walks as models of animal foraging. *Journal of the Royal Society Interface* **8**, 1233–1247. [119]

Jansen, M.J.W. (1998) Prediction error through modelling concepts and uncertainty from basic data. *Nutrient Cycling in Agroecosystems* **50**, 247–253. [205, 216]

Jettestuen, E., Nermoen, A., Hestmark, G., Timdal, E. and Mathiesen, J. (2010) Competition on the rocks: community growth and tessellation. *PLoS ONE* **5**, e12820. [168]

Jongman, R.H.G., ter Braak, C.J.F. and van Tongeren, O.F.R. (1999) *Data Analysis in Community and Landscape Ecology.* Cambridge University Press. [253]

JTS Topology Suite (2012). http://tsusiatsoftware.net/jts/main.html. [188]

Jungnickel, D. (1999) *Graphs, Networks and Algorithms.* Springer. [183]

Kansky, K.J. (1963) *Structure of Transportation Networks: Relationships Between Network Geometry and Regional Characteristics.* University of Chicago. [183]

Kardar, M., Parisi, G. and Zhang, Y-C. (1986) Dynamic scaling of growing interfaces. *Physical Review Letters* **56**, 889–892. [148, 149]

Kawasaki, K., Takasu, F., Caswell, H. and Shigesada, N. (2006) How does stochasticity in colonization accelerate the speed of invasion in a cellular automaton model? *Ecological Research* **21**, 334–345. [155]

Keeling, M.J. and Eames, K.T.D. (2005) Networks and epidemic models. *Journal of the Royal Society Interface* **2**, 295–307. [168, 184]

Keeling, M.J. and Rohani, P. (2008) *Modeling Infection Diseases in Humans and Animals.* Princeton University Press. [163]

Kelly, R.L. (1983) Hunter-gatherer mobility strategies. *Journal of Anthropological Research* **39**, 277–306. [128]

Kesten, H. (2006) What is ... percolation? *Notices of the American Mathematical Society* **53**, 572–573. [141]

Kimura, M. (1953) '*Stepping stone' model of population.* Annual Report, National Institute of Genetics of Japan. [78]

Kimura, M. (1968) Evolutionary rate at the molecular level. *Nature* **217**, 624–626. [82]

Kimura, M. (1983) *The Neutral Theory of Molecular Evolution.* Cambridge University Press. [82]

Kirch, P.V. (2005) Archaeology and global change: the Holocene record. *Annual Review of Environment and Resources* **30**, 409–440. [231]

Kitzberger, T., Aráoz, E., Gowda, J.N., Mermoz, M. and Morales, J. (2012) Decreases in fire spread probability with forest age promotes alternative community states, reduced resilience to climate variability and large fire regime shifts. *Ecosystems* **15**, 97–112. [165]

Klafter, J. and Sokolov, I.M. (2005) Anomalous diffusion spreads its wings. *Physics World* **18**, 29–32. [110, 143]

Klausmeier, C.A. (1999) Regular and irregular patterns in semiarid vegetation. *Science* **284**, 1826–1828. [91]

Kleindorfer, G.B., O'Neill, L. and Ganeshan, R. (1998) Validation in simulation: various positions in the philosophy of science. *Management Science* **44**, 1087–1099. [52, 212, 228]

Klüpfel, H., Schreckenberg, M. and Meyer-König, T. (2005) Models for crowd movement and egress simulation. In *Traffic and Granular Flow 03* (ed. Hoogendoorn, S.P., Luding, S., Bovy, P.H.L., Schreckenberg, M. and Wolf, D.E.) Springer pp. 357–372. [130]

Knackstedt, M. and Paterson, L. (2009) Invasion percolation. In *Encyclopedia of Complexity and Systems Science* (ed. Meyers, R.A.) Springer. [145, 146]

Koch, A.J. and Meinhardt, H. (1994) Biological pattern formation: from basic mechanisms to complex structures. *Reviews of Modern Physics* **66**, 1481–1507. [90]

Koch, A.L. (1966) The logarithm in biology 1. Mechanisms generating the log-normal distribution exactly. *Journal of Theoretical Biology* **12**, 276–290. [26]

Kreps, D.M., Milgrom, P., Roberts, J. and Wilson, R. (1982) Rational cooperation in the finitely repeated prisoners' dilemma. *Journal of Economic Theory* **27**, 245–252. [92]

Lake, M.W. (2000) MAGICAL Computer simulation of Mesolithic foraging. In *Dynamics in Human and Primate Societies: Agent-Based Modelling of Social and Spatial Processes* (ed. Kohler, T.A. and Gumerman, G.J.) Oxford Univesity Press pp. 107–143. [128]

Lake, M.W. (2001) The use of pedestrian modelling in archaeology, with an example from the study of cultural learning. *Environment and Planning B: Planning and Design* **28**, 385–403. [128]

Lande, R., Engen, S. and Sæther, B-E. (2003) *Stochastic Population Dynamics in Ecology and Conservation.* Oxford University Press. [14]

Laplace, P.S. (1814) *Essai philosophique sur les probabilités.* Gauthiers-Villars. [195]

Larkin, P.A. (1977) An epitaph for the concept of Maximum Sustained Yield. *Transactions of the American Fisheries Society* **106**, 1–11. [198]

Laube, P. and Purves, R.S. (2006) An approach to evaluating motion pattern detection techniques in spatio-temporal data. *Computers, Environment and Urban Systems* **30**, 347–374. [126]

Laurie, A. and Jaggi, N. (2003) Role of 'vision' in neighborhood racial segregation: a variant of the Schelling segregation model. *Urban Studies* **40**, 2687–2704. [85]

Lawler, G.F. and Limic, V. (2010) *Random Walk: A Modern Introduction*. Cambridge University Press. [99]

Lawton, J.H. (1999) Are there general laws in ecology? *Oikos* **84**, 177–192. [30, 50]

Lazar, N. (2010) Ockham's Razor. *Wiley Interdisciplinary Reviews: Computational Statistics* **2**, 243–246. [216]

Lazer, D., Pentland, A., Adamic, L., Aral, S., Barabási, A-L., Brewer, D., Christakis, N., Contractor, N., Fowler, J., Gutmann, M., Jebara, T., King, G., Macy, M., Roy, D. and van Alstyne, M. (2009) Computational social science. *Science* **323**, 721–723. [126]

Levin, S.A. (1992) The problem of pattern and scale in ecology: the Robert H. MacArthur award lecture. *Ecology* **73**, 1943–1967. [32, 34]

Levins, R. (1966) The strategy of model building in population biology. *American Scientist* **54**, 421–431. [23, 27, 199, 206]

Liggett, T.M. (1985) *Interacting Particle Systems*. Springer. [70]

Liggett, T.M. (1999) *Stochastic Interacting Systems: Contact, Voter and Exclusion Processes*. Springer. [70, 71, 78, 83]

Liggett, T.M. (2010) Stochastic models for large interacting systems and related correlation inequalities. *Proceedings of the National Academy of Sciences of the United States of America* **107**, 16413–16419. [70]

Limpert, E. and Stahel, W.A. (2011) Problems with using the normal distribution—and ways to improve quality and efficiency of data analysis. *PLoS ONE* **6**, e21403. [26]

Limpert, E., Stahel, W.A. and Abbt, M. (2001) Log-normal distributions across the sciences: keys and clues. *BioScience* **51**, 341–352. [24, 26]

Link, W.A. and Barker, R.J. (2010) *Bayesian Inference: With Ecological Applications*. Elsevier/Academic Press. [205, 222]

Lloyd-Smith, J.O., Schreiber, S.J., Kopp, P.E. and Getz, W.M. (2005) Superspreading and the effect of individual variation on disease emergence. *Nature* **438**, 355–359. [135]

Lock, G. and Molyneaux, B. (eds) (2006) *Confronting Scale in Archaeology*. Springer. [32, 33]

Loosmore, N.B. and Ford, E.D. (2006) Statistical inference using the *G* or *K* point pattern spatial statistics. *Ecology* **87**, 1925–1931. [44]

MacArthur, R.H. (1957) On the relative abundance of bird species. *Proceedings of the National Academy of Sciences of the United States of America* **43**, 293–295. [87]

MacArthur, R.H. and Pianka, E.R. (1966) On optimal use of a patchy environment. *American Naturalist* **100**, 603–609. [118]

MacArthur, R.H. and Wilson, E.O. (1967) *The Theory of Island Biogeography*. Princeton University Press. [82]

MacEachren, A.M., Robinson, A., Hopper, S., Gardner, S., Murray, R., Gahegan, M. and Hetzler, E. (2005) Visualizing geospatial information uncertainty: what we know and what we need to know. *Cartography and Geographic Information Science* **32**, 139–160. [200]

Macy, M.W. and Willer, R. (2002) From factors to actors: computational sociology and agent-based modeling. *Annual Review of Sociology* **28**, 143–166. [51]

Madras, N. and Slade, G. (1996) *The Self-Avoiding Walk*. Birkhäuser. [108]

Makse, H.A., Andrade, J.S., Batty, M., Havlin, S. and Stanley, H.E. (1998) Modeling urban growth patterns with correlated percolation. *Physical Review. E* **58**, 7054–7062. [157]

Makse, H.A., Havlin, S. and Stanley, H.E. (1995) Modelling urban growth patterns. *Nature* **377**, 608–612. [157]

Malamud, B.D., Morein, G. and Turcotte, D.L. (1998) Forest fires: an example of self-organized critical behaviour. *Science* **281**, 1840–1842. [162, 164]

Malanson, G.P. (2003) Dispersal across continuous and binary representations of landscapes. *Ecological Modelling* **169**, 17–24. [161]

Malanson, G.P. and Cramer, B.E. (1999a) Ants in labyrinths: lessons for critical landscapes. *The Professional Geographer* **51**, 155–170. [161]

Malanson, G.P. and Cramer, B.E. (1999b) Landscape heterogeneity, connectivity, and critical landscapes for conservation. *Diversity and Distributions* **5**, 27–39. [161]

Malarz, K. and Galam, S. (2005) Square-lattice site percolation at increasing ranges of neighbor bonds. *Physical Review E* **71**, 016125. [138, 141]

Malpezzi, S. (2008) Hedonic pricing models: A selective and applied review. In *Housing Economics and Public Policy* (ed. O'Sullivan, T. and Gibb, K.) Blackwell Science Ltd pp. 67–89. [9]

Mandelbrot, B.B. (1967) How long is the coast of Britain? Statistical self-similarity and fractional dimension. *Science* **156**, 636–638. [34]

Manson, S.M. (2001) Simplifying complexity: a review of complexity theory. *Geoforum* **32**, 405–414. [22]

Mardia, K.V. (1972) *Statistics of Directional Data*. Academic Press. [105]

Mardia, K.V. and Jupp, P.E. (1999) *Directional Statistics*. John Wiley & Sons. [107]

Marsh, L.M. and Jones, R.E. (1988) The form and consequences of random walk movement models. *Journal of Theoretical Biology* **133**, 113–131. [103]

Martínez, I., Wiegand, T., Camarero, J.J., Batllori, E. and Gutiérrez, E. (2011) Disentangling the formation of contrasting tree-line physiognomies combining model selection and Bayesian parameterization for simulation models. *American Naturalist* **177**, E136–E152. [222, 224, 225]

Martínez, V.J., Jones, B.J.T., Domínguez-Tenreiro, R. and van de Weygaert, R. (1990) Clustering paradigms and multifractal measures. *Astrophysical Journal* **357**, 50–61. [88]

Matott, L.S., Babendreier, J.E. and Purucker, S.T. (2009) Evaluating uncertainty in integrated environmental models: a review of concepts and tools. *Water Resources Research* **45**, W06421. [193, 194, 198, 200, 202, 215]

Matthews, R., Gilbert, N., Roach, A., Polhill, J.G. and Gotts, N. (2007) Agent-based land-use models: a review of applications. *Landscape Ecology* **22**, 1447–1459. [94, 95]

May, R.M., Levin, S.A. and Sugihara, G. (2008) Complex systems: ecology for bankers. *Nature* **451**, 893–895. [12]

Mayer, D.G. and Butler, D.G. (1993) Statistical validation. *Ecological Modelling* **68**, 21–32. [193, 199, 212, 213]

McCarthy, M.A., Possingham, H.P., Day, J.R. and Tyre, A.J. (2001) Testing the accuracy of population viability analysis. *Conservation Biology* **15**, 1030–1038. [212, 215]

McGill, B.J. (2003) A test of the Unified Neutral Theory of Biodiversity. *Nature* **422**, 881–885. [82]

McIntire, E.J.B. and Fajardo, A. (2009) Beyond description: the active and effective way to infer processes from spatial patterns. *Ecology* **90**, 46–56. [267, 269]

McKay, M.D., Beckman, R.J. and Conover, W.J. (1979) A comparison of three methods for selecting values of input variables in the analysis of output from a computer code. *Technometrics* **21**, 239–245. [204]

McMahon, S.M., Dietze, M.C., Hersh, M.H., Moran, E.V. and Clark, J.S. (2009) A predictive framework to understand forest responses to global change. *Annals of the New York Academy of Sciences* **1162**, 221–236. [193]

Meakin, P. (1983). Diffusion-controlled cluster formation in 2–6-dimensional space. *Physical Review A* **27**, 1495–1507. [156, 157]

Meakin, P. (1990) Fractal structures. *Progress in Solid State Chemistry* **20**, 135–233. [34]

Meentemeyer, V. (1989) Geographical perspectives of space, time, and scale. *Landscape Ecology* **3**, 163–173. [29, 32, 33, 34, 37]

Mercader, R.J., Siegert, N.W., Liebhold, A.M. and McCullough, D.G. (2011) Influence of foraging behavior and host spatial distribution on the localized spread of the emerald ash borer, *Agrilus planipennis*. *Population Ecology* **53**, 271–285. [131]

Metzler, R. and Klafter, J. (2000) The random walk's guide to anomalous diffusion: a fractional dynamics approach. *Physics Reports* **339**, 1–77. [143]

Miller, H.J. (2004) Tobler's first law and spatial analysis. *Annals of the Association of American Geographers* **94**, 284–289. [57]

Millington, J.D.A. and Perry, G.L.W. (2011) Multi-model inference in biogeography. *Geography Compass* **5**, 448–463. [222]

Millington, J.D.A., Demeritt, D. and Romero-Calcerrada, R. (2011) Participatory evaluation of agent-based land-use models. *Journal of Land Use Science* **6**, 195–210. [54, 199, 228, 267]

Millington, J.D.A., O'Sullivan, D. and Perry, G.L.W. (2012) Model histories: Narrative explanation in generative simulation modelling. *Geoforum* **43**, 1025–1034. [269]

Millington, J.D.A., Perry, G.L.W. and Malamud, B.D. (2006) Models, data and mechanisms: quantifying wildfire regimes. *Geological Society, London, special Publications* **261**, 155–167. [162]

Milne, B.T., Johnson, A.R., Keitt, T.H., Hatfield, C.A., David, J.L. and Hraber, P.T. (1996) Detection of critical densities associated with piñon-juniper woodland ecotones. *Ecology* **77**, 805–821. [168]

Mitchell, M. (2008) *Complexity: A Guided Tour*. Oxford University Press. [22]

Mithen, S. and Reed, M. (2002) Stepping out: a computer simulation of hominid dispersal from Africa. *Journal of Human Evolution* **43**, 433–462. [128]

Mitzenmacher, M. (2004) A brief history of generative models for power law and lognormal distributions. *Internet Mathematics* **1**, 226–251. [26]

Mooij, W.M. and DeAngelis, D.L. (1999) Error propagation in spatially explicit population models: a reassessment. *Conservation Biology* **13**, 930–933. [207]

Morales, J.M. and Carlo, T.A. (2008) The effects of plant distribution and frugivore density on the scale and shape of dispersal kernels. *Ecology* **87**, 1489–1496. [116]

Morgan, F.J. (2011) *Residential Property Developers in Urban Agent-Based Models: Competition, Behaviour and the Resulting Spatial Landscape*. Unpublished Ph.D. thesis University of Auckland, New Zealand. [87]

Morgenthaler, S. (2009) Exploratory data analysis. *Wiley Interdisciplinary Reviews: Computational Statistics* **1**, 33–44. [212]

Moritz, M.A., Hessburg, P.F. and Povak, N.A. (2011) Native fire regimes and landscape resilience. In *The Landscape Ecology of Fire* (ed. McKenzie, D., Miller, C. and Falk, D.A.) vol. 213 of Ecological Studies, Springer Netherlands pp. 51–86. [162, 164]

Morrison, D.W. (1978) On the optimal searching strategy for refuging predators. *American Naturalist* **112**, 925–934. [103]

Morrison, M. (2009) Models, measurement and computer simulation: the changing face of experimentation. *Philosophical Studies* **143**, 33–57. [14, 222]

Moussaïd, M., Helbing, D. and Theraulaz, G. (2011) How simple rules determine pedestrian behavior and crowd disasters. *Proceedings of the National Academy of Sciences of the United States of America* **108**, 6884–6888. [130]

Mueller, T., Fagan, W.F. and Grimm, V. (2010) Integrating individual search and navigation behaviors in mechanistic movement models. *Theoretical Ecology* **4**, 341–355. [118]

Muetzelfeldt, R.I. and Massheder, J. (2003) The Simile visual modelling environment. *European Journal of Agronomy* **18**, 345–358. [175]

Murray, J.D. (1981) A pre-pattern formation mechanism for animal coat markings. *Journal of Theoretical Biology* **88**, 161–199. [90]

Murray, J.D. (2002) *Mathematical Biology: I. An Introduction* Interdisciplinary applied mathematics, 3rd edn. Springer. [91]

Murray, J.D. (2003) *Mathematical Biology II: Spatial Models and Biomedical Applications*, Interdisciplinary applied mathematics, 3rd edn. Springer. [136]

Murray-McIntosh, R.P., Scrimshaw, B.J., Hatfield, P.J. and Penny, D. (1998) Testing migration patterns and estimating founding population size in Polynesia by using human mtDNA sequences. *Proceedings of the National Academy of Sciences of the United States of America* **95**, 9047–9052. [230]

Nathan, R., Getz, W.M., Revilla, E., Holyoak, M., Kadmon, R., Saltz, D. and Smouse, P.E. (2008) A movement ecology paradigm for unifying organismal movement research. *Proceedings of the National Academy of Sciences of the United States of America* **105**, 19052–19059. [125]

Newman, M.E.J. (2010) *Networks: An Introduction*. Oxford University Press. [168, 183, 185]

Newman, M.E.J. and Ziff, R.M. (2001) Fast Monte Carlo algorithm for site or bond percolation. *Physical Review E* **64**, 016706. [141]

Nikitas, P. and Nikita, E. (2005) A study of hominin dispersal out of Africa using computer simulations. *Journal of Human Evolution* **49**, 602–617. [128]

Norte Pinto, N. and Pais Antunes, A. (2010) A cellular automata model based on irregular cells: application to small urban areas. *Environment and Planning B: Planning and Design* **37**, 1095–1114. [95, 184]

Nowak, M.A. and May, R.M. (1992) Evolutionary games and spatial chaos. *Nature* **359**, 826–829. [92, 93]

O'Brien, W.J., Browman, H.I. and Evans, B.I. (1990) Search strategies of foraging animals. *American Scientist* **78**, 152–160. [126]

Okabe, A., Boots, B. and Sugihara, K. (1994) Nearest neighbourhood operations with generalised Voronoi diagrams: a review. *International Journal of Geographical Information Systems* **8**, 43–71. [87]

Okabe, A., Boots, B., Sugihara, K. and Chiu, S.N. (2000) *Spatial Tessellations: Concepts and Applications of Voronoi Diagrams*, 2nd edn. John Wiley & Sons. [87]

Okabe, A., Satoh, T., Furuta, T., Suzuki, A. and Okano, K. (2008) Generalized network Voronoi diagrams: concepts, computational methods, and applications. *International Journal of Geographical Information Science* **22**, 965–994. [87]

O'Kelly, M.E., Kim, H. and Kim, C. (2006) Internet reliability with realistic peering. *Environment and Planning B: Planning and Design* **33**, 325–343. [13]

Okubo, A. (1980) *Diffusion and Ecological Problems: Mathematical Models* Biomathematics 10. Springer-Verlag. [136]

Okubo, A. (1986) Dynamical aspects of animal grouping: swarms, schools, flocks, and herds. *Advances in Biophysics* **22**, 1–94. [122, 128]

Okuyama, T. (2009) Local interactions between predators and prey call into question commonly used functional responses. *Ecological Modelling* **220**, 1182–1188. [206]

O'Malley, L., Korniss, G. and Caraco, T. (2009) Ecological invasion, roughened fronts, and a competitor's extreme advance: integrating stochastic spatial-growth models. *Bulletin of Mathematical Biology* **71**, 1160–1188. [151, 152]

O'Neill, R.V., Gardner, R.H. and Turner, M.G. (1992a) A hierarchical neutral model for landscape analysis. *Landscape Ecology* **7**, 55–61. [158, 160]

O'Neill, R.V., Gardner, R.H., Turner, M.G. and Romme, W.H. (1992b) Epidemiology theory and disturbance spread on landscapes. *Landscape Ecology* **7**, 19–26. [168]

Oreskes, N., Shrader-Frechette, K. and Belitz, K. (1994) Verification, validation, and confirmation of numerical models in the earth sciences. *Science* **263**, 641–646. [26, 52, 226, 227, 228]

O'Sullivan, D. (2001) Graph cellular automata: a generalised discrete urban and regional model. *Environment and Planning B: Planning and Design* **28**, 687–705. [95, 184]

O'Sullivan, D. (2002) Toward micro-scale spatial modeling of gentrification. *Journal of Geographical Systems* **4**, 251–274. [95]

O'Sullivan, D. (2004) Complexity science and human geography. *Transactions of the Institute of British Geographers* **29**, 282–295. [22, 267]

O'Sullivan, D. (2005) Geographical information science: time changes everything. *Progress in Human Geography* **29**, 749–756. [176]

O'Sullivan, D. (2008) Geographical information science: agent-based models. *Progress in Human Geography* **32**, 541–550. [51]

O'Sullivan, D. and Unwin, D.J. (2010) *Geographic Information Analysis*, 2nd edn. John Wiley & Sons. [31, 40, 41, 44, 115, 181]

O'Sullivan, D., Millington, J.D.A., Perry, G.L.W. and Wainwright, J. (2012) Agent-based models: because they're worth it? In *Agent-Based Models for Geographical Systems*. (ed. Heppenstall, A., Crooks, A., See, L.M. and Batty, M.) Springer pp. 109–123. [51]

Packard, N.H. and Wolfram, S. (1985) Two-dimensional cellular automata. *Journal of Statistical Physics* **38**, 901–946. [69]

Painter, K.J., Hunt, G.S., Wells, K.L., Johansson, J.A. and Headon, D.J. (2012) Towards an integrated experimental-theoretical approach for assessing the mechanistic basis of hair and feather morphogenesis. *Interface Focus* **2**, 433–450. [90]

Pan, X., Han, C.S., Dauber, K. and Law, K.H. (2007) A multi-agent based framework for the simulation of human and social behaviors during emergency evacuations. *AI & Society* **22**, 113–132. [130]

Parker, D.C., Manson, S.M., Janssen, M.A., Hoffmann, M.J. and Deadman, P. (2003) Multi-agent systems for the simulation of land-use and land-cover change: a review. *Annals of the Association of American Geographers* **93**, 314–337. [51, 94, 95]

Parker, J. and Epstein, J.M. (2011) A distributed platform for global-scale agent-based models of disease transmission. *ACM Transactions on Modeling and Computer Simulation* **22**, 1–25. [33, 185]

Parker, W.S. (2008) Computer simulation through an error-statistical lens. *Synthese* **163**, 371–384. [193, 228, 269]

Parysow, P., Gertner, G. and Westervelt, J. (2000) Efficient approximation for building error budgets for process models. *Ecological Modelling* **135**, 111–125. [200]

Pascual, M. and Guichard, F. (2005) Criticality and disturbance in spatial ecological systems. *Trends in Ecology & Evolution* **20**, 88–95. [163]

Patlak, C.S. (1953) Random walk with persistence and external bias. *Bulletin of Mathematical Biology* **15**, 311–338. [104, 108]

Pauly, D., Christensen, V., Guénette, S., Pitcher, T.J., Sumaila, U.R., Walters, C.J., Watson, R. and Zeller, D. (2002) Towards sustainability in world fisheries. *Nature* **418**, 689–695. [198]

Pearson, K. (1905a) The problem of the random walk. *Nature* **72**, 294. [102]

Pearson, K. (1905b) The problem of the random walk. *Nature* **72**, 342. [102]

Pearson, K. (1906) *Mathematical Contributions to the Theory of Evolution XV: A Mathematical Theory of Random Migration* vol. III of Drapers Company Research Memoirs Biometric Series. Dulau & Co. [103]

Pegman, A.P.McK. (2012) *Reconstruction of Seed Dispersal via Modeling, Seedling Recruitment, and Dispersal Efficiency of* Hemiphaga novaeseelandiae *in* Vitex lucens *and* Prumnopitys ferruginea *in New Zealand*. Unpublished Ph.D. thesis University of Auckland, New Zealand. [174]

Pemantle, R. (1992) Vertex-reinforced random walk. *Probability Theory and Related Fields* **92**, 117–136. [108, 129]

Pemantle, R. (2007) A survey of stochastic processes with reinforcement. *Probability Surveys* **4**, 1–79. [108]

Pereira, M.G., Malamud, B.D., Trigo, R.M. and Alves, P.I. (2011) The history and characteristics of the 1980-2005 Portuguese rural fire database. *Natural Hazards and Earth System Sciences* **11**, 3343–3358. [163]

Perry, G.L.W. (2009) Modelling and simulation. In *A Companion to Environmental Geography* (ed. Castree, N., Demeritt, D., Liverman, D. and Rhoads, B.) John Wiley & Sons pp. 336–357. [23, 199]

Perry, G.L.W. and Millington, J.D.A. (2008) Spatial modelling of succession-disturbance dynamics in forest ecosystems: concepts and examples. *Perspectives in Plant Ecology, Evolution and Systematics* **9**, 191–210. [51]

Perry, G.L.W., Miller, B.P. and Enright, N.J. (2006) A comparison of methods for the statistical analysis of spatial point patterns in plant ecology. *Plant Ecology* **187**, 59–82. [39, 43, 44, 52]

Perry, G.L.W., Wilmshurst, J.M., McGlone, M.S., McWethy, D.B. and Whitlock, C. (2012) Explaining fire-driven landscape transformation during the Initial Burning Period of New Zealand's prehistory. *Global Change Biology* **18**, 1609–1621. [166]

Pesavento, U. (1995) An implementation of von Neumann's self-reproducing machine. *Artificial Life* **2**, 337–354. [18]

Peterson, G.D. (2002) Contagious disturbance, ecological memory, and the emergence of landscape pattern. *Ecosystems* **5**, 329–338. [32, 165]

Petrovskii, S. and Morozov, A. (2009) Dispersal in a statistically structured population: fat tails revisited. *American Naturalist* **173**, 278–289. [111]

Phillips, D.L. and Marks, D.G. (1996) Spatial uncertainty analysis: propagation of interpolation errors in spatially distributed models. *Ecological Modelling* **91**, 213–229. [207]

Phillips, J.D. (2004) Doing justice to the law. *Annals of the Association of American Geographers* **94**, 290–293. [57]

Phipps, M. (1989) Dynamical behaviour of cellular automata under the constraint of neighborhood coherence. *Geographical Analysis* **21**, 197–215. [180]

Pidd, M. (2010) *Tools for Thinking: Modelling in Management Science*, 3rd edn. John Wiley & Sons. [175]

Pidgeon, N.F. and Fischhoff, B. (2011) The role of social and decision sciences in communicating uncertain climate risks. *Nature Climate Change* **1**, 35–41. [194]

Pilkey, O.H. and Pilkey-Jarvis, L. (2007) *Useless Arithmetic: Why Environmental Scientists Can't Predict The Future*. Columbia University Press. [14]

Piou, C., Berger, U. and Grimm, V. (2009) Proposing an information criterion for individual-based models developed in a pattern-oriented modelling framework. *Ecological Modelling* **220**, 1957–1967. [222]

Polasky, S., Carpenter, S.R., Folke, C. and Keeler, B. (2011) Decision-making under great uncertainty: environmental management in an era of global change. *Trends in Ecology & Evolution* **26**, 398–404. [194]

Pontius Jr., R.G. (2000) Quantification error versus location error in comparison of categorical maps. *Photogrammetric Engineering and Remote Sensing* **66**, 1011–1016. [214]

Pontius Jr., R.G. (2002) Statistical methods to partition effects of quantity and location during comparison of categorical maps at multiple resolutions. *Photogrammetric Engineering and Remote Sensing* **68**, 1041–1049. [207, 214]

Pontius Jr., R.G. and Millones, M. (2011) Death to Kappa: birth of quantity disagreement and allocation disagreement for accuracy assessment. *International Journal of Remote Sensing* **32**, 4407–4429. [95]

Pontius Jr., R.G., Boersma, W., Castella, J-C., Clarke, K.C., de Nijs, T., Dietzel, C., Duan, Z., Fotsing, E., Goldstein, N., Kok, K., Koomen, E., Lippitt, C.D., McConnell, W., Mohd Sood, A., Pijanowski, B., Pithadia, S., Sweeney, S., Trung, T.N., Veldkamp, A.T. and Verburg, P.H. (2008) Comparing the input, output, and validation maps for several models of land change. *Annals of Regional Science* **42**, 11–37. [95]

Portugali, J. (2000) *Self-Organization and the City* Springer Series in Synergetics. Springer. [86]

Poundstone, W. (1992) *Prisoner's Dilemma*. Anchor Books. [91, 92]

Pyke, G.H., Pulliam, H.R. and Charnov, E.L. (1977) Optimal foraging: a selective review of theory and tests. *The Quarterly Review of Biology* **52**, 137–154. [118]

Quinn, G.P. and Keough, M.R. (2001) *Experimental Design and Data Analysis for Biologists*. Cambridge University Press. [253]

R-Development-Core-Team (2012) R: A language and environment for statistical computing. http://www.r-project.org/. [xiv, xxiv]

Railsback, S.F. and Grimm, V. (2012) *Agent-Based and Individual-Based Modeling: A Practical Introduction*. Princeton University Press. [198, 199, 202, 203, 205, 206, 223, 229, 268]

Ramos-Fernández, G., Mateos, J.L., Miramontes, O., Cocho, G., Larralde, H. and Ayala-Orozco, B. (2004) Lévy walk patterns in the foraging movements of spider monkeys (*Ateles geoffroyi*). *Behavioral Ecology and Sociobiology* **55**, 223–230. [116]

Ratz, A. (1995) Long-term spatial patterns created by fire: a model oriented towards boreal forests. *International Journal of Wildland Fire* **5**, 25–34. [165]

Rawlence, N.J., Metcalf, J.L., Wood, J.R., Worthy, T.H., Austin, J.J. and Cooper, A. (2012) The effect of climate and environmental change on the megafaunal moa of New Zealand in the absence of humans. *Quaternary Science Reviews*. **50**, 141–153. [230]

Rayleigh, J.W. (1905) The problem of the random walk. *Nature* **72**, 318. [102]

Refsgaard, J.C., van der Sluijs, J.P., Højberg, A.L. and Vanrolleghem, P.A. (2007) Uncertainty in the environmental modelling process—a framework and guidance. *Environmental Modelling & Software* **22**, 1543–1556. [193, 194]

Regan, H.M., Colyvan, M. and Burgman, M.A. (2002) A taxonomy and treatment of uncertainty for ecology and conservation biology. *Ecological Applications* **12**, 618–628. [193]

Révész, P. (2005) *Random Walks in Random and Non-Random Environments*. World Scientific. [99]

Reynolds, C.W. (1987) Flocks, herds, and schools: A distributed behavioral model. *Computer Graphics (SIGGRAPH '87 Conference Proceedings)* **21**, 25–34. [121, 124]

Rice, S.P., Lancaster, J. and Kemp, P. (2010) Experimentation at the interface of fluvial geomorphology, stream ecology and hydraulic engineering and the development of an effective, interdisciplinary river science. *Earth Surface Processes and Landforms* **35**, 64–77. [7]

Ripley, B.D. (1977) Modelling spatial patterns. *Journal of the Royal Statistical Society, Series B*. **39**, 172–212. [43]

Rogers, T. and McKane, A.J. (2011) A unified framework for Schelling's model of segregation. *Journal of Statistical Mechanics: Theory and Experiment*, P07006. [85]

Rolett, B. and Diamond, J. (2004) Environmental predictors of pre-European deforestation on Pacific islands. *Nature* **431**, 443–446. [230]

Rosenstock, H.B. and Marquardt, C.L. (1980) Cluster formation in two-dimensional random walks: application to photolysis of silver halides. *Physical Review B* **22**, 5797–5809. [155]

Rosindell, J., Hubbell, S.P., He, F., Harmon, L.J. and Etienne, R.S. (2012) The case for ecological neutral theory. *Trends in Ecology & Evolution* **27**, 203–208. [82, 266]

Rossmanith, E., Blaum, N., Grimm, V. and Jeltsch, F. (2007) Pattern-oriented modelling for estimating unknown pre-breeding survival rates: the case of the Lesser Spotted Woodpecker (*Picoides minor*). *Biological Conservation* **135**, 555–564. [223]

Rosvall, M. and Bergstrom, C.T. (2008) Maps of random walks on complex networks reveal community structure. *Proceedings of the National Academy of Sciences of the United States of America* **105**, 1118–1123. [185]

Rouchier, J., Cioffi-Revilla, C., Polhill, J.G. and Takadama, K. (2008) Progress in model-to-model analysis. *Journal of Artificial Societies and Social Simulation* **11**, http://jasss.soc.surrey.ac.uk/11/2/8.html. [232]

Roux, S. and Guyon, E. (1989) Temporal development of invasion percolation. *Journal of Physics A: Mathematical and General* **22**, 3693–3705. [146]

Ruckelshaus, M., Hartway, C. and Kareiva, P.M. (1997) Assessing the data requirements of spatially explicit dispersal models. *Conservation Biology* **11**, 1298–1306. [207, 208, 236]

Ruckelshaus, M., Hartway, C. and Kareiva, P.M. (1999) Dispersal and landscape errors in spatially explicit population models: a reply. *Conservation Biology* **13**, 1223–1224. [207]

Rudnick, J. and Gaspari, G. (2004) *Elements of the Random Walk: An Introduction for Advanced Students and Researchers*. Cambridge University Press. [99]

Runions, A., Fuhrer, M., Lane, B., Federl, P., Rolland-Lagan, A.G. and Prusinkiewicz, P. (2005) Modeling and visualization of leaf venation patterns. *ACM Transactions on Graphics* **24**, 702–711. [87]

Russell, E.S. (1931) Some theoretical considerations on the 'overfishing' problem. *ICES Journal of Marine Science: Journal du Conseil* **6**, 3–20. [197]

Ruxton, G.D. (1996) Effects of the spatial and temporal ordering of events on the behaviour of a simple cellular automata model. *Ecological Modelling* **84**, 311–314. [171]

Ruxton, G.D. and Saravia, L.A. (1998) The need for biological realism in the updating of cellular automata models. *Ecological Modelling* **107**, 105–112. [171]

Rykiel, E.J. (1996) Testing ecological models: the meaning of validation. *Ecological Modelling* **90**, 229–244. [212]

Saltelli, A., Chan, K. and Scott, E.M. (2000) *Sensitivity Analysis*. John Wiley & Sons. [202]

Samet, H. (1990) *The Design and Analysis of Spatial Data Structures*. Addison-Wesley. [86, 187]

Sander, L.M. (2000) Diffusion-limited aggregation: a kinetic critical phenomenon? *Contemporary Physics* **41**, 203–218. [155, 156, 157]

Saura, S. and Martínez-Millán, J. (2000) Landscape pattern simulation with a modified random clusters method. *Landscape Ecology* **15**, 661–678. [159, 160, 233, 236]

Savakis, A.E. and Maggelakis, S.A. (1997) Models of shrinking clusters with applications to epidermal wound healing. *Mathematical and Computer Modelling* **25**, 1–6. [149]

Sawada, Y., Ohta, S., Yamazaki, M. and Honjo, H. (1982) Self-similarity and a phase-transition-like behavior of a random growing structure governed by a nonequilibrium parameter. *Physics Review A* **26**, 3557–3563. [149]

Sawyer, S. (1977) Rates of consolidation in a selectively neutral migration model. *The Annals of Probability* **5**, 486–493. [80]

Sayer, A. (1992) *Method in Social Science: A Realist Approach*. Routledge. [9]

Schellinck, J. and White, T. (2011) A review of attraction and repulsion models of aggregation: methods, findings and a discussion of model validation. *Ecological Modelling* **222**, 1897–1911. [125]

Schelling, T.C. (1969) Models of segregation. *American Economic Review* **59**, 488–493. [83]

Schelling, T.C. (1971) Dynamic models of segregation. *Journal of Mathematical Sociology* **1**, 143–186. [83]

Schelling, T.C. (1978) *Micromotives and Macrobehavior*. Norton. [83]

Scheucher, M. and Spohn, H. (1988) A soluble kinetic model for spinodal decomposition. *Journal of Statistical Physics* **53**, 279–294. [80]

Schmolke, A., Thorbek, P., DeAngelis, D.L. and Grimm, V. (2010) Ecological models supporting environmental decision making: a strategy for the future. *Trends in Ecology & Evolution* **25**, 479–486. [231, 247]

Schneider, D.C. (2001) The rise of the concept of scale in ecology. *BioScience* **51**, 545–553. [36]

Schneider, D.C. (2009) *Quantitative Ecology: Measurement, Models, and Scaling*. Elsevier/Academic Press. [32, 33, 36, 37]

Schönfisch, B. (1997) Anisotropy in cellular automata. *Biosystems* **41**, 29–41. [181]

Schönfisch, B. and de Roos, A. (1999) Synchronous and asynchronous updating in cellular automata. *BioSystems* **51**, 123–143. [171]

Schroeder, M. (1991) *Fractals, Chaos, Power Laws: Minutes from an Infinite Paradise*. W. H. Freeman. [22]

Semboloni, F. (2000) The growth of an urban cluster into a dynamic self-modifying spatial pattern. *Environment and Planning B: Planning and Design* **27**, 549–564. [95, 184]

Shao, J., Havlin, S. and Stanley, H.E. (2009) Dynamic opinion model and invasion percolation. *Physical Review Letters* **103**, 018701. [168]

Shi, W. and Pang, M.Y.C. (2000) Development of Voronoi-based cellular automata–an integrated dynamic model for geographical information systems. *International Journal of Geographical Information Science* **14**, 455–474. [95, 184]

Shiwakoti, N., Sarvi, M., Rose, G. and Burd, M. (2009) Enhancing the safety of pedestrians during emergency egress. *Transportation Research Record: Journal of the Transportation Research Board* **2137**, 31–37. [130]

Shlesinger, M.F. and Klafter, J. (1986) Lévy walk versus Lévy flight. In *On Growth and Form* (ed. Stanley, H.E. and Ostrowsky, N.) Martinus Nijhoff pp. 279–283. [110]

Sigmund, K. (2009) *The Calculus of Selfishness* Princeton Series in Theoretical and Computational Biology. Princeton University Press. [91, 92]

Simon, H.A. (1955) A behavioral model of rational choice. *The Quarterly Journal of Economics* **69**, 99–118. [112]

Simon, H.A. (1956) Rational choice and the structure of the environment. *Psychological Review* **63**, 129–138. [112, 114]

Simon, H.A. (1996) *The Sciences of the Artificial*, 3rd edn. MIT Press. [22, 111]

Simon, H.A. and Bonini, C.P. (1958) The size distribution of business firms. *The American Economic Review* **48**, 607–617. [26]

Sims, D.W., Humphries, N.E., Bradford, R.W. and Bruce, B.D. (2012) Lévy flight and Brownian search patterns of a free-ranging predator reflect different prey field characteristics. *Journal of Animal Ecology* **81**, 432–442. [126]

Skellam, J.G. (1951) Random dispersal in theoretical populations. *Biometrika* **38**, 196–218. [136]

Smith, E. (2001) Uncertainty analysis. In *Encyclopedia of Environmetrics* (ed. El-Shaarawi, A.H. and Piegorsch, W.W.) John Wiley & Sons, Ltd. [194, 202, 203, 204, 205]

Smith, J.M. (2004) Unlawful relations and verbal inflation. *Annals of the Association of American Geographers* **94**, 294–299. [57]

Smith, N. (1979) Toward a theory of gentrification: a back to the city movement by capital not people. *Journal of the American Planning Association* **45**, 538–548. [32]

Smith, N. (1996) *The New Urban Frontier: Gentrification and the Revanchist City.* Routledge. [32]

Soetaert, K. and Herman, P.M.J. (2009) *A Practical Guide To Ecological Modelling: Using R As a Simulation Platform.* Springer. [133, 201, 229]

Solé, R.V. (2011) *Phase Transitions.* Princeton University Press. [141, 143]

Solé, R.V. and Bascompte, J. (2006) *Self-Organization in Complex Ecosystems.* Princeton University Press. [22, 138]

Solé, R.V. and Goodwin, B. (2000) *Signs of Life: How Complexity Pervades Biology.* Basic Books. [22]

Solow, A.R. (2005) Power laws without complexity. *Ecology Letters* **8**, 361–363. [164]

South, A. (1999) Dispersal in spatially explicit population models. *Conservation Biology* **13**, 1039–1046. [207, 211]

Spiegelhalter, D., Pearson, M. and Short, I. (2011) Visualizing uncertainty about the future. *Science* **333**, 1393–1400. [213]

Spitzer, F. (1970) Interaction of Markov processes. *Advances in Mathematics* **5**, 246–290. [70]

Spitzer, F. (2001) *Principles of Random Walk.* Springer. [99]

Stainforth, D.A., Aina, T., Christensen, C., Collins, M., Faull, N., Frame, D.J., Kettleborough, J.A., Knight, S., Martin, A., Murphy, J.M., Piani, C., Sexton, D., Smith, L.A., Spicer, R.A., Thorpe, A.J. and Allen, M.R. (2005) Uncertainty in predictions of the climate response to rising levels of greenhouse gases. *Nature* **433**, 403–406. [194]

Stanford, K. (2009) Underdetermination of scientific theory. In *The Stanford Encyclopedia of Philosophy* (ed. Zalta, E.N.), Winter 2009 edn. [26, 52, 266]

Starfield, A.M. and Bleloch, A.L. (1986) *Building Models for Conservation and Wildlife Management.* MacMillan. [12]

Stark, C.P. (1991) An invasion percolation model of drainage network evolution. *Nature* **352**, 423–425. [146, 147, 148]

Stark, C.P. (1994) Cluster growth modeling of plateau erosion. *Journal of Geophysical Research* **99**, 13957–13969. [166, 167, 229]

Stauffer, D. and Aharony, A. (1994) *Introduction to Percolation Theory*, 2nd edn. Taylor & Francis. [137, 141, 142, 145]

Stein, M. (1987) Large sample properties of simulations using Latin hypercube sampling. *Technometrics* **29**, 143–151. [204, 258]

Stein, R.A. (2011) Super-spreaders in infectious diseases. *International Journal of Infectious Diseases* **15**, e510–e513. [135]

Sterman, J. (2000) *Business Dynamics.* Irwin/McGraw-Hill. [175]

Stevens, D. and Dragićević, S. (2007) A GIS-based irregular cellular automata model of land-use change. *Environment and Planning B: Planning and Design* **34**, 708–724. [95]

Stocki, R. (2005) A method to improve design reliability using optimal Latin hypercube sampling. *Computer Assisted Mechanics and Engineering Sciences* **12**, 393–411. [258]

Stoll, P. and Prati, D. (2001) Intraspecific aggregation alters competitive interactions in experimental plant communities. *Ecology* **82**, 319–327. [79]

Stoyan, D. and Penttinen, A. (2000) Recent applications of point process methods in forestry statistics. *Statistical Science* **15**, 61–78. [42, 43]

Stumpf, M.P.H. and Porter, M.A. (2012) Critical truths about power laws. *Science* **335**, 665–666. [26, 164]

Sugden, A. and Pennisi, E. (2006) When to go, where to stop. *Science* **313**, 775. [126]

Sui, D.Z. (2004) Tobler's first law of geography: a big idea for a small world? *Annals of the Association of American Geographers* **94**, 269–277. [57]

Takeyama, M. (1996) Geo-Algebra: *A Mathematical Approach to Integrate Spatial Modelling and GIS* Unpublished Ph.D. thesis University of California, Santa Barbara, CA. [95]

Takeyama, M. and Couclelis, H. (1997) Map dynamics: integrating cellular automata and GIS through Geo-Algebra. *International Journal of Geographical Information Science* **11**, 73–91. [95]

Tchen, C.M. (1952) Random flight with multiple partial correlations. *The Journal of Chemical Physics* **20**, 214–217. [105]

Tesfatsion, L. and Judd, K. (eds) (2006) *Agent-Based Computational Economics* vol. 2 of *Handbook of Computational Economics*. Elsevier/North-Holland. [51]

Theraulaz, G., Bonabeau, E., Nicolis, S.C., Solé, R. V. RV, Fourcassié, V., Blanco, S., Fournier, R., Joly, J.-L., Fernández, P., Grimal, A., Dalle, P. and Deneubourg, J.-L. (2002) Spatial patterns in ant colonies. *Proceedings of the National Academy of Sciences of the United States of America* **99**, 9645–9649. [91]

Thiele, J.C. and Grimm, V. (2010) NetLogo meets R: Linking agent-based models with a toolbox for their analysis. *Environmental Modelling & Software* **25**, 972–974. [xiv, xxiv]

Thiele, J.C., Kurth, W. and Grimm, V. (2012) RNetLogo: an R package for running and exploring individual-based models implemented in NetLogo. *Methods in Ecology and Evolution* **3**, 480–483. [xiv, xxiv]

Thiéry, J.M., D'Herbès, J.M. and Valentin, C. (1995) A model simulating the genesis of banded vegetation patterns in Niger. *Journal of Ecology* **83**, 497–507. [90]

Thompson, J.R., Wiek, A., Swanson, F.J., Carpenter, S.R., Fresco, N., Hollingsworth, T., Spies, T.A. and Foster, D.R. (2012) Scenario studies as a synthetic and integrative research activity for long-term ecological research. *BioScience* **62**, 367–376. [13, 269]

Tobler, W.R. (1970) A computer movie simulating urban growth in the Detroit region. *Economic Geography* **46**, 234–240. [57, 93]

Tobler, W.R. (1979) Cellular geography In *Philosophy in Geography* (ed. Gale, S. and Olsson, G.) D. Reidel Publishing Company pp. 379–386. [93]

Tobler, W.R. (2004) On the first law of geography: a reply. *Annals of the Association of American Geographers* **94**, 304–310. [57]

Tomlin, D. (1990) *Geographical Information Systems and Cartographic Modeling*. Prentice-Hall. [176]

Torrens, P.M. (2012) Moving agent pedestrians through space and time. *Annals of the Association of American Geographers* **102**, 35–66. [125, 130]

Travis, D.J., Carleton, A.M. and Lauritsen, R.G. (2002) Contrails reduce daily temperature range. *Nature* **418**, 601. [17]

Turchin, P. (1991) Translating foraging movements in heterogeneous environments into the spatial distribution of foragers. *Ecology* **72**, 1253–1266. [108]

Turcotte, D.L. (1999) Applications of statistical mechanics to natural hazards and landforms. *Physica A: Statistical Mechanics and its Applications* **274**, 294–299. [164]

Turing, A.M. (1952) The chemical basis of morphogenesis. *Philosophical Transactions of the Royal Society of London. Series B.* **237**, 37–72. [89, 90]

Turner, M.G. (1989) Landscape ecology: the effect of pattern on process. *Annual Review of Ecology and Systematics* **20**, 171–197. [32]

Turner, M.G., Wu, Y., Romme, W.H. and Wallace, L.L. (1993) A landscape simulation model of winter foraging by large ungulates. *Ecological Modelling* **69**, 163–184. [125]

Unwin, D.J. (1996) GIS, spatial analysis and spatial statistics. *Progress in Human Geography* **20**, 540–551. [30, 32]

Urban, D.L. (2005) Modeling ecological processes across scales. *Ecology* **86**, 1996–2006. [32, 33, 34, 36, 37]

Uriarte, M., Anciães, M., da Silva, M.T.B., Rubim, P., Johnson, E. and Bruna, E.M. (2011) Disentangling the drivers of reduced long-distance seed dispersal by birds in an experimentally fragmented landscape. *Ecology* **92**, 924–937. [116]

van den Berg, M., Bolthausen, E. and den Hollander, F. (2001) Moderate deviations for the volume of the Wiener sausage. *Annals of Mathematics* **153**, 355–406. [115]

van Vliet, J., White, R. and Dragićević, S. (2009) Modeling urban growth using a variable grid cellular automaton. *Computers, Environment and Urban Systems* **33**, 35–43. [95]

Vasconcelos, M.J. and Zeigler, B.P. (1993) Discrete-event simulation of forest landscape response to fire disturbances. *Ecological Modelling* **65**, 177–198. [174]

Vasconcelos, M.J., Zeigler, B.P. and Graham, L.A. (1993) Modeling multi-scale spatial ecological processes under the discrete event systems paradigm. *Landscape Ecology* **8**, 273–286. [174]

Vellend, M. (2010) Conceptual synthesis in community ecology. *The Quarterly Review of Biology* **85**, 183–206. [267]

Vichniac, G.Y. (1984) Simulating physics with cellular automata. *Physica D: Nonlinear Phenomena* **10**, 96–116. [68]

Vicsek, T. and Szalay, A.S. (1987) Fractal distribution of galaxies modeled by a cellular-automaton-type stochastic process. *Physical Review Letters* **58**, 2818–2821. [63, 64]

Vicsek, T., Czirók, A., Ben-Jacob, E., Cohen, I. and Shochet, O. (1995) Novel type of phase transition in a system of self-driven particles. *Physical Review Letters* **75**, 1226–1229. [122, 123]

Vincenot, C.E., Giannino, F., Rietkerk, M., Moriya, K. and Mazzoleni, S. (2011) Theoretical considerations on the combined use of system dynamics and individual-based modeling in ecology. *Ecological Modelling* **222**, 210–218. [52, 175]

Vinković, D. and Kirman, A. (2006) A physical analogue of the Schelling model. *Proceedings of the National Academy of Sciences of the United States of America* **103**, 19261–19265. [85]

Viswanathan, G.M., Afanasyev, V., Buldyrev, S.V., Murphy, E.J., Prince, P.A. and Stanley, H.E. (1996) Lévy flight search patterns of wandering albatrosses. *Nature* **381**, 413–415. [119]

Viswanathan, G.M., Buldyrev, S.V., Havlin, S., da Luz, M.G.E., Raposo, E.P. and Stanley, H.E. (1999) Optimizing the success of random searches. *Nature* **401**, 911–914. [119]

Viswanathan, G.M., da Luz, M.G.E., Raposo, E.P. and Stanley, H.E. (2011) *The Physics of Foraging: An Introduction to Random Searches and Biological Encounters*. Cambridge University Press. [109, 119]

von Bertalanffy, L. (1950) An outline of general system theory. *The British Journal for the Philosophy of Science* **1**, 134–165. [175]

Waddington, C.H. (1977) *Tools for Thought: How to Understand and Apply the Latest Scientific Techniques of Problem Solving*. Basic Books. [14]

Wainwright, J. (2008) Can modelling enable us to understand the rôle of humans in landscape evolution. *Geoforum* **39**, 659–674. [263]

Waldrop, M. (1992) *Complexity*. Simon and Schuster. [21]

Walker, W.E., Harremoës, P., Rotmans, J., van der Sluijs, J.P., van Asselt, M.B.A., Janssen, P. and von Krauss, M.P.K. (2003) Defining uncertainty: a conceptual basis for uncertainty management in model-based decision support. *Integrated Assessment* **4**, 5–17. [193]

Wang, Q. and Malanson, G.P. (2008) Neutral landscapes: bases for exploration in landscape ecology. *Geography Compass* **2**, 319–339. [158]

Wassermann, S. and Faust, K. (1994) *Social Network Analysis: Methods and Applications.* Cambridge University Press. [183, 184]

Watts, D.J. and Strogatz, S.H. (1998) Collective dynamics of 'small-world' networks. *Nature* **393**, 440–442. [183]

Weaver, W. (1948) Science and complexity. *American Scientist* **36**, 536–544. [21, 22]

Weidlich, W. (1971) The statistical description of polarization phenomena in society. *British Journal of Mathematical and Statistical Psychology* **24**, 251–266. [78]

Weisberg, M. (2006) Robustness analysis. *Philosophy of Science* **73**, 730–742. [206]

Werner, B.T. (1995) Eolian dunes: computer simulations and attractor interpretation. *Geology* **23**, 1107–1110. [91]

Werner, B.T. and Fink, T.M. (1993) Beach cusps as self-organized patterns. *Science* **260**, 968–971. [91]

Wesselung, C.G., Karssenberg, D.-J., Burrough, P.A. and van Deursen, W.P.A. (1996) Integrating dynamic environmental models in GIS: the development of a Dynamic Modelling language. *Transactions in GIS* **1**, 40–48. [176]

White, R. and Engelen, G. (1994) Urban systems dynamics and cellular automata: fractal structures between order and chaos. *Chaos, Solitons and Fractals* **4**, 563–583. [94]

White, R. and Engelen, G. (1997) Cellular automata as the basis of integrated dynamic regional modelling. *Environment and Planning B: Planning and Design* **24**, 235–246. [94]

White, R., Engelen, G. and Uljee, I. (1997) The use of constrained cellular automata for high-resolution modelling of urban land-use dynamics. *Environment and Planning B: Planning and Design* **24**, 323–343. [94]

Whitlock, C., Higuera, P.E., McWethy, D.B. and Briles, C.E. (2010) Paleoecological perspectives on fire ecology: revisiting the fire-regime concept. *Open Ecology* **3**, 6–23. [135]

Wiegand, K., Jeltsch, F. and Ward, D. (1999) Analysis of the population dynamics of *Acacia* trees in the Negev desert, Israel with a spatially-explicit computer simulation model. *Ecological Modelling* **117**, 203–224. [186]

Wiegand, T., Gunatilleke, S., Gunatilleke, N. and Okuda, T. (2007) Analyzing the spatial structure of a Sri Lankan tree species with multiple scales of clustering. *Ecology* **88**, 3088–3102. [49]

Wiegand, T., Huth, A., Getzin, S., Wang, X., Hao, Z., Gunatilleke, S. and Gunatilleke, N. (2012) Testing the independent species' arrangement assertion made by theories of stochastic geometry of biodiversity. *Proceedings of the Royal Society of London. Series B*, **279**, 3312–3320. [56]

Wiegand, T., Jeltsch, F., Hanski, I. and Grimm, V. (2003) Using pattern-oriented modeling for revealing hidden information: a key for reconciling ecological theory and application. *Oikos* **100**, 209–222. [199, 222]

Wiens, J.A. (1989) Spatial scaling in ecology. *Functional Ecology* **3**, 385–397. [37]

Wilensky, U. (1999) Netlogo. http://ccl.northwestern.edu/netlogo/. Center for Connected Learning and Computer-Based Modeling, Northwestern University. Evanston, IL. [xiv, xv, xxiv]

Wilkinson, D. and Willemsen, J.F. (1983) Invasion percolation: a new form of percolation theory. *Journal of Physics A: Mathematical and General* **16**, 3365–3376. [145, 146]

Wilkinson, D.H. (1952) The random element in bird 'navigation'. *Journal of Experimental Biology* **29**, 532–560. [103]

Williams, T. and Bjerknes, R. (1972) Stochastic model for abnormal clone spread through epithelial basal layer. *Nature* **236**, 19–21. [149]

Wilson, A.G. (2006) Ecological and urban systems models: some explorations of similarities in the context of complexity theory. *Environment and Planning A* **38**, 633–646. [136]

Wilson, R.J. (1996) *Introduction to Graph Theory*. Longman. [183]

Winsberg, E. (2009a) Computer simulation and the philosophy of science. *Philosophy Compass* **4**, 835–845. [11]

Winsberg, E. (2009b) A tale of two methods. *Synthese* **169**, 575–592. [222]

With, K.A. (2002) The landscape ecology of invasive spread. *Conservation Biology* **16**, 1192–1203. [136]

Witten, T.A. and Sander, L.M. (1981) Diffusion-limited aggregation, a kinetic critical phenomenon. *Physical Review Letters* **47**, 1400–1403. [155, 156]

Wolfram, S. (1984) Universality and complexity in cellular automata. *Physica D: Nonlinear Phenomena* **10**, 1–35. [94]

Wolfram, S. (1986) *Theory and Applications of Cellular Automata*. World Scientific. [20]

Wood, S.N. (2010) Statistical inference for noisy nonlinear ecological dynamic systems. *Nature* **466**(7310), 1102–1104. [45, 220, 223]

Worboys, M. and Duckham, M. (2004) *GIS: A Computing Perspective*, 2nd edn. CRC Press. [175, 179]

Wu, H., Li, B-L., Springer, T.A. and Neill, W.H. (2000) Modelling animal movement as a persistent random walk in two dimensions: expected magnitude of net displacement. *Ecological Modelling* **132**, 115–124. [107]

Wu, H.C. and Sun, C.T. (2005) What should we do before running a social simulation? The importance of model analysis *Social Science Computer Review* **23**, 221–234. [93]

Xie, Y. (1996) A generalized model for cellular urban dynamics. *Geographical Analysis* **28**, 350–373. [94]

Young, D.A. (1984) A local activator-inhibitor model of vertebrate skin patterns. *Mathematical Biosciences* **72**, 51–58. [90]

Zeng, Y. and Malanson, G.P. (2006) Endogenous fractal dynamics at alpine treeline ecotones. *Geographical Analysis* **38**, 271–287. [168]

Zhang, J. (2009) Tipping and residential segregation: a unified Schelling model. *Journal of Regional Science* **51**, 167–193. [85]

Zinck, R.D. and Grimm, V. (2008) More realistic than anticipated: a classical forest-fire model from statistical physics captures real fire shapes. *The Open Ecology Journal* **1**, 8–13. [162, 164]

Zinck, R.D., Pascual, M. and Grimm, V. (2011) Understanding shifts in wildfire regimes as emergent threshold phenomena. *American Naturalist* **178**, E149–E161. [166]

Index

Spatial Simulation: Exploring Pattern and Process, First Edition.
David O'Sullivan and George L.W. Perry.
© 2013 John Wiley & Sons, Ltd. Published 2013 by John Wiley & Sons, Ltd.

Plate 1 An example of the heterogeneity caused by processes such as wildfire where pattern and process interact in complicated ways. Here, a wildfire in the Front Range of the Colorado Rockies has produced a mosaic of burned (black) and unburned (green) areas that will act to constrain and amplify future wildfire events. *Source*: authors' collection.

Plate 2 Snapshot of a 200 × 200 toroidally wrapped voter model, initialised with a unique state at each of the 40 000 sites, after 32 million events (800 generations). There are now 129 distinct states in clusters of varying sizes.

Spatial Simulation: Exploring Pattern and Process, First Edition.
David O'Sullivan and George L.W. Perry.
© 2013 John Wiley & Sons, Ltd. Published 2013 by John Wiley & Sons, Ltd.

Plate 3 Snapshot of the model of Plate 2 after 100 million events (2500 generations). Only 54 distinct states remain, although considerable continuity from the previous snapshot is evident.

Plate 4 Snapshot of the model in previous plates after 200 million events (5000 generations). Twenty-seven distinct states now remain.

Plate 5 A spotted eagle ray with markings resembling those produced by a simple reaction-diffusion model. *Source*: authors' collection.

Plate 6 Lichens slowly spreading across a rock surface in Lamar Valley, Yellowstone National Park, Wyoming. The patterns here are reminiscent of those produced by simple models such as diffusion-limited aggregation and Eden growth processes. *Source*: authors' collection.

Plate 7 Example of a spanning cluster produced by the invasion percolation process on an uncorrelated 720 × 360 lattice with trapping (sideways view with invasion starting from bottom of image). Sites in darker colours were invaded earlier in the process and repeated changes in the main 'front' of the invasion are evident. This cluster includes 24 459 invaded sites.

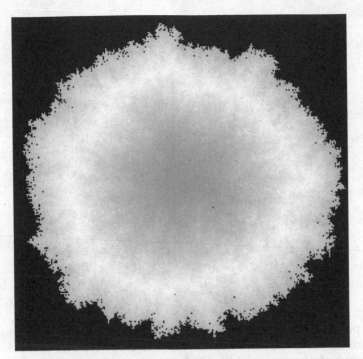

Plate 8 Evolution of the Eden growth process over time shown by colouring on a red (oldest) to blue (newest) colour ramp. The space is a 256 × 256 lattice and this cluster evolved over 42 063 steps.

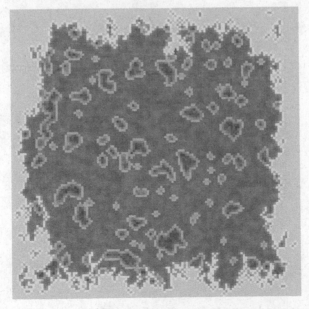

Plate 9 Initial conditions in the model run shown in Figure 8.7. One group has arrived on the island (the orange icon to the north). Low-value resource is green and high-value resource in shades of grey from light (low levels) to dark (high levels).

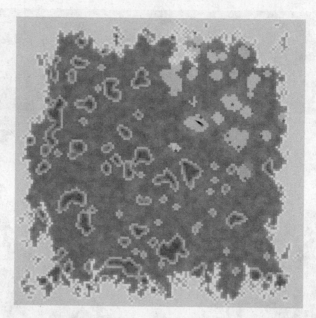

Plate 10   The model run of Figure 8.7 after 25 years. Exploitation of the high-value resource has only occurred to the north-east, although the home camp has moved to the centre of the island. Current hunting spots are indicated by red stars.

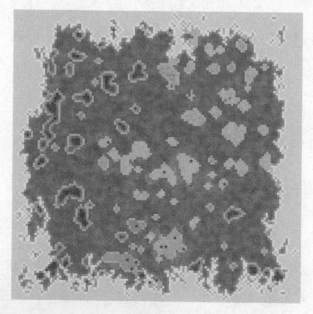

Plate 11   The same model run at 50 years. Exploitation of high-value resource has continued, but is now concentrated in the south. Meanwhile there has been some recovery of the high-value resource in the north-east.

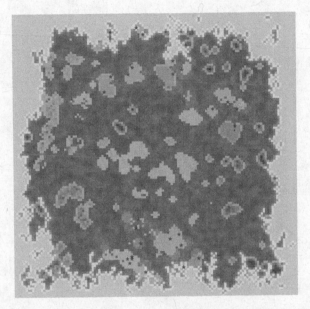

Plate 12 Rapid population growth from around the 130th until the 150th year sees the human population split into two groups (in the south and north-east) and more far-ranging high-value exploitation is evident.

Plate 13   At 200 years, there are three groups on the island and the high-value resource has been almost completely 'hunted out'. The island colonisation persists for 32 more years, with declining human population and sees the high-value resource almost completely eliminated.

Printed in the United States
By Bookmasters